PREFACE

The object of this book is to provide advanced students, and especially students of geography, with a reasoned account of the world's climatic types. It is not intended to exempt the student from the reading of original works, but to give a foundation on to which he or she may build. Without entering into the physics of meteorological processes, with which the student is expected to be already familiar, effect is as far as possible related to cause, and, since the book is intended primarily for geographers, prominence is given to the human aspect and the practical application. In short, the book attempts to be reasoned and not merely descriptive, hence the arrangement adopted is not regional but is based on climatic types, with a view to emphasizing the essential similarity of environment in regions similarly situated and climatically allied. Particular attention is paid to the normal type, the regional peculiarities of the more important areas being dealt with separately after the general description. The classification followed departs only in matters of detail from those in general use, but the boundary lines adopted are not always coincident with those generally recognized; the reasons for these departures are set out in some detail in Chapter V.

Except in matters of treatment and presentation the book makes no other claim to originality, and the author acknowledges his indebtedness to a mass of climatological literature much too large and varied to be listed in detail. Generally this is acknowledged in the text or in footnotes, but since the book is intended only as a text-book for students, chapter and verse are not always given in the references. The object in view in quoting references is rather to allow the student to follow up a particular line of inquiry should he wish to do so, and this purpose is effected by means of a short guide to further reading at the end of each chapter; but it sometimes happens that an interesting side-track is exposed in the text which it is impossible to pursue further, and in such cases the way is pointed out in a footnote. In general only those books and journals are quoted which would be readily accessible to a student in the library of the university or college.

For the sake of uniformity the same general order has been followed in the description of each climatic type, but different types lend themselves to most satisfactory treatment in different ways; thus the arrangement is sometimes by seasons and sometimes by elements.

When controversial topics arise it is not always feasible to give due consideration to all views held, and generally one explanation only, the one that appears most satisfactory to the author, is given. The last chapter, in particular, dealing with the evolution of climate, covers such an enormous subject in such a short space that it is impossible to

v

do more than provide a bald outline of the sequence of events. In the present inadequate state of our knowledge there are inevitably numerous apparent contradictions of fact and divergences of opinion among authors as to the explanations. Limitation of space has forbidden full treatment of these and it has been necessary to ignore the difficulties in attempting to present a consecutive account.

The Fahrenheit scale of temperature is used throughout and rainfall is expressed in inches, since students are generally more familiar with these units. Often the decimal points are ignored, temperature and rainfall figures being rounded off to the nearest degree or inch, for except when dealing with very small rainfall amounts or very small temperature variations these are of little significance; moreover, they give a misleading impression of exactitude which, in point of fact, does not exist, especially when dealing with mean values.

It is assumed that the student is equipped with a good modern atlas containing the usual maps of annual and seasonal temperature, pressure and rainfall, vegetation maps, etc., which are not, therefore, reproduced in the book.

The author wishes to express his gratitude to Mr. L. C. W. Bonacina for kindly reading Chapter V and for offering valuable suggestions and criticisms, and to Dr. H. A. Matthews for generous assistance and for undertaking the reading of the proofs.

Reading, 1931   A. A. M.

PREFACE TO NINTH EDITION

Climatology has been changing continuously since this book was first written. It was heavily revised in 1943 (third edition), when a new chapter on Air Masses was included. In the eighth edition the text was again reset and opportunity was taken to refine on the old system of climatic classifications with the introduction of a more realistic desert boundary and the redefinition of some temperature zones, now uniformly defined by the duration of temperate seasons. The subject remains descriptive and explanatory, but the findings of physical, biological, agricultural and medical sciences contribute to the proper understanding of the environment. The contributions of these sciences are usually precise in their treatment of experiments under controlled conditions, but climate deals generally with the 'mean' conditions of an unreliable environment, so only the relevant geographical aspects of this knowledge are included and they are brought up to date.

CLIMATOLOGY

by

A. AUSTIN MILLER, D.SC.

Professor of Geography in the
University of Reading

WITH 83 MAPS

AND DIAGRAMS

METHUEN & CO. LTD
11 New Fetter Lane London E.C. 4

First published April 30th 1931
Third Edition, revised, January 1944
Eighth Edition, revised and reset, 1953
Ninth Edition 1961
Reprinted four times
Reprinted 1971

SBN 416 31230 6

Printed in Great Britain by Butler & Tanner Ltd, Frome and London
Bound by Leighton-Straker Bookbinding Co. Ltd
Standard Road N.W.10

NOTE TO 1965 REPRINT

The Meteorological Office has decided to use degrees Centigrade in all publications, and this practice is now adopted in the Weather Forecasts, though for the time being, temperatures are announced in both units. Furthermore the School Examining Bodies are insisting that degrees Centigrade shall be used in the answers to all questions at 'O' and 'A' level. No decision has yet been announced about rainfall, which is now given in inches, but may in future be announced in millimetres, to bring this country in line with the figures applied to rainfall in many countries using these units. For the time being scales are provided for the quick conversion from degrees Fahrenheit to degrees Centigrade and of inches into millimetres.

CONTENTS

MAPS AND DIAGRAMS

I

THE MEANING AND SCOPE OF CLIMATOLOGY

The subject of climatology is intimately interwoven with the affairs of everyday life. This industrial era, in which a large percentage of the population does its daily work under cover from the elements, certainly feels the controlling hand of weather less than an earlier agricultural age, but the climatic control of its daily life and habits is probably as great today as it ever was. The agriculturist is still almost entirely at the mercy of the weather and climate, but the industrialist's dependence, though less direct, is not less real, and often the site-area of an industrial concentration itself originally depended on local climatic advantages, as Lancashire well knows. California's rapid increase of population is due to the attractiveness of its climate as well as to its industries, and the peninsula of Florida has little else to recommend it but the sublimity of its winter warmth.

Climate and Trade. Climate limits the choice of crops and therefore the local production of food; and climate determines the site for the cultivation of those other foodstuffs and raw materials of industry which modern life demands; this climatic control of production and requirements is one of the bases of the world's trade. Furthermore climates, in general, change from north to south, and the products of different climates move, therefore, chiefly along meridional routes, although much deflected by economic factors; climate therefore controls the direction as well as the existence of trade routes. The great wind belts, too, are determinants of ocean routes, for steam power has diminished but not removed the obstacle of a head wind and current, and the same clockwise swirl of air and water in the North Atlantic which led Columbus to America on the trades and back on the westerlies still aids much of the steam shipping in these waters; round the world with the 'Roaring Forties' is almost as universal for the steamship today as it was for the 'windjammers' of a hundred years ago. The development of air lines is insisting on the importance of air currents at all levels, and a dependence even greater than that of the sailing ship is forced on the aeroplane despite its high operating speeds.

Climate and Daily Life. In every climate the life habits of the natives have been regulated, often after many bitter experiences, in accordance with prevailing conditions. Nature enforces obedience to certain rules of diet, clothing and behaviour by native and visitor alike, the neglect of which is fraught with serious consequences. The adoption of many native methods and habits has always been a wise procedure for immigrant or conqueror, and this has ultimately led to his absorption into the life of the zone. Thus, in the long run, climate triumphs over extraneous introductions and a certain continuity of character is assured which reflects the individuality of the zone.

Climate and Race. Controversy centres round the discussion of the extent to which the physical and psychological characters of the inhabitants of different climates can be attributed directly to the climate. Is the dark pigment of the negro's skin the result of centuries spent in an environment of torrid heat? Can the dim light of the equatorial forests be held responsible for the dwarfing of the pygmies? There is no doubt that many of these characters are racial and hereditary, for the 'black man' does not become noticeably less black in a temperate environment, and the pygmy does not grow appreciably taller when transplanted from his native forest. The 'yellow' race shows little or no climatic control and is found from the Arctic (Esquimaux) to the Equator (Malays) and from arid climates (Gobi) to humid (Java).

The relative importance of heredity and environment provide a fruitful subject for inconclusive debate, but it must at least be admitted that there is a certain selective influence at work. For the 'black' races are limited in the natural state to tropical climates, and in general they do not thrive beyond their limits, while the gradual paling of complexion and hair polewards in Europe is an incontrovertible fact which cannot be devoid of significance.

Climate and Character. Psychologically each climate tends to have its own mentality, innate in its inhabitants and grafted on its immigrants. There appears to be a direct relationship between mental vigour and changeability of climates; all the world's great civilizations are now in regions experiencing abrupt and often unexpected changes of weather; the temperate zone governs the tropical zone by virtue of its infinitely greater energy and initiative. In U.S.A., where other factors are presumably uniform, there is a remarkable difference in the output of able and prominent men between the northern (temperate) and southern (sub-tropical) states. The enervating monotonous climates of much of the tropical zone, together with the abundant and easily obtained food-supply, produce a lazy and indolent people, indisposed to labour for hire and therefore in the past subjected to coercion culminating in slavery.

Climate and Colonization. It is increasingly clear that colonization, as distinct from occupation and exploitation, can only succeed when the climate of the colony is somewhat similar to that of the parent country. North Europeans succeed best in Canada, South Europeans in Brazil; Spaniards colonized the Argentine but Scots and Welshmen are needed in Patagonia and the bleak Falkland Islands. This is a general principle which has held true throughout history; the 'Barbarians' of the Steppes, accustomed to a continental climate, were rapidly converted to the ways of India on reaching the plains through the north-western gates and lost their individuality, just as the 'Shepherd Kings' soon became Egyptian in an Egyptian climate. The Roman Empire exceeded its climatic limits in the colonization of Britain and Germany, and was glad to retire, leaving little in language or literature to bear witness of its occupation. France, especially southern France, Italy and Spain, on the other hand, were climatically somewhat similar to Rome, and it was here that Roman culture left a lasting mark. Conversely the Teutonic peoples found a climatically suitable outlet for their colonizing activity in Britain, but were beyond their climatic range along the Mediterranean in Italy and Spain.

This aspect of colonization is today one of the most interesting and important applications of climatic study. Three well-known examples, briefly enumerated, will serve to illustrate this:

1. The climatic difficulties, particularly the impossibility of sustained manual labour in the moist heat of Queensland, which provide the most formidable obstacle to the 'White Australia' policy (see p. 13).

2. The moderation of an equatorial climate by altitude on the East African plateau, which has made possible the establishment of real colonies here (see p. 37).

3. India, with its excessive heat and moisture from the tropical monsoon which was always one of the weak spots in the British Empire, because it could not be settled by Europeans, but had to be ruled by a transitory autocracy of British officials whose real home was elsewhere.

Climate and Health. The governing and policing of dependencies and the prosecution of trade and industry call for the presence of white overseers in climates which are unsuitable for white occupation. Of such pioneers climate has taken a heavy toll of life and health, while even among the native population the death-rate in certain climates is extremely high. Climate was at one time accorded full blame for such appalling death-rates, but it is now clear that germ-borne and insect-borne diseases are potent influences, and that insanitary conditions and overcrowding of a native population, together with an ill-balanced dietary, are factors predisposing to ill-health. Consequently the tendency has been to rebound to the other extreme and to exonerate climate from all blame. But it is clear that by encouraging decay, by nourishing swarms of insect life, by lowering man's resistance to disease, and in a host of other ways climate is at the bottom of this ill-health; and it is clear that the maintenance of health can only be accomplished by an enormous expenditure of labour and money on hygienic precautions and the treatment of disease. There is, of course, another side to the picture, and certain climates possess undoubted pathological virtues. Mountain air, because of its clear, dry, rarefied nature, is frequently recommended for diseases of the lungs; desert air, equally clear and dry, but less rarefied and therefore not so trying to the heart, is beneficial for heart complaints. The tonic effect of a sea-voyage is not entirely imaginary, and sunlight, which may be sought with more chance of success in some climates than others, is credited with considerable medicinal virtues.

The Data of Climatology. It is a common definition of climatology that it deals with average weather conditions; and it is clear that averages, to be of value, must be based on careful observations over a considerable period of time. These are the raw materials of climatology, these observations of temperature, rainfall and other climatic elements which have been collected, mostly during the last hundred years, in nearly all the civilized and many other parts of the world. This miscellaneous collection of data is of very unequal value, especially to the geographer; many of them have little or no biological significance, while others, such as rainfall, temperature, humidity and sunshine, are geographical influences of the first magnitude.

The climatologist demands accuracy and reliability in climatic figures, a reliability which is only achieved by records over a long period of time; for this purpose thirty-five years is considered the minimum. In thirty-five years a station may be considered to have experienced all types of weather which are likely to occur there, and the mean of the observations may be taken to reflect average conditions. But owing to the incompleteness, as yet, of meteorological organization, there are huge areas on the face of the earth where records are not available for anything approaching this length of time, and in fact there remain wide expanses for which no data exist at all. Incomplete data, extending over a few years only, may serve a useful purpose as affording samples of the climate, though they should always be treated with caution, but where no data exist at all the geographer must be able, by applying the general principles of the science, to deduce the climate of the place from a variety of observed responses. The nature of the relief, the vegetation, the habits of the people, their architecture and their occupations will be attuned to the prevailing rainfall and temperature conditions, and it is the response rather than the cause which the geographer wishes to know.

Climatology and Geography. The attitude of the geographer herein differs somewhat from that of the meteorologist, for the latter will try to convert these observations into average figures for temperature, precipitation, etc., while the geographer is more usually concerned with the reverse process, the translation of average figures into certain biological responses. He will develop the habit of thinking in such terms as these, and a wet-bulb temperature of 80°, for example, will not be a figure merely, but will convey a picture of stifling heat with all work in the open air at a standstill.

Figures may be looked upon as a convenient shorthand method of describing climatic environment, and are, therefore, of incalculable value to meteorologist and geographer alike; but valuable as such average figures are, to rely too fully on them is to miss much that is important in climatic study, especially when climate is being considered as a geographical influence.

Accurate and detailed figures frequently have a real value, because they represent conditions which are critical, as, for example, the 15-inch isohyet in U.S.A. which in general defines the limit of cereal cultivations. But such values are not always of universal application; to attempt cereal cultivation with 15 inches of rain on the South African plateau would be to court disaster.

Climatology and Meteorology. Mean values, such as mean monthly temperatures and even mean daily maxima and minima, tend to obscure those extremes of heat and cold which may have such disastrous sequels for plants, animals and even man himself, especially when these are existing near the limits of their natural range. Hot and cold waves, thunderstorms and tornadoes, droughts and floods are events which may occur rarely in the annals of any area, but their effects are so far-reaching that they deserve careful mention among the more prosaic facts of mean temperatures, rainfall and humidities.

A very significant difference between the climatic belts which lie equatorwards and polewards of the mid-latitude high-pressure belt is to be sought in the striking differences in the nature of those two great air currents; the steady uneventful progression of the trades provides a striking contrast with the swirling, turbulent, eddying rush of the westerlies; climate is the control in the former, weather can wreak havoc in the latter.

There are, then, certain climates in which a proper appreciation of life conditions can only be attained by a study of weather types; not only the normal succession of everyday weather, but also those rarer, but more moving types which may be infrequent, but which may have consequences of vital importance. Moreover, in some climates certain weather types occur with such regularity as to be valuable elements of climate, whose frequency and distribution are worthy to be recorded in the same way as rainfall and sunshine amount. Yet in the description of the climate of a country, weather types are usually split up into their component parts and recorded as isolated data of each element. Weather types are the integrals which go to make up the climatic whole, and there is a danger of their losing their individuality unless climate is carefully examined, as it were microscopically, to appreciate its texture. Average figures create an illusion of steadiness and uniformity which is seldom justified by the facts; the study of weather types provides the corrective.

Need for Scientific Treatment. While it is true that climatology is primarily descriptive and lacks, or rather dispenses with, the careful analysis of causes which is essential in meteorology, there is much to be said for an analytical method of approaching the subject. Such a method has two practical recommendations: in the first place the understanding of a phenomenon is a valuable aid to memory and in the second it enables intelligent anticipation of similar results where similar causes are at work. The structure of the science being revealed, its outward form is the easier to remember, and the parts unseen may the more accurately be supplied from one's knowledge of the general plan.

Factors and Elements of Climate. The climate of a place is defined by a number of *elements*, or component parts, such as the temperature and humidity of the air, the rainfall, the wind velocity, the duration of sunshine and a host of others of less importance and less significance to man. These elements are the results of the interaction of a number of *factors*, or determining causes, such as latitude, altitude, wind direction, distance from the sea, relief, soil type, vegetation, etc. A clear distinction should be made between elements and factors, as defined above, and the use of the terms should be carefully restricted. The distinction is not always easy; for example, wind direction is a factor of great importance in determining climate, yet in some respects it is an element of climate; wind velocity is undoubtedly an element, yet may also act as a factor, for it may control precipitation by the rate at which it brings up supplies of moisture from the sea (see p. 146). The length of day, i.e. the duration for which the sun is above the horizon, is a factor since it helps to determine temperature; but the duration of actual sunshine is an element with far-reaching effects on plant and animal life.

Factors and elements of climate are clearly of two kinds, the first mathematically determined and therefore constant, the second variable and unreliable. Among the former is latitude, a factor which determines the length of day throughout the year, a mathematically controlled and therefore reliable element. Latitude also determines the intensity of insolation, and this, together with the length of day, influences the duration of sunshine and the temperature, elements which, however, being also influenced by other and variable factors, such as prevailing wind, marine influence, etc., are not constant. Ocean currents provide an example of the other group of variable factors which exert fluctuating influences on the temperature, rainfall and sunshine of adjacent coasts.

Technological Progress. The great strides now being made in science and technology have profoundly altered the impact of climate on our way of life. In the first place they have made possible a much fuller and more economical use of existing resources of climate in the heating and cooling of homes and in the protection of the body from the discomfort of extreme cold and heat. New varieties of crops can be grown in much better harmony with the climatic environment, which is now much more fully understood: climatic power services, such as hydro-electricity and solar energy can be used more efficiently and are gradually ousting the use of fossil fuels, such as coal and oil for certain purposes. On the other hand the grip of climatic control on man's activities has been loosened and the climatic environment can be by-passed by airways, by 'snow-cats', refrigeration, greenhouses, irrigation and by atomic energy. A stage has been reached at which climatic hazards exist only to be overcome. Yet the climate remains the same; apart from minor fluctuations all that has changed, and is rapidly changing, is man's use of it. We still need to know the climate before we can take full advantage of it or master the problems it presents.

SUGGESTIONS FOR FURTHER READING

For the relation of climate to health and human activities reference should be made to R. de C. Ward's *Climate, considered especially in Relation to Man*, 2nd ed., 1917; E. Huntington's *Civilization and Climate*, 3rd ed., 1924; and *The Human Habitat*, 1928, by the same author. See also G. Taylor, 'The Frontiers of Settlement in Australia', *Geog. Rev.*, 1926; G. T. Trewartha, 'Recent Thought on White Acclimatization in the Tropics', *Geog. Rev.*, 1926; S. F. Markham, *Climate and the Energy of Nations*; C. F. Mills, 1946, *Climate Makes the Man*. *Climate and Man* (the 1941 Year book of Agriculture), published by the U.S. Dept. of Agriculture, provides over 1,000 pages of stimulating essays on numerous aspects of applied climatology as well as very full statistical summaries of the climate of each State of the Union. In *Climate in Every Day Life*, C. E. P. Brooks describes a number of practical problems in applied climatology. The matter of bodily comfort has been the subject of research in the Services during the war; see 'Clothing for Global Man', *Geog. Rev.*, 1949 and 'Environmental Warmth and its Measurement', *Medical Research Council, War Memorandum*, No. 17, T. Bedford, H.M.S.O., 1946.

Some suggestions for future developments in Climatology are given in 'The Use and Misuse of Climatic Resources' by the author in his Presidential Address to Section E of the British Association (*Advancement of Science*, September 1956), and 'Climatology applied in the service of man' in the *Advancement of Science*, March 1957.

The publications of climatic consultants contain material of practical value to climatologists as well as to the farmer, e.g. The Drexel Institute, Laboratory of Climatology, Centerton, New Jersey.

II

THE ELEMENTS OF CLIMATE

Solar Radiation. Radiation from the sun consists of rays of three different natures according to their wavelength, heat rays, light rays and actinic rays, each of which, intercepted by solid bodies, produces its peculiar effects in varying degrees according to the nature of the surface on which it falls. The light rays are, of course, responsible for the phenomenon of daylight: light rays and actinic rays are necessary for the life processes of plants, but from a climatological point of view the heat rays are the most important of the three, and temperature is the most important manifestation of solar energy.

Temperatures in the Sun. Temperature figures quoted in climatology are generally 'shade' temperatures, i.e. the temperature of the air measured with due precautions taken to exclude the influence of the direct rays of the sun, but it is a common experience that it is much hotter in the sunshine, wherefore it is frequently of interest and importance to know the value of 'temperatures in the sun'. These are generally measured by means of a thermometer with the bulb coated with lamp-black mounted in a glass tube exhausted of air (the black bulb *in vacuo*), and the element which they measure is the intensity of the sun's radiant heat.

Many mountain resorts frequented by invalids have air temperatures often in the neighbourhood of zero in winter, yet the bright sunshine produces a feeling of warmth and comfort and allows light clothing to be worn without any sensation of cold. On days such as this the black bulb *in vacuo* may be recording over 100°, which indicates the power of the sun's rays. Apart from their warming capacity these rays have beneficial tonic effects, stimulating the system and imparting a general feeling of well-being.

Radiant energy, like all forms of vibration, can be reflected from solid surfaces and intensified, or rather waste can be diminished, by the reflection. Reflection from walls is a frequent device for the ripening of peaches and pears; reflection from bare ground also assists in the ripening of melons and other creeping plants; while reflection from water surfaces enhances the climatic reputation of water-side resorts and is a powerful agent in producing sunburn.

Shade Temperatures. Of all the meteorological elements there is none of such vital importance to living things as temperature, and there is none which exercises so profound a control over human distribution. Waterless deserts may be made habitable and productive by irrigation, but clothing and artificial heat are inadequate to overcome

8

the handicap of the low temperatures of the polar caps and high altitude, since these remain unproductive of vegetation. The temperature requirements of plants cannot be artificially supplied on a large scale as can their moisture requirements; man can convert tropical desert into tropical cultivation but he cannot change tropical cultivation to temperate, or vice versa. Temperature is therefore recognized as the element of chief classificatory value and its accurate measurement and statement is a matter requiring careful attention. By temperature at any given time is meant the temperature of the air measured under standardized conditions and with certain recognized precautions against errors introduced by radiation from the sun or other heated body.[1]

The mean daily temperature is, strictly speaking, the mean of twenty-four readings taken at hourly intervals throughout the day and night; but except where self-recording instruments are used twenty-four hourly observations are not usually available and the mean daily temperature is calculated from readings taken at certain times of the day, usually, morning, afternoon and evening. For this purpose various combinations are recommended,[2] such as 7 a.m. +2 p.m. +9 p.m. ÷3, 7 a.m. +2 p.m. +9 p.m. +9 p.m. ÷4, 6 a.m. +2 p.m. +10 p.m. ÷3, etc., all of which give a fairly satisfactory mean differing little from the mean of twenty-four hourly observations. Often the mean is calculated by dividing the sum of the maximum and minimum temperatures by 2, a practice which saves trouble as these thermometers need only be inspected once a day. This, however, gives a result which is usually rather high, but is good enough for generalizations.

The mean monthly temperature is the total of the daily means for the month divided by the number of days in the month.

The mean annual temperature should strictly be calculated from the daily means added together and divided by 365, but the usual practice is to divide the total of the monthly means by 12, which gives practically the same result. The mean annual can be considered reliable if based on thirty-five years' observations, but this is not always necessary; in fact, it is found that in equatorial climates the mean annual temperature based on two or three years' observations is reliable within a small margin of error. But in any case the figure is of little value outside equatorial climates, since the monthly temperatures all depart from it by considerable amounts. Extremes mutually cancel out to give a mean which may be the same for climates of very different type, e.g.:

	Hottest Month	Coldest Month	Mean Annual	Range
Peking	78·8	23·5	53·1	55·3
Scilly Isles	60·8	45·3	52·2	15·5

[1] See *Observer's Handbook*, published by the Meteorological Office.
[2] See A. McAdie, *Mean Temperatures and their Corrections in the United States*, Washington, 1891; and W. Ellis, *Q. J. Roy. Met. Soc.*, 1890.

The difference between the means of the warmest and coldest months is the *mean annual range*.[1]

The mean of the maxima recorded in a given period is the *mean maximum*[1] and the mean of the minima *the mean minimum*.[1]

The difference between the mean maximum and the mean minimum for the month is the *mean diurnal range* for that month.

The difference between the maxima and minima recorded in a particular month throughout the whole period of observation is the *absolute monthly range*.

The means of the highest and lowest temperatures which occur during any month of the year are the *mean monthly and mean annual extremes*.

The difference between the mean extremes of the hottest and coldest months is the *annual extreme range*.[1]

Diurnal and Annual Range of Temperature. The inclination of the axis of the earth's rotation to the plane of revolution round the sun is the cause of the seasonal contrasts, especially of temperature. At the equator there is no difference throughout the year in the length of daylight, but it increases to a summer day $13\frac{1}{2}$ hours long at the tropics and diminishes to $10\frac{1}{2}$ hours at midwinter (see p. 34). The consequence is, of course, the variation in length and in temperature of the seasons and of the length of day. The midnight sun can be seen on midsummer day at the polar circles. At the poles the daylight lasts for six months and the winter night for the other six; there is one maximum of insolation (at midsummer) and the winter night receives none at all, the temperature continuing to fall until just before the sun reappears skirting the horizon at the equinox and climbing spirally until midsummer. There is, of course, no vegetation at the poles, but within the arctic tundra lands there is a long summer and a long winter and the variation is seasonal, but within the tropics there is little difference between summer and winter; the night being the coldest period the variation is diurnal (see p. 275).

Accumulated Temperatures. Another aspect of temperature with well-marked botanical controls is the duration of temperatures above a certain minimum amount, e.g. the minimum for growth; 200 frostless days are required for the successful cultivation of cotton and 150 for maize. Experiments at Rothamsted[2] showed that 42° may be considered the basal temperature for wheat; each degree excess of the mean daily temperature over this critical value may be called a 'day degree', and if they are added together for the period of growth they give a measure of the accumulated temperature. The experiments showed that wheat required 1,960 day degrees between germination and ripening at Rothamsted, but in Canada the requirements are considerably less on account of the longer duration of daylight.[3]

[1] See Maps in Bartholomew's *Atlas of Meteorology*.
[2] Summarized by R. H. Curtis in Symon's *Meteorological Magazine*, 1905.
[3] J. F. Unstead, 'Climatic Limits of Wheat Cultivation', *Geog. Journ.*, 1912.

Isotherms. The distribution of temperatures can be expressed graphically by means of isotherms, lines joining places with the same temperature, but in the great majority of isothermal maps it must be borne in mind that the effect of altitude has been eliminated by reduction of all temperatures to corresponding sea-level temperatures by means of a formula. These maps therefore show the combined effects of the other factors, latitude, continentality, etc., and are of great value in the study of climate; but to obtain the approximate temperature of any place from the map an allowance must be made for its altitude. For biological distributions the actual temperature is of greater significance than the sea-level figure, which has only a theoretical value; but maps constructed on this basis are little more than relief maps and become extremely complicated and confused unless drawn on a large scale.

A much clearer conception of isothermal lines is obtained if they are considered as the lines of intersection of isothermal planes with the surface of the sphere. For example, the July isotherm of 80° passes near Mombasa and is met again at Cairo, while between the two is Port Sudan where occurs the isotherm of 90°. Somewhere about 3,000 feet above this latter station there must exist a surface with mean July temperatures of 80°, in fact the isothermal surface whose edges meet the ground at Cairo and Mombasa. This isothermal surface is continuous and could be flown along by an aeroplane taking off at Mombasa, passing over Port Sudan at a height of about 3,000 feet and landing at Cairo. About one-third of the way along this route the Abyssinian highlands rise into layers of air with mean July temperatures below 80° and thus intersect the isothermal surface, but since isotherms are reduced to sea-level values the 80° isotherm does not appear at this point on the map.

As will be shown later the decrease of temperature with altitude, although variable, is approximately 1°F. for each 330 feet of ascent, while the decrease with latitude is still more variable but, except in the vicinity of the Equator, averages about 1°F. for each degree of latitude (approximately 328,000 feet). The vertical decrease of temperature in mid-latitudes is therefore about 1,000 times that of the horizontal; that is to say, the inclination of the isothermal surfaces is, on an average, about 1 in 1,000.

Isanomalous Lines. Another interesting and informative method of representing temperature conditions is by the method of differences. The difference between the mean temperature of a place (reduced to sea-level) and the mean temperature along its latitude is its temperature anomaly, and lines joining places with equal and similar anomalies are known as isanomalous lines. They indicate with great clearness those places which are too warm for their latitude (having a positive anomaly) and those which are too cold (having a negative anomaly).

Sensible Temperature. The temperature recorded by the thermometer does not always agree with the sensations of heat recorded
by the human body. The sensation of heat depends on other air
conditions besides temperature, the chief of these being air movement

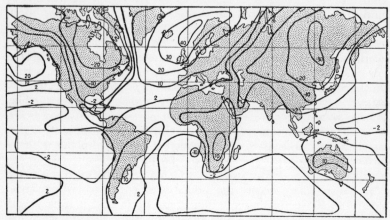

Fig. 1.—January Isanomalous Lines (Batchelder), differences in degrees Fahrenheit

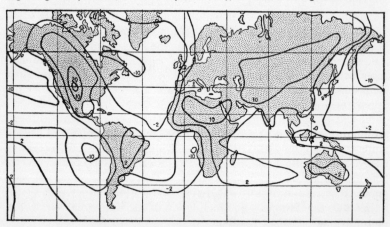

Fig. 2.—July Isanomalous Lines (Batchelder)

and humidity. For the human body is cooled by a double process of
radiation and evaporation (from the sweat glands), and any condition
which produces activity of either gives an impression of cold. Temperatures of 60° below zero are easily endured in Siberia where, under
the influence of the winter anticyclone, the air is practically still, but
blizzards with temperatures of 60° higher than this are insufferable.
On the other hand, the comfort of a cooling breeze on a hot day is
universally recognized, it accelerates both radiation and evaporation

from the skin. Again, 80° in the equatorial zone is distinctly more uncomfortable than 100° in the desert because of the humidity, while at the other extreme the dry cold of continental interiors is less trying than the raw cold of moister climates. Both excessive heat and excessive cold are, in fact, more trying in wet than in dry air, for the dry air mitigates heat by accelerating evaporation from the skin, while the damp air, being a better conductor of heat than the dry, allows escape of heat from the body in cold weather. These factors, it is true, work in opposite directions, but evaporation is more important than

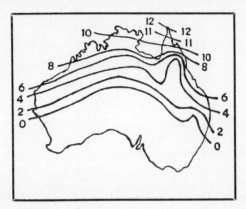

Fig. 3.—Climatic Control of Settlement (Griffith Taylor)

conduction at high temperatures, while the reverse obtains at low temperatures and evaporation practically ceases.

In moist climates the blood becomes diluted because evaporation is checked, and conversely in dry climates (i.e. deserts and mountains) the blood is more concentrated. These physiological conditions have interesting psychological sequels; the consequences of wet climates are nervous depression and lethargy, while dry climates breed nervous energy, excitability and sleeplessness; these characters are very pronounced in visitors before acclimatization has been attained.

Wet-Bulb Temperatures. The combination of heat and high humidity is one which the human body is least able to endure and is a very important control of human activity. A measure of this combination is given by the wet-bulb thermometer, which, like the human body, records the value of air temperature, decreased by evaporation from a moist surface. Wet-bulb temperatures of 75° or 80° conduce to heat-stroke, although dry-bulb temperatures of 90° or 100° can be safely endured. Dr. Griffith Taylor[1] has shown that continued wet-bulb temperatures of 70° place a limit on white colonization by rendering impossible sustained manual labour and by prohibiting even

[1] 'Control of Settlement by Temperature, etc.', 1916, *Weather Bulletin*, **14,** Melbourne.

sedentary and domestic occupations from being followed with health and comfort.

In Fig. 3 isopleths are plotted to show the number of months with mean wet-bulb temperatures above this amount. The coastal regions north of the tropic are clearly unfavourable for white settlement.

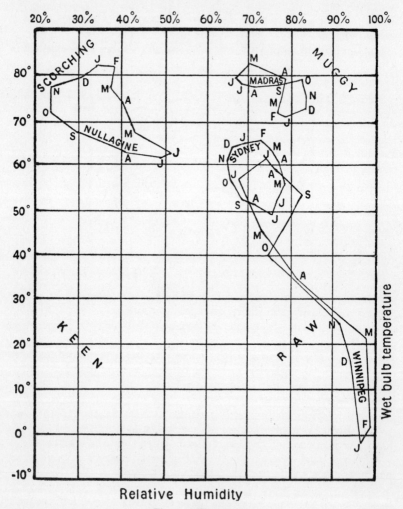

Fig. 4.—Climographs
(*After* Griffith Taylor)

Climographs and Homoclimes. The same author has described an interesting graphical method of depicting climates which emphasize these physiological reactions. With wet-bulb temperatures as ordinates

and relative humidity as abscissae the twelve monthly figures are
plotted, giving a twelve-sided polygon (the climograph). According
to the position of the polygon on the graph the nature of the climate
is seen at a glance. The N.E. corner with high wet-bulb temperatures
and high relative humidity is described as 'muggy'; the N.W. corner
with low relative humidity but high wet-bulb temperatures as 'scorching', the S.W. as 'keen', and the S.E. as 'raw'. Climographs for four
contrasted stations are shown in Fig. 4.

Stations having similar climographs are described as 'homoclimes'.
Alice Springs is the homoclime of Biskra (Algeria), Perth of Cape
Town, Brisbane of Durban, etc.

Humidity. The depression of the wet-bulb temperature below that
of the dry-bulb provides one of the means employed for estimating
the humidity of the air,[1] i.e. the water vapour content, a most important
element, both because of its biological effects, described above, and
its influence on rainfall. But its value is, in many climates, subject
to rapid fluctuations under the influence of wind direction, especially
in relation to moisture supply, and this makes average figures of little
significance. Maps of the distribution of relative humidity are therefore
seldom used, but the resultants, namely, cloudiness and rainfall, are
shown in all atlases of meteorology.

Rainfall. After temperature, rainfall is the most important of the
climatic elements. The agricultural or pastoral utilization of the land
is the only real and lasting source of wealth, and both of these are in a
large measure dependent on rainfall. The yield of wheat, sugar, maize
and numerous other crops varies with the rainfall of the year in such
a way as to leave no doubt that the rainfall has been the real determinant, and the carrying power of grazing land in head of stock per
square mile emphasizes the same truth. As early as 1874, Sir W.
Rawson[2] showed that the yield of sugar in Barbados was intimately
related to the rainfall of the preceding year, and was even able to forecast the total production of the island, whose sugar acreage had been
constant for many years, by means of a formula, viz. that every inch
of rain in the year would result in a yield of 800 hogsheads of sugar
from the whole island in the following year. Sir Napier Shaw[3] devised
a formula by which the wheat yield in England could be forecast from
the rainfall of the preceding autumn as follows: yield $= 39.5$ bushels
per acre $- \frac{3}{4}$ (previous autumn rainfall in inches). From which it
appears that the autumn rainfall is excessive in the British Isles, and
a reduced autumn rainfall augurs an increased yield. In India the
opposite obtains, rainfall during germination is deficient and an

[1] For the method, see *Observer's Handbook.*
[2] *Reports upon Rainfall of Barbados and upon its Influence on the Sugar Crops,* 1847–1871,
by Governor Rawson, C.B., Barbados, 1874.
[3] *Proc. Roy. Soc.* (A), 1905.

increased fall results in an increased yield. Here, however, such factors as the moisture storage capacity of the soil and the evaporation must be allowed for. Wallen[1] has carefully worked out the relationship between rainfall and the yield of cereals (wheat, barley, oats and rye) in Sweden.

Effective Rainfall. Run-off and Evaporation. Rainfall amount is stated in inches or millimetres (1 mm. $= \frac{1}{25}$ inch) for each month of the year and includes the total depth of water resulting from all forms of condensation, whether rain, dew, fog, frost, hail or snow. It should be borne in mind that the months are not of equal duration; thus, unless allowance is made by weighting the monthly totals, January, with 31 days, may be expected to have a rainfall some 10 per cent higher than February with 28 days, other things being equal. But in addition to the monthly and annual totals some information is desirable on the nature of the rainfall, its persistence and its intensity.

This is provided by data concerning:

1. The number of rainy days (a rainy day is one on which an appreciable amount of rain, say 0·01 inch, falls). The average rainfall per day can, of course, be obtained by dividing the mean annual rainfall by the number of rainy days; it gives a valuable measure of the intensity of the rain.

2. Maximum precipitation per day, per hour, or for shorter periods. In general the rainfall of low latitudes is much more torrential than that of high, but over a short period of time the temperate zones can show an intensity almost equal to that of the tropics. $1\frac{1}{4}$ inches fell in five minutes at Preston on 10th August, 1893, and nearly 4 inches in an hour at Maidenhead on 12th July, 1901; but such 'cloudbursts' rapidly exhaust the available water vapour and are short-lived. In the tropics, on the other hand, where the warmer air holds infinitely greater stores of moisture, these intensities can be maintained for hours on end; 30 inches in 24 hours is by no means uncommon from typhoons, 40 inches has been recorded from many stations, and Baguio, in the Philippines, has recorded 45 inches.

The intensity affects run-off and evaporation, thus qualifying the effectiveness of the rain. For example, 30 inches of rain at Pretoria[2] is inadequate for agriculture and the country round is chiefly engaged in pastoral pursuits. The explanation is supplied by the nature of the rain, which occurs generally in the form of heavy downpours compacting the surface soil into a relatively impervious layer over which the rain runs without penetrating; and under the cloudless blue skies, which are such a typical feature of the climate, evaporation is extremely rapid.

The rate of evaporation depends primarily on the dryness of the

[1] 'Sur la Corrélation entre les récoltes et les variations de la température et de l'eau tombée en Suède', Stockholm, K. Svenska, *Vet. Ak. Handl.*, 57, No. 8, 1917.

[2] *Rainfall and Farming in Transvaal*, F. E. Plummer and H. D. Leppan, Pretoria, 1927.

air, but it is also affected by numerous other influences such as air movement and plant cover. Steady winds, by importing fresh supplies of air before saturation is reached, enormously accelerate the process, especially if the winds are coming from an arid area, e.g. the sirocco coming off the Sahara, or if it is descending and therefore being adiabatically warmed, e.g. the Foehn and Chinook winds (see pp. 268 and 235).

The nature of the ground surface, too, has an effect on the rate of evaporation. A wet soil surface in the Transvaal, because of its irregular surface and consequent large area exposed,[1] has been shown to evaporate 4·75 inches a week compared with only 1·88 inches from a free water surface. Plant cover also checks the direct loss by evaporation, and a loss of 52 per cent was found[2] on a summer tilled plot as compared with only 14 per cent on land under grass sod. On the other hand, the plant cover causes a loss by transpiration from the leaf surface which generally exceeds the loss by evaporation; in the case of forests this loss, or use, by transpiration is enormous.

Evapotranspiration. The importance of evapotranspiration was introduced by Thornthwaite in his proposed classification of climates. Though it is difficult to express accurately there is no doubt that air temperature, especially the daily maximum, is one of the conditions that affects the availability of rain and its effects on soil moisture. It has been shown, especially by H. L. Penman, that the east of England is liable to temporary drought when the level of water in the soil falls and checks the growth of crops; a device has been suggested for calculating the supplemental irrigation required to maintain maximum yield. The extra evaporation in the heat of summer is responsible for the fall of rivers even if summer is the season of heaviest rainfall. The discharge of the River Thames falls from an average of 3,910 cubic feet per second in winter, to 1,823 in summer. The rainfall in the Thames basin (above Teddington) averages about 16 inches from October to March and the run-off (2,102 million gallons a day) averages nearly 7 inches (44 per cent) but from April to September the rainfall is nearly 14 inches, but the run-off (980 million gallons a day) is the equivalent of only 3·2 inches (about 23 per cent).

Seasonal Distribution of Rain. It has been shown that effective rainfall differs materially from the recorded amount because of evaporation and run-off. Seasonal régime is a further qualification which profoundly affects its utility and which therefore deserves careful attention in climatic descriptions. The seasonal incidence may render much of the rainfall useless and wasteful as in parts of Bombay which receives the unnecessary amount of 75 inches in four months (June to September) and has drought throughout the rest of the year; or scanty rain may be concentrated into a short growing season so that optimum

[1] *Op. supra cit.*   [2] *Op. supra cit.*

utilization is obtained. For example, wheat is grown in parts of Western Australia with less than 10 inches of rain, but only because it comes at just the right time; elsewhere 30 inches may not be enough.

Rainfall Reliability. Another feature of great significance, especially in areas where the yearly or seasonal mean is barely adequate, is the reliability of the rain, generally expressed as a mean or extreme percentage departure from the normal (see pp. 137 and 167). Regions with high variability will be victims of drought in bad years and their crop production will fluctuate widely. In pastoral areas the results are still more serious, for flocks and herds are sadly depleted by the failure of the pastures and the process of recovery may take years.

A reliable dry period at harvest time is a valuable asset, in which respect the Mediterranean climates are especially fortunate, while the unreliability of such a period in Norway reduces the harvest to a gamble. By restricting crop rotation and by compelling seasonal fallow, the seasonal incidence of rain enforces a seasonal régime of labour on the farm.

Types of Rainfall and their Seasonal Distribution. If the different types of rainfall are examined it will be found that each tends to have maxima and minima at certain fairly definite seasons.

Relief rain will occur chiefly when the supply of moisture is greatest. This may be: (1) when the seas are warm and the land is growing cooler, i.e. in autumn or early winter in intermediate zones. Seathwaite in the Lake District receives 41 inches out of 130 inches (32 per cent) in October, November and December, Ben Nevis 52 inches out of 171 inches (34 per cent) in the same three months.

Or (2) on the arrival of a strong off-sea wind as in monsoon climates. Calicut has 90 inches out of 119 inches (75 per cent) in June, July and August; Cape York (Queensland) has 58 inches out of 82 inches (71 per cent) in January, February and March.

Convectional rainfall will tend to occur:

(1) When the annual swing of the climatic belts brings the region under the influence of the doldrum zone of ascending currents of air: e.g. a summer maximum of rain, as in tropical climates—Timbuktu has 7 inches out of 9 inches (78 per cent) in June, July and August.

(2) When high temperatures set up convectional currents, often initiating thunderstorms, i.e. again a summer maximum, as in continental interiors: e.g. Moscow has 8 inches out of 21 inches (38 per cent) in June, July and August.

Cyclonic rain will tend to occur when cyclonic activity is greatest, i.e.:

(1) When the swing of the climatic belts brings the region into the belt of the stormy westerlies, i.e. in winter on the equatorial side of this belt: e.g. Algiers has 14 inches out of 30 inches (47 per cent) in November, December and January.

(2) When the shrinkage of the polar continental air masses in summer permits air of maritime origin to invade regions which are firmly under the influence of continental air in winter, or when marked temperature differences exist between arctic and continental air, giving rise to frontogenesis. In both cases a late summer maximum occurs: e.g. Hebron has 9 inches out of 19·8 inches (46 per cent) in July, August and September.

Isohyets. The cartographic representation of rainfall amount is by means of lines drawn through points having equal rainfall during any given period and known as isohyets. Unlike isothermal and isobaric maps these require no correction for altitude, indeed such would be impossible as the relation between rainfall and altitude is subject to no fixed laws (see p. 36). Higher land generally has a heavier rainfall than lower levels, especially where the precipitation is in the form of relief rain, and to some extent isohyetal maps function as relief maps, but the influence is not sufficient to obscure the other controls of rainfall.

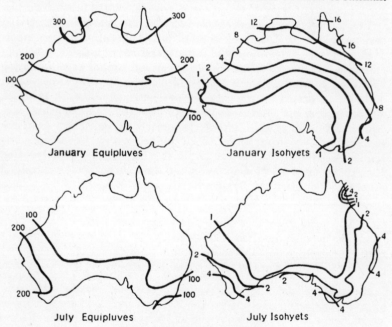

Fig. 5.—Equipluves and Isohyets

Equipluves.[1] The method of differences (cf. p. 11) may be conveniently applied to the study of rainfall, giving an informative impression of the raininess of a particular month, or other period. The amount of rainfall which would be precipitated at a given place, assuming that

[1] For a full description of this method, see B. C. Wallis, 'Geographical Aspects of Climatological Investigation', *Scot. Geog. Mag.*, 1914.

such rain were perfectly evenly distributed through the year, is taken as
a 'norm,' and the actual mean amount of rain is stated as a percentage
of this (the pluviometric coefficient). Lines joining places with equal
pluviometric coefficients, and therefore with equal and similar departures from the norm, are known as equipluves.

Rainfall amount and seasonal régime may be instructively presented
together by a combination of yearly isohyets with graphs of the monthly
values placed on the site of the station (see Fig. 5).

Snow. Snowfall is included in rainfall figures, but its measurement
presents considerable difficulties.[1] It is sometimes stated that 10 inches
of snow are equivalent to 1 inch of rain, but the nature of the snow,
the size of the flakes (largely dependent on temperature), the degree
of compactness, etc., introduce considerable errors, and it is to be
recommended that the snow be either weighed directly or melted and
the resultant water measured. But when these precautions are taken
the result is not necessarily accurate, as a still greater difficulty is
experienced in making a correct catch, especially if the snow is drifting.

The ways in which snow affects everyday life are so numerous and
so far-reaching that data of snowfall are of the utmost importance.
It interferes with the working of railway and road transport, involving
enormous yearly expenditure in clearing, and capital expense during
construction for snow-sheds and snow-ploughs. On the other hand,
the snow provides a natural highway for sledges, on which, for example,
the success of the lumbering industry in eastern Canada and elsewhere largely depends. It protects underlying vegetation from the
ill-effects of severe frost; winter wheat can be grown in Ontario because
snow comes before frost (see p. 50); its late melting may delay sowing;
it holds up the winter precipitation and then discharges it suddenly in
what may be disastrous floods when the thaw comes.

For reasons such as these it is of interest to know:

1. The average number of days on which snow falls.
2. The average duration of snow cover.
3. The average depth of snow cover.
4. The average dates of the first and last snows.

Dew, Mist and Fog. Condensation of water vapour may occur
without necessarily resulting in rain; fog, mist and cloud are such forms
of condensation whose frequency and seasonal distribution deserve
mention in the climatic data of a station. They make very little
difference to the recorded precipitation, since they are very small in
amount and much of this is lost by evaporation, but they have quite
an appreciable effect on vegetation in certain climates. The regular
dense fogs of the Kalahari coastal strip nourish a scanty vegetation,

[1] For a discussion of these (with references) see R. de C. Ward's *Climate of the
United States*, pp. 234-6.

and the dew and mist of the cold season in the Central Provinces of India play an important role in nourishing the wheat crop of this great producing area.

Mist or fog will occur wherever air in contact with the ground is cooled below the dew-point. This may happen in one of two ways:

1. The warm moist air may be cooled *in situ* (radiation fog) either from direct loss of heat by radiation, such as commonly occurs at night under anticyclonic conditions, or from sinking of cold air into valley bottoms, chilling the moist air which lies there, especially near the stream.

2. The moist air may be cooled by drifting into contact with a chilling surface (advection fog) as when a current of air passes from a warm sea on to cold land, or from a warm current to a cold (e.g. along the 'Cold Wall' where the Gulf Stream meets the Labrador current), or in front of a feeble depression where warm air drifts over colder.

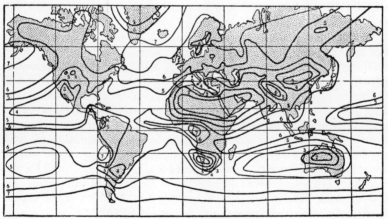

Fig. 6.—Mean Annual Cloudiness
Figures indicate tenths of the sky covered

Cloudiness and Sunshine. Above ground-level the cooling of air below dew-point results in clouds, which to a greater or less extent obscure the sun. Cloudiness is stated in percentage (or more usually tenths) of the sky covered with clouds. Stations with equal cloud amount may be plotted on a map and joined by lines called *Isonephs*. The map of mean annual cloudiness (Fig. 6) shows two belts of high cloud amount corresponding to the equatorial and circumpolar low-pressure belts. Within the zone of the trade winds the areas of clear sky are strikingly extended westward of the deserts as the dry air is carried seawards. Westwards from the desert shores of Australia and

California these are shown extending through some 80° of longitude. In fact, the only places in this belt where high cloud amounts are recorded is where these winds blow on-shore on the eastern margins of continents. There is, however, an interruption of this general cloudlessness in the China Seas, the result of the monsoonal interruption of the trades. In addition to the mean annual cloudiness the monthly and daily distributions of cloud are often important. There may be high cloudiness in the morning, disappearing as the day advances, or clear mornings may turn cloudy later on the arrival, for example, of a sea-breeze.

The duration of sunshine (shown on maps by lines of equal duration known as *isohels*) is not the converse of cloudiness, for the possible hours of sunshine in each month vary according to latitude and, to a less extent, according to altitude. Thus within the Arctic Circle on 21st June the sun is above the horizon for twenty-four hours, twice as long as at the Equator; the Pole could therefore record the same sunshine hours as the maximum at the Equator though the sky was covered with cloud for half the day. The nature of the cloud (whether cirrus, cumulus or stratus) and its position in the sky also affect the sunshine hours.

Cloudiness is one of the characteristics of marine climates and it is one of the means by which, in these climates, temperatures are kept uniform; for clouds check solar radiation by day and terrestrial radiation by night. They afford protection from the sun, which is sometimes an advantage but more often the reverse. Coffee and certain kinds of tobacco are crops requiring shade, which often has to be artificially supplied at considerable expense. On the other hand, most fruits require direct sunlight for ripening; South Africa, California and the Mediterranean lands, the chief fruit-growing areas in the world, are some of the sunniest. In the British Isles, Kent, Norfolk and Devon have the highest sunshine hours and produce the most fruit, while the fruit-growing district of Evesham benefits from the loop of the 1,400-hours isohel which spreads up the Severn-Avon valley.

Pressure and Winds. Pressure only becomes an element of climate at high altitudes when it is decreased sufficiently to produce physiological effects (see p. 36), but as a factor it is directly responsible for winds and storms which are elements of great importance. To separate those circumstances in which wind functions as an element from those in which it is more properly considered as a factor would lead to much repetition and inconvenience, and the whole subject of pressure, wind and atmospheric circulation will be discussed here. The effect of winds on sensible temperature has already been discussed, they cool the body both by conduction and evaporation; but while windy climates are generally more stimulating to animals and man than those with persistent calms, the winds are often injurious to plants on account of the accelerated evaporation; a windy day can be as desiccating as a hot day. Winds are more constant and their velocity is greater over the sea than on land, since free air movement is less interfered with; even the trade winds are not entirely reliable on land, especially on land of considerable relief. The nearest approach to the conditions at sea is found on flat, featureless plains such as the prairies and the pampas where the strength and reliability of the winds have been turned to account in the use of numerous windmills for pumping water and for other purposes. Windmills are familiar features of the landscape in the flat lands of Holland and the Norfolk Broads.

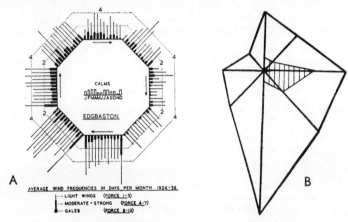

Fig. 7.—Wind Roses for Edgbaston (monthly) and Mendoza (yearly)
The shaded inner rose at Mendoza shows percentage of thunderstorms according to wind direction

The measurement of wind velocity has not received the same attention as that of other elements. The anemometer, by means of which wind velocity can be measured and expressed (usually in miles per hour or metres per second), is not altogether satisfactory, since its exposure, in spite of precautions, is nearly always subjected to eddies and gusts which introduce errors. But at many stations wind velocity is merely estimated by its effect on smoke, trees, etc., and stated according to Beaufort's Scale.[1]

Wind direction can be represented diagrammatically by 'wind roses', in which the percentage frequency of winds from each point of the compass is indicated by the length of a line drawn in that direction radially from a centre. They may be instructively combined with the frequency of other elements, and in this way the dependence of such elements as rainfall on wind direction may be demonstrated (see Fig. 7).

Circulation of the Atmosphere. The broad essentials of the atmospheric circulation in each hemisphere are the three belts of trade

[1] See *Observer's Handbook.*

winds, the westerlies and the polar winds; but it should be borne in mind that these only represent the surface winds, often only shallow currents, and that the direction is generally different and the force always greater at higher altitudes. The importance of this will doubtless be great in an era of commercial flying, since, by careful selection of flying level, favourable winds may often be found in both directions along a route. The important air route to Australia, for example, passes over Queensland where strong and steady trade winds, i.e. following winds on the northward journey, blow for months at a time.

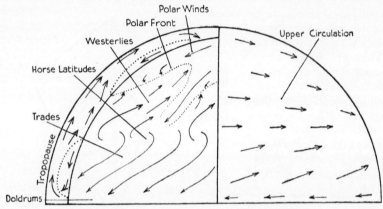

Fig. 8.—Generalized Circulation of the Atmosphere

But above these is a N.W. return current, i.e. a following wind on the southward journey, which occurs at 12,000 feet near the tropic but is found at lower and lower levels polewards until it is only 4,000 feet up at Melbourne. The existence of a return current of air, flowing from S.W. to N.E. above the northern hemisphere trades had long been recognized from the drift of high clouds and the movements of dust thrown to considerable heights by volcanic eruptions. It now appears that this eastward drift is universal at least as far polewards as the areas of permanent low pressure where the polar winds and the westerlies meet, and probably as far as the poles themselves, but the return current above the westerlies is apparently directed towards the south of east so that there occurs an accumulation of air over the Horse Latitude highs which finds relief by descent to renew the surface trades and westerlies. The general circulation of the troposphere therefore appears to be as shown in Fig. 8.

The 'Horse Latitude' Front of Divergence. The belt of high pressure known as the 'Horse Latitudes' from which diverge, at the earth's surface, the trade winds and the westerlies, is a front of divergence, and the two currents, being divergent, do not interfere. It is a zone of calms, a zone of stability, a region of descending, and there-fore dry, air currents. On land it is a belt of deserts, widest in Africa and Asia where there is most land, scarcely existent in North America where the place of the Sahara is taken by the Gulf of Mexico, i.e. where a great desiccating agent is replaced by one of the greatest sources of atmospheric moisture.

This zone of high pressure is not, however, continuous thoughout the year. During the summer in each hemisphere the belt is interrupted by the low pressures which develop over continents, but during winter it again becomes continuous, the pressure over continents being now even higher than over the seas. Five areas remain, however, permanent centres of high pressure, isolated during summer into independent anticyclones; two are in the northern, three in the southern, hemisphere. They occur over the oceans whose relatively low temperatures in summer maintain heavy air above them. These permanent 'centres of action' are of vital concern to the climates of the surrounding lands, and, combined with the waxing and waning of the continental pressure systems, are responsible for the seasonal changes in wind direction.

The Equatorial Front of Convergence. The trade winds origi-nating in these high-pressure areas of the northern and southern hemispheres meet in the equatorial belt of low pressure, the 'Doldrum' belt, which is directly due to the intense heating which it receives. This belt of calms is a front of convergence where two air currents meet and escape by upward movement, and this causes it to be turbu-lent and stormy in great contrast with the Horse Latitude calms. But the air currents which meet here are of similar temperature and humidity, and thus it is less unreliable and less squally than the other front of convergence, which exists between the westerlies and the polar winds. Since seasonal temperature changes in these low latitudes are very slight, there are virtually no seasonal pressure changes; low pressure is continuous round the globe throughout the year.

The Swing of the Pressure Belts. The primary cause of the pressure belts is the unequal distribution of insolation; the low pressure of the equatorial belt results from the excessive insolation received and tends to coincide with the heat equator. But the position of the heat equator is subject to a seasonal migration, lagging somewhat behind the zenithal position of the sun, and the low-pressure belt also swings north and south, lagging somewhat behind the heat equator. The other pressure belts, intimately connected with this, move north and south also, so that certain areas lying at or about the junction of two zones experience a season in each; the essence of the tropical climates, for example, is its alternation of equatorial calms and trade wind condi-tions. On land the wind belts are interrupted to such an extent and so modified by continental influences that they are often rendered entirely unrecognizable, but except in the Indian and West Pacific

Oceans, where they are displaced by monsoons, they retain their characteristics over the sea. Their limits here can be traced with some degree of reliability and are stated in the table below. The lag of the wind belts behind the sun is such that the limits of the migrations are not reached until two or three months after the solstices.

		March		September	
		Atlantic	Pacific	Atlantic	Pacific
N.E. Trades {	Northern limit	26° N.	25° N.	35° N.	30° N.
	Southern limit }	3° N.	5° N.	11° N.	10° N.
Equatorial {	Northern limit }				
Calms {	Southern limit }	0°	3° N.	3° N.	7° N.
S.E. Trades {	Northern limit }				
	Southern limit	25° S.	28° S.	25° S.	20° S.

It will be noticed that all the belts are north of their natural position, a fact which is explained by the greater amount of land in the northern hemisphere, and whose consequences are far-reaching. It results in warm air and water (from currents) finding their way across the Equator into the northern hemisphere, thus increasing its temperature at the expense of the southern, and strengthening the Gulf Stream and the Kuro Siwo at the expense of the Brazilian and East Australian currents (see p. 45). Other important consequences of this will be described later, especially in connection with the climates of West Africa and Central America.

The Polar Anticyclones. The permanent ice caps of the polar regions tend to maintain high pressure in these areas from which winds are radially dispersed. The conditions are simplest in Antarctica where a powerful katabatic effect is added by the high plateau nature of the continent, giving rise to extremely unpleasant high winds in what Sir Douglas Mawson described as the 'Home of the Blizzard'. In the north polar regions matters are much complicated by the existence in high latitudes of the high plateau of Greenland and the great continental anticyclones of Siberia and Canada grouped around a polar ocean. These two great high-pressure areas seem to be joined by a ridge of high pressure almost across the pole, so that the dispersal of cold air, though extremely irregular, is, in broad outline, outwards from the polar region. These polar highs are quite shallow systems, and seem to be replaced above about 6,000 feet by enormous low-pressure areas which are the foci of the upper air circulation as shown in Fig. 8.

The Polar Front of Convergence. By contrast with the equatorial front there meet, along this line, air currents of very different natures,

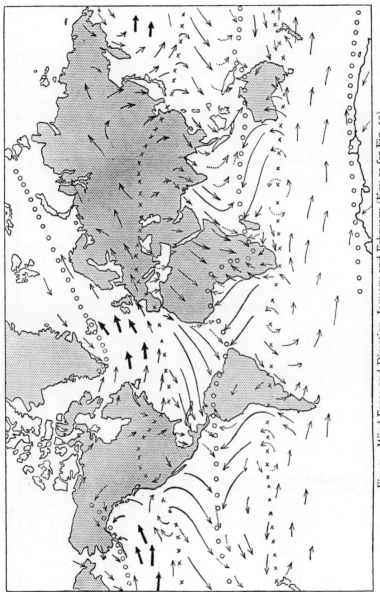

Fig. 9.—Wind Force and Direction, January and February (Key as for Fig. 10)

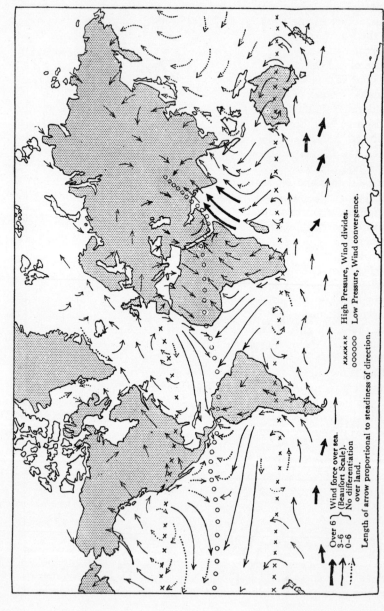

Fig. 10.—Wind Force and Direction, July and August

the warm wet westerlies of equatorial origin and the cold heavy air gravitating out from the poles. Along the plane of contact eddies are set up, and patches of the warmer, lighter westerly circulation penetrate into the heavier polar air and so rise swirling upwards. Thus are born the greater number of cyclonic storms which form such a characteristic feature of the climates of the 'temperate' zones.

Owing to the constant procession of cyclones, the polar front is normally a zone of low pressure, but being in a zone subject to a considerable annual temperature range, it is interrupted in the northern hemisphere in winter by the growth of the continental highs of Eurasia and North America. Thus there are left two areas of low pressure in winter, the Icelandic and the Aleutian, which owe their existence to the pronounced temperature contrasts between the warm waters of the North Atlantic and North Pacific Drifts and the cold land surfaces of Greenland-Iceland and Alaska-Siberia. Now the Greenland ice cap is permanent, enduring through the summer, consequently the temperature contrast endures and the Icelandic low persists throughout summer as a still lower pressure in a low-pressure belt. But Alaska and Siberia have no permanent snow cover and temperatures here rise considerably in summer; the temperature contrast therefore ceases to exist and the Aleutian low is overshadowed by the extension of the continental lows of North America and Siberia, and ceases to have a separate existence.

In the southern hemisphere there is practically no land in these latitudes and the low-pressure belt is permanent, continuous and uniform.

The Wind Belts. The diagrammatic representation of the atmospheric circulation given in Fig. 8 ignores the influence of land masses and the individual pressure systems to which they give rise. It has already been shown that these interrupt the continuity of the pressure belts and it must be obvious that they complicate the theoretical wind 'belts' to such an extent that they are often quite unrecognizable. Figs. 9 and 10 show, in a generalized way, the mean positions of the pressure systems and the resultant mean wind directions for January and July, which are now seen to take the form of a series of swirls round individual highs and lows. Only over mid-oceans are the 'belts' recognizable as such; where sea meets land the winds swing round towards a meridional direction, especially in the region of the horse latitude highs. Thus the 'Mediterranean climates' which are described in many text-books as 'coming under the trade winds in summer' are seen to have prevailing north-westerly or northerly winds in the Mediterranean and California, and south-westerly or southerly winds in Chile, Cape Province and Swanland.

The Trade Winds. These are supplied mainly from the eastern and equatorward sides of the sub-tropical highs, from whence a steady stream of air pours towards the equatorial low. Over the oceans they live up to their name and are the world's most constant winds in force and direction; 70 to 80 per cent of occasions show wind from S.S.E., S.E., or E.S.E. (in the southern hemisphere), the force being generally from 10 to 20 m.p.h., and in some places shown on the map (Fig. 9) 15 to 25 m.p.h.; calms are rare. Their deflection from the N.E. or S.E. direction by the land masses should be noted. In the Indian Ocean and among the islands of the south-west Pacific, they are reversed in the summer hemisphere by the monsoons.

The Westerlies. These constitute the air flow from the sub-tropical highs to the sub-arctic lows, but they are strongly contrasted with the trades in regularity. On the whole their force is much greater; winds of gale force are quite frequent, but on the other hand they may dwindle away or even be reversed in direction. Some idea of the extent of their variability can be obtained from the following data of percentage frequency of wind direction at Valentia (Ireland):

	N.	N.E.	E.	S.E.	S.	S.W.	W.	N.W.	Calm
January .	6	6	10	13	21	16	12	7	9
February .	6	9	9	13	19	14	11	7	12
March . .	9	10	8	8	15	15	11	10	14
April . .	11	10	8	9	15	13	12	9	13
May . .	13	7	7	10	15	12	10	11	15
June . .	12	5	5	10	15	13	13	12	15
July . .	11	3	2	6	15	16	16	17	14
August . .	8	4	3	8	17	16	15	13	16
September .	9	6	8	11	15	11	12	10	18
October .	9	10	9	10	14	11	12	9	15
November .	8	9	10	10	15	14	13	9	12
December .	6	5	9	11	19	17	13	9	11

Whereas the goal of the trades is a more or less continuous belt of low pressure, the goal of the westerlies at any particular moment is a travelling low-pressure system; the wind direction and force thus vary from day to day and from hour to hour, according to the position of the low-pressure centre at the time. In the northern hemisphere they are most constant in summer when the continents have developed their own low-pressure systems which give the sub-arctic lows their nearest approximation to a belt form. They are most variable in winter when the continents have become high-pressure areas and the pressure gradient between these and the Icelandic and Aleutian lows is at its greatest. At this season travelling depressions are most numerous and most intense. Winds approaching gale force are usual and winds exceeding gale force are then not uncommon, but there is no certainty that they will be from the S.W.; in fact the worst gales are often from a point north of west in rear of depressions.

In the southern hemisphere the almost continuous water in these latitudes encourages a more regular flow and a higher average wind velocity in the 'roaring forties'. But here also these are day-to-day variations introduced by travelling centres of very low pressure. The meteorological phenomena in these belts are extremely complicated; they are dealt with more fully in Chapter IV on Air Masses.

The limits of these winds, and of the meteorological phenomena associated with them, swing north and south with the seasons; no precise latitudinal limits can be set down; but reference should be made to Figs. 9 and 10, and to the chapters on warm temperate climates.

Polar Winds. These consist of the cold air travelling outwards from the shallow polar anticyclones. In the southern hemisphere the trough of lowest pressure is found at about the Antarctic Circle, and this trough becomes the goal of strong winds blowing off the high plateau of Antarctica. Southerly winds prevail from here to the pole and most expeditions have reported them to be monotonously strong; blizzards, except in a few localities, favourably situated with regard to topography, are unpleasantly frequent.

In the Arctic regions the pressure distribution is highly complicated by the presence of large land masses surrounding the Arctic Ocean and the outflow of polar air is similarly complicated. In view of the important influence of this air on the weather of high and intermediate latitudes the system merits fuller consideration, which it receives in Chapter IV on Air Masses.

SUGGESTIONS FOR FURTHER READING

For an account of the elements of climate the standard work is J. Hann' *Handbook of Climatology*, vol. 1, translated by Ward, 1903. A. J. Herbertson's *The Distribution of Rainfall over the Land*, London, 1901, is still valuable; also M. Jefferson, 'A New Map of World Rainfall', *Geog. Rev.*, 1926. B. Franze's monograph on the rainfall of South America is reviewed and summarized by H. A. Matthews in the *Geog. Journ.*, 1929.

For the relation of climate and agriculture the following articles may be consulted. H. Mellish, 'The Relations of Meteorology with Agriculture', *Q.J. Roy. Met. Soc.*, 1910; B. C. Wallis, 'Rainfall and Agriculture in the United States of America', *Monthly Weather Review*, 1915; R. de C. Ward, 'The Larger Relation of Climate and Crops in the United States', *Q.J. Roy. Met. Soc.*, 1919; G. Taylor, 'Agricultural Climatology of Australia', *Q.J. Roy. Met. Soc.*, 1920; R. H. Hooker, 'Forecasting Crops from the Weather', *Q.J. Roy. Met. Soc.*, 1921; H. M. Leake, 'The Agricultural Value of Rainfall in the Tropics', *Proc. Roy. Soc.*, 1928; H. L. Manning, 'The Statistical Assessment of Rainfall Probability and its Application in Uganda Agriculture', *Empire Cotton Growing Association*, 1956.

The wind belts and atmospheric circulation are well described in the *Admiralty Weather Manual* and in most text-books on Meteorology, such as Petterssen's *Introduction to Meteorology*. The climates of the oceans and their shores can be studied from the following publications, available from H.M. Stationery Office, Kingsway. *Weather in the Indian Ocean* (M.O. 451), 1943, *Monthly Meteorological Charts of the Indian Ocean* (M.O. 519), 1949, *Monthly Meteorological Charts of the Atlantic Ocean* (M.O. 483), 1948, *Monthly Meteorological Charts of the Western Pacific* (M.O. 484), 1947, *Monthly Meteorological Charts of the Eastern Pacific* (M.O. 518), 1950. The influence of climate on river discharge is dealt with in E. E. Foster, *Rainfall and run-off*.

The standard atlas is Bartholomew's *Atlas of Meteorology*, being vol. III of Bartholomew's *Physical Atlas*. The *Klimadiagramm Weltatlas* reproduces hundreds of climatic diagrams all over the world, including maxima and minima of temperature and rainfall. The *Atlas of American Agriculture* is very useful; the new *Oxford Economic Atlas* contains plates of crop distribution with helpful explanations. Reference may be made to the *Russian Climatological Atlas*, *Klima Atlas von Deutschland*, the *Climatological Atlas of India*, *Climatological Atlas of Canada*, *Climatological Atlas of Japan*, *Climatological Atlas of the United States*, etc. Many national atlases have valuable climatic plates, e.g. the *Atlas de France*, *Atlas of Finland*, *Atlas Niedersachsen*, *Atlas van Tropisch Nederland* (with English translation of titles).

Any investigation into regional climatology will call for abundant data for illustrations and analysis. Collections of the mean monthly values of temperature and rainfall are readily available in most text-books such as this one and Kendrew's *Climates of the Continents*. Some 500 stations, spread over the world, are described in *World Weather Records* published by the Smithsonian Institute; Mean Pressure, Temperature and Rainfall are there given for each month of each year and the records are kept up to date by a supplementary publication every ten years. The *Réseau Mondial*, published annually by the Meteorological Office, London, between 1910 and 1932, then discontinued, provided data for about 1,500 stations (two for each 10° square of latitude and longitude). It supplied maximum, minimum, mean and extreme temperatures, monthly rainfall totals and mean pressure for each month of the particular year, and, as departures from the normal were given, the 'normal' value could easily be determined if required.

Other data, sunshine, cloud, rain-days, snowfall, thunder, wind direction and force, extremes of temperature, etc., can be requested from the Climatology Section of the Meteorological Office. The best available collections of such data are to be found in the various regional volumes of the *Handbuch der Klimatologie* edited by W. Köppen and R. Geiger. This monumental series, written by world authorities, edited and published in Germany, was not quite complete when interrupted by war in 1939; there are still a few irritating gaps. Some volumes are written in English, some in German, but the full and varied tables can be profitably used with a minimum acquaintance with either language.

Conrad and Pollak *Methods in Climatology* deals with the mathematical treatment and analysis of climatic statistics. The simpler methods are explained in *The Skin of the Earth* by A. Austin Miller.

III

THE FACTORS OF CLIMATE

Solar and Physical Climate. The latitude of a place, together with its altitude and relation to surrounding relief, determines once and for all the possible duration and intensity of the light and heat received from the sun. But this theoretically possible amount will not be attained because of interference by clouds and because of heat imported or exported by air currents. These interferences are subject to no rigid laws and cannot be evaluated with accuracy in estimating the climate of a place. But the climate, in so far as it is governed by the amount of insolation received, is capable of accurate estimation and is known as the 'solar climate'. The actual climate, resulting from the interference by the other factors with the solar climate, is known as the 'physical climate'.

Latitude is the prime factor in determining climatic zones, for the only considerable source of warmth is the sun, and the heating effect is greatest where the sun's rays are most nearly vertical. The amount of heat received by a given area depends both on the intensity and duration of sunlight, both of which depend on latitude. The intensity of insolation is greatest where the sun's rays fall vertically on the surface of the earth, both because here a bundle of rays of given width is spread over the minimum area and because the rays are penetrating a minimum thickness of atmosphere and therefore absorption will be a minimum. The duration of sunlight increases in summer with increasing latitude and decreases in winter; thus in summer the low angle of the sun in high latitudes is partially offset by the greater length of day. This is a fact of great significance in cereal cultivation and allows, for example, the ripening of wheat in Canada in places where the growing season is less than 100 days and where the accumulated temperatures are less than 1,400 day degrees (see p. 10). Furthermore, the long days of weak sunshine in high latitudes seem to produce better grain than the shorter hours of more intense sunshine of lower latitudes; Canadian wheat is better quality than Egyptian, Italian rice is superior to Indian. In winter, on the other hand, the feeble sun, low in the sky, has little power in the short hours of daylight to warm up the ground, and the heat received by day is entirely dissipated by radiation in the long winter night.

Insolation at each Latitude. At the Equator the value of insolation varies little throughout the year, for the day is 12 hours long, and the sun never departs far from the vertical (see Fig. 11). There will, however, be two maxima at the equinoxes and two minima at the solstices. Polewards the length of day in summer increases, until at the tropic the midsummer day is $13\frac{1}{2}$ hours long. It is clear that insolation at midsummer is greater here than it can ever be at the Equator, for the duration is longer and the intensity is just as great, the sun being directly overhead. At midwinter, on the other hand, the day is only $10\frac{1}{2}$ hours long and the altitude of the midday sun is only 43°; insolation at this season is clearly much less than it can ever be at the Equator. The single maximum and minimum characteristic of the tropic is, in point of fact, reached at about 12° N. and S.; for beyond this point the gain of insolation by increased length of day outweighs the loss of insolation consequent on the rather low angles

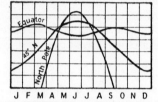

Fig. 11.—Annual Variation
of Insolation

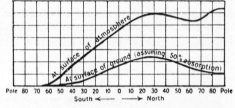

Fig. 12.—Insolation at Different Latitudes on
21st June

of the sun. Polewards again the length of day increases steadily while the midsummer altitude of the sun steadily declines. For a while the former more than compensates for the latter and midsummer insolation increases polewards until about $43\frac{1}{2}$°, where a maximum is reached. Winter insolation here is, of course, much less than at the tropic, for days are short and the midday angle of the sun is only 23° (Fig. 11). Further polewards still the relative values of the two factors are reversed, the increasing length of day fails to counteract the low angle of the sun, and the value of insolation decreases to a minimum at 62° N. Beyond this, however, the length of day increases rapidly until at the Arctic Circle it is 24 hours long; the values of the factors concerned are again reversed and insolation rises again. It might be thought that beyond the Arctic Circle the lower elevation of the midday sun, without the compensation of increased length of day, would cause insolation to decline, but actually the opposite is the case and the curve continues to rise until a final maximum is reached at the Pole, exceeding all previous maxima. This is accounted for by the increased altitude of the sun at midnight, which is $23\frac{1}{2}$° at the Pole, while at the Arctic Circle the midnight sun at midsummer only skims the horizon. At midwinter within the polar circle there is, of course, no insolation as the sun does not rise; insolation at the Pole ceases at the autumnal equinox and begins again at the vernal equinox (see Fig. 11).

Effect of the Atmosphere. Up to this point we have considered only the theoretical value of insolation arriving at the surface of the sphere, but to arrive at the insolation value at the surface of the earth a correction must be applied for the loss of heat in the passage of the sun's rays through the atmosphere. The amount which penetrates will be determined by two factors:

(1) The thickness of air to be traversed, a fixed and calculable value, and (2) the degree of transparency of the air which varies with conditions of cloud, dust, etc. Other things being equal, the absorption of the sun's rays will be greatest in high latitudes where the oblique rays have to pass through a greater depth of atmosphere; the theoretical high values of insolation here are therefore reduced considerably when the correction is applied. If the coefficient of transmission is taken as 0·50, an average amount in fairly clear weather, only 18 per cent of the total insolation reaches ground-level at the Pole. The lower curve in Fig. 12 is drawn on this assumption and the upper curve shows the value of insolation received by the outer layers of the atmosphere. It should be remembered that absorption is much greater in the cloudy belts of the equatorial calms and the stormy westerlies than under the clear skies of the high-pressure belts of the horse latitudes, so that the lower curve, based on a fixed coefficient of transmission, has little more than a theoretical value.

The value of insolation, thus corrected, supplies an approximate measure of the sun's heating effect, but this solar control of climate is effective in broad outline only. Certainly it determines the latitudinal arrangement of the main climatic zones but in detail it is profoundly modified by such factors as altitude, marine influence, physical features, plant cover, etc., the resultant being the *Physical Climate*.

Effect of Altitude. Height above sea-level has a profound influence on climate, in many respects imitating the effects of increased latitude. The peculiarities of climates of high altitudes will be dealt with in detail in a later chapter, but it will be necessary here to examine some of the more important effects. Briefly these are: (1) a decreased pressure, (2) a decreased mean temperature, (3) an increased precipitation.

Pressure and Altitude. Although the decrease of pressure with altitude is slightly more rapid in cold climates than in hot, the difference is not significant, and as a rough approximation it may be assumed that:

From sea-level to 2,000 feet the decrease is 4   per cent per 1,000 feet
   ,,   2,000 feet to 5,000 ,,   ,,   ,, 3   ,,   ,,
   ,,   5,000 feet to 10,000 ,,   ,,   ,, 2·5   ,,   ,,

Or, stated in another way:

The barometer stands at 30 inches at sea-level.

"	"	"	29	"	"	830 feet
"	"	"	28	"	"	1,800 "
"	"	"	26	"	"	3,800 "
"	"	"	24	"	"	5,900 "
"	"	"	20	"	"	10,600 "
"	"	"	16	"	"	16,000 "

At 18,500 feet the pressure of the atmosphere is only half that at sea-level.

The direct physiological results of diminished pressure are breathlessness and a feeling of lethargy culminating in 'mountain sickness'. The height at which this malady appears varies with the individual and with conditions, but is usually about 15,000 feet. It is an anaemic condition, a purely physical result of decreased oxygenation of the blood in the rarefied atmosphere. That it is not the result of exertion is proved by the fact that passengers on mountain railways suffer in the same way as climbers on foot, but the symptoms naturally become more acute in proportion as the system demands more oxygen for greater exertion. A high hæmoglobin content, which increases the capacity for absorbing oxygen, is characteristic of the blood of all native animals and inhabitants of these high altitudes, and acclimatization of visitors is slowly achieved by increasing the number of red corpuscles, often by more than 50 per cent.

The decrease in density of the air brings about a marked reduction in the absorption of insolation, an effect which is further increased by the virtual absence, at these high altitudes, of water vapour, which acts as a powerful absorbent at lower levels. The result is a marked difference between sunshine and shade temperatures and other peculiarities which will be dealt with later (pp. 269 and 279).

Temperatures and Altitude. The rate of decrease of temperature with altitude is subject to considerable variations for local reasons, and there may be actual inversions, but it is usually in the neighbourhood of 1°F. for every 300 feet of ascent. It is usually less in winter (410 feet in the British Isles) than in summer (270 feet), less at night than during the day, less on plateaux (290 feet) than on mountains (265 feet) and still less on plains (365 feet). There are numerous exceptions to these generalizations, for example the winter lapse rate in eastern Brazil is 1°F. in 226 feet and the summer lapse rate only 1°F. in 500 feet. This is a result of the high humidity and ascending air currents of the belt of calms centred here in summer, for in the condensation which results heat is liberated in great quantities and delays the fall in temperature. The chief cause of the normal temperature lapse is the rarity of the air at high altitudes and its relative poverty in moisture and carbon dioxide, whereby its power of absorbing heat is much diminished. The sun's rays are allowed to pass through without warming the air, and free radiation from the land surface is promoted at night or whenever the sun ceases to shine. At lower levels the denser air, and more particularly the clouds, have a blanketing effect which checks this loss by radiation. Again, in mountain districts the large land surface exposed facilitates radiation, while much of the surface, being in shadow, receives little or no heat from the sun. On the other hand, although air temperatures are low the sun's rays,

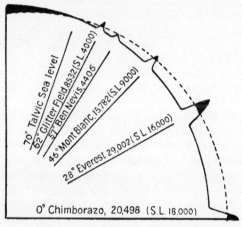

Fig. 13.—Altitude of the Snow-line

passing freely through the clear thin air, beat fiercely on solid objects such as rock surfaces, heating them intensely and burning the exposed skin of hands and face. Mountains and plateaux thus enjoy climates characterized, *inter alia*, by lower temperatures, and thus appear as 'islands' of, perhaps, tundra in the midst of forest. The benefits which this sometimes confers may be seen in Africa and South America, both of which have enormous areas of high land in the equatorial belt; but for their altitude these would be sheer jungle, as impenetrable as the Amazon and Congo forests, and as unhealthy; actually these plateaux are among the most prosperous and promising parts of the continents.

Snowfall on Mountains. High mountains, even on the Equator, experience regular temperatures below freezing-point and are, subject to adequate precipitation, capped by perpetual snow. It is important, however, to realize that precipitation, as well as temperature, helps to define the snow-line, for the loss by melting and evaporation must be made good by fresh falls; the Pamirs, although rising far above the snow-lines as determined by temperature, have very little snow cover because of their aridity, placed as they are out of the course of moisture-bearing currents; the snow-line on the windward side of the Andes and Himalayas is hundreds of feet lower than on the leeward side. Fig. 13 shows diagrammatically the average altitude of the snow-line at various latitudes.

Rainfall and Altitude. More far-reaching than the effects on pressure and temperature is the profound modification of rainfall distribution by high land. Mountains in general enjoy a higher rainfall than lowlands similarly placed, as can be clearly seen from almost any isohyetal map. On the mean annual rainfall map of the United States, for example, the Cordilleras and the Appalachians are clearly brought out by ridges of heavy precipitation, and even such small relief features as the Ozarks and the Black Hills of Dakota stand out conspicuously. The latter receive 25 inches of rain in the midst of an area with only 15 inches, and the vegetation responds in no uncertain manner. Their name, in fact, refers to the dark green forests, the basis of a local lumber industry, which make such a striking feature in the surrounding prairie. The higher land of the Sahara traps the rain-bearing winds to make altitude oases such as Dar-Fur which gets its water from the Jebel Maria. The increase of rainfall is explained by the cooling of the air by: (1) forced ascent up a slope and (2) by contact with the cold surfaces of higher altitudes; consequently the effect is most pronounced where the high land opposes a barrier to rain-bearing winds, i.e. where the rainfall is orographic, and least marked in regions of calms where the rainfall is convectional. Further, the moisture capacity of air is not a simple function of temperature, and the change of capacity of warm air on cooling is, degree for degree, greater than that of cool air; hence the ascent of air in tropical latitudes results in heavier rain than a similar ascent in temperate climates. The temperature lapse, also, is steeper in low latitudes than in high, so that a given ascent brings about a greater cooling. The heaviest rainfalls in the world are recorded under such conditions, Kauai,[1] in the Hawaiian Islands, has an average rainfall of 476 inches and Cherrapunji, in Assam, has 450 inches.

Rain Shadow. Hawaii lies in the path of the N.E. trades, which, coming across a warm sea, bring more than 140 inches to Hilo on the N.E. coast and perhaps 180 inches to the higher slopes. But this excessive rain must be compensated elsewhere and there is a complementary zone of deficient rainfall in lee of the high land. Here the conditions are reversed, the air is being warmed by descent and, in addition, is passing from the cold hill-tops to the warmer plains, its capacity for moisture is increased and little rain falls. Hilea, on the lee side of Hawaii, has only 35 inches and the result is the desert of

[1] Climatological data, Hawaiian Section, 1922, quoted by S. S. Visher in *Climatic Laws*, 1924.

Kau. The Western Ghats, opposed for five months to the constant south-west monsoon, have 250 inches in places on their western slope, while many stations on the eastern slope have 25 inches or less. The effect of mountains is thus to bring about not so much an actual increase of rainfall as a redistribution.

Rate of Increase of Rainfall with Altitude. The increase becomes noticeable before the mountains are actually reached, as is shown by the figures for the following stations, all at about 65 feet O.D. on the Gangetic plain:[1]

Dacca	100 miles from Khasi Hills	.	78 inches
Bogra	60 ,,	,,	,,	,,	.	92	,,
Mymensing	30 ,,	,,	,,	,,	.	110	,,
Sylhet	20 ,,	,,	,,	,,	.	150	,,

This fact is a warning against attempts to estimate accurately the rate of increase with altitude, for it is clear that another factor is in operation. Air is stowed against the slope of the mountain, thus compelling upper air currents to rise before the mountain is reached, causing stations to windward to have a rainfall out of all proportion to its height above sea-level. Again, distance from the sea, temperature of the rain-bearing currents, the temperature of the land surface, the steepness of the slope and the presence or absence of gaps in the crest all have an influence on the precipitation at different levels. On the South Downs behind Bognor and Brighton,[2] where the chalk ridge is unbroken, the rate of increase is about 2 per cent per 100 feet, but on the discontinuous green-sand hills behind Cranleigh it is only about 1 per cent; the air currents go round, not over the hills. On the steeper slopes of the Welsh mountains the figure is considerably exceeded (4 or 5 per cent) and in tropical climates it is much higher still.

Zone of Maximum Precipitation. The increase of precipitation is not maintained indefinitely up a slope; there comes a level at which the rate of increase slackens and eventually ceases, while higher still there is an actual decrease. There is, then, a zone of maximum precipitation whose altitude varies slightly from place to place, is lower in tropic than temperate zones, in humid climates than in arid, in the cold season than the hot, in the wet season than in the dry. In Java it appears to occur at about 3,300 feet, in the Western Ghats about 5,000 feet, in the Sierra Nevada behind Los Angeles about 5,000 feet, and in the Alps about 7,000 feet, while in the British Isles this zone is probably never reached. The explanation of the decrease above the zone of maximum precipitation is to be found in the decreased absolute humidity at high altitudes; temperature is lower and the moisture capacity of the air is less, consequently there is less available

---

[1] H. F. Blandford, *The Rainfall of India*. Quoted by Hann.
[2] M. de. C. Salter, *Rainfall of the British Isles*, 1921.

for condensation. An interesting example of seasonal variation of this zone is quoted by Hann from Swerzow. In the Tian Shan the zone of maximum winter snowfall is about 7,000 to 9,000 feet, and here are the coniferous forests, while the zone below is devoid of forests because of its aridity. In summer the zone of maximum precipitation rises above the coniferous forest and the plentiful summer rain nourishes a rich grass which provides pasture for the flocks and herds which the Khirgiz maintain here in winter, safely above the zone of heavy snow.

Mountains as Climatic Divides. By interfering with the free flow of air currents and by influencing the distribution of rainfall and temperature, mountains tend to coincide with important climatic divides. The Dinaric Alps separate the extreme continental climate of the Hungarian plain, with January temperatures below freezing, from the warm Adriatic littoral with January temperatures of 50°. It would never be imagined at Fiume that such low temperatures could exist so near at hand, but unpleasant confirmation is supplied when the passage of a depression to the south draws off some of the cold air over the lip of the Hungarian basin down into the Adriatic as the icy 'Bora'. The Alpine-Himalayan mountain system throughout its length provides a climatic divide of the first magnitude, excluding polar influences from the lands to the south and tropical influences from the lands to the north. The winters of Siberia, Turan and China are disproportionately cold, those of India are disproportionately warm; Multan (January temperature 54°) and Shanghai (January temperature 38°) are on the same parallel. The Sacramento Valley, cut off by the coast ranges from the cooling influence of the sea, has mean July temperatures in the neighbourhood of 90°, while San Francisco records only 57°. The contrasts in rainfall are just as marked, 100 inches and 10 inches occur within 200 miles of each other on the west and east slopes of the Andes respectively in southern Chile and Argentina.

The north-and-south grain of America and the east-and-west grain of Eurasia offer an interesting comparison of climatic results. In North America there is no barrier to the free meridional movement of air, for the Gulf-Hudson Bay divide is below 1,000 feet and the plains offer no obstruction. The low pressure of cyclones over the southern states can therefore draw huge stores of icy air from the great Canadian reservoir down to the very shores of a tropical sea. In Asia, on the other hand, the warm tropical air of India is kept separate from the Siberian cold by a well-nigh impassable barrier. The contrast does not stop there; for the direct result of the great Siberian well of cold is to generate a huge anticyclone, the source of the winter monsoon. In America this cannot exist to the same degree because of mixture with warm air from the Gulf. In summer the heated south of North

America is tempered by northern influences, while in the enclosed basin of India the unmitigated heat sets up an intense low pressure which is the focus of the summer monsoon. Thus it is the disposition of the relief, rather than the relative sizes of the continents, which accounts for the feeble development of the monsoon in America as compared with Asia.

Influence of Land and Sea on Climate. Next to the variation of insolation with latitude, the distribution of land and water over the surface of the earth is the most important control of climate. A variety of physical effects combine to make water much more conservative of heat than land; slower to warm up, slower to cool down, it has a moderating influence on temperature which may be felt at considerable distances inland. The distance to which marine influence penetrates will depend on the prevailing wind direction coupled with the facilities which the relief offers for free entry of ocean winds. Canada and the northern United States, for example, lie in the belt of the prevailing westerly winds, but marine influence is restricted by the Cordilleras to a narrow coastal strip in British Columbia, Washington, Oregon and California. The East coast is a lee shore and the climate is continental right up to the seaboard. Marine and continental climates in this part of North America are clearly and sharply differentiated, the coast-line being simple and uniform, but in Western Europe the intimate intermingling of gulf, bay and inland sea with islands and peninsulas, together with the absence of marked relief features athwart the path of the winds, allows the penetration of marine influences for hundreds of miles into the continent. The complexity of the coast-line and of continental relief in Europe has its sequel in a complexity of climates, which contrasts strangely with such a compact continent of uniform relief as, for example, Africa. The differences between marine and continental climates may conveniently be grouped under three headings: (1) Rainfall, (2) Temperature, and (3) Pressure and Winds.

Continental and Marine Rainfall. The oceans are the chief sources of atmospheric moisture, and it is to be expected that humidity and rainfall will be greatest where the direction of the prevailing wind is on-shore. The sudden decrease of rainfall inland along the western mountain divide of both Americas in the zones of the westerly winds is an indication of the narrow fringe of oceanic climates there, just as the gradual eastward diminution of rainfall in Europe in these latitudes shows its great extension here. Rainfall in marine climates is chiefly orographic and usually fairly adequately distributed throughout the year, rainfall in continental climates is chiefly summer rain resulting from convectional overturning set up by the high summer temperatures.

Continental and Marine Temperatures. The diurnal variation of temperature of sea surfaces is almost a negligible amount, and even the annual variation is extremely small; within the tropics it seldom exceeds 10°, in mid-latitudes it reaches 20° or 25°, but these figures are only exceeded where the oscillation of boundaries between currents bring a certain place now under the influence of a warm current, now of a cold. Fluctuations of the Gulf Stream and the Labrador current are responsible for a yearly range of surface temperature of over 50° off the coast of Maine, but this is local and climatically ineffective, especially as the wind is here off-shore. The variations of air temperature over the water surface are thus kept very small and present strong contrasts with the ranges recorded over the land. The uniform temperatures of winds blowing off the sea, together with the greater

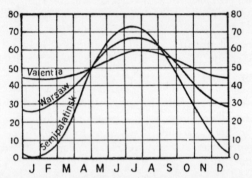

Fig. 14.—Yearly March of Temperature in Continental
and Marine Climates

humidity of the air which impedes insolation by day and radiation by night, results in:

1. A diminution of the daily range of temperature.
2. A diminution of the annual range of temperature.
3. A delay in the attainment of the daily and yearly maxima and minima. These are features characteristic of marine climates and are illustrated in Fig. 14, which shows temperature curves for three stations, all situated in approximately the same latitude. It will be noticed that the annual range at Valentia is less than 15°, that the hottest month is August and the coldest February. At Semipalatinsk the range is over 70°, the maximum is reached early in July and the minimum early in January.

It is interesting to compare the northern and southern hemispheres in respect of the continentality of their climates; the northern hemisphere contains most of the land, the southern is essentially a sea hemisphere. In the northern hemisphere the 32° isotherm for January comes down to latitude 35°N. in China, while nearly all of Asia north of 40°N. has a mean January temperature below freezing. North

America is slightly more favoured, but even here the 32° isotherm extends south of 40°N. in the Mississippi basin.

In the southern hemisphere, none of the three southern continents reach the 32° July isotherm, although the tip of Tierra del Fuego reaches 55°S. At the other extreme the summer temperature of Tierra del Fuego is only 50°, but in Asia the 50° July isotherm reaches 70°N. In the absence of a great continental mass in high latitudes there can be no great reservoir of cold air in the southern hemisphere such as exists in Siberia and Canada to supply the 'Cold Waves' and 'Northers' of the United States and the 'Buran' of Siberia. Compared

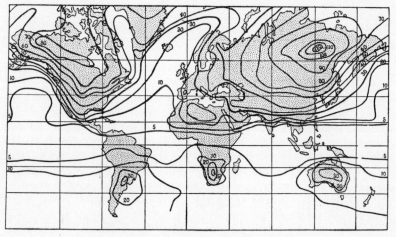

Fig. 15.—Mean Annual Range of Temperature

with these the 'Southerly bursters' of Australia and New Zealand and the 'Pampero' of the Argentine are relatively mild.

Fig. 15 shows the mean annual range of temperature and also emphasizes the contrasts between the two hemispheres. Nowhere in the southern hemisphere does the range exceed 40°, but in the northern hemisphere huge areas in Canada have more than double that figure and in Siberia more than treble.

Continental and Marine Pressure and Winds: Land and Sea Breezes. The difference in behaviour of land and water towards diurnal and annual temperature changes begets a difference of pressure which results in periodic diurnal and seasonal winds known as land and sea breezes and monsoons. The heating of the land during the day causes an ascent of air over the land and an indraught of oceanic air; the descent of air cooled by radiation over the land at night causes an expulsion of land air out to sea. Land and sea breezes are most noticeable and regular where temperature changes are most regular and particularly in equatorial climates. In Java the native fishing industry depends on its regular occurrence, the boats start out at night with the land breeze and return about noon with the sea breeze.

In the belt of the westerlies the cyclonic disturbances frequently mask this diurnal periodicity, but in settled weather it is usually well marked. In the tropics the sea breeze sets in at any time between 8 a.m. and 1 p.m., but its appearance is usually punctual at any given spot. Its arrival is anxiously awaited since it is both refreshing and healthy. Houses are set to catch the breeze and the best residential districts are those which are most favourably situated for its reception. In Senegambia[1] a fall in temperature of 20° and a rise of humidity of 42 per cent have been noted within 15 minutes of the arrival of the sea breeze. In Senegal the inland temperatures in the hot season reach 110°, but along the coast, thanks to the sea breeze, they seldom rise above 85°.

At about sunset the sea breeze gradually dies away to a calm and the land breeze springs up; frequently it is as unpleasant and unhealthy as the sea breeze is refreshing. It is not usually so marked as the sea breeze since the land is hotter than the sea throughout the year in tropical latitudes and in the summer in temperate latitudes; the diurnal temperature gradient is therefore stronger than the nocturnal. In addition, the sea breeze is not retarded by friction to the same extent as the land breeze. The pressure gradients which give rise to these breezes are not very great and are easily modified or even reversed by other influences. In the trade wind belt they are most strongly felt when they reinforce the prevailing wind, the opposite breeze only serving to moderate or modify the trades. In California, for example, the sea breeze reinforces the prevailing west wind during the day and is welcome as moderating the day temperatures, especially as in summer the breeze is further strengthened by a seasonal low pressure inland due to the high temperatures of the interior.

Monsoons. Just as the diurnal march of temperature gives rise to land and sea breezes, so the annual march of temperature produces land winds in winter and sea winds in summer, known as monsoons. Almost every large land mass which is subject to a considerable annual variation of temperature generates a monsoon, but it is only in the case of the largest land masses that the effect is sufficiently powerful to overcome the normal planetary circulation, and in practice the name is restricted to the most conspicuous of these. They will be dealt with fully on a regional basis in later chapters.

Effect of Ocean Currents. There are many aspects of continental and marine climates which are not adequately explained by the contrasts in behaviour of land and sea, and these are usually the results of ocean currents. A map of the ocean currents shows, as the main

---

[1] *Comptes Rendus*, 1894, quoted by Hann.

features of the circulation, a great east to west movement of water in the equatorial zone compensated by a west to east movement in the temperate zone, the swirls being completed by a poleward movement along the western shores of the oceans (eastern margins of continents) and an equatorward return along eastern shores (western margins). These currents carry tropical temperatures towards temperate seas on the eastern margins and temperate waters towards tropical seas on the western margins. We expect, as a result of these movements, a warming effect along eastern margins in low latitudes (e.g. the North Equatorial current) and a cooling on western margins (e.g. Californian current), while in higher latitudes it is the western margins which will be warmed (e.g. the North Pacific Drift) and the east cooled (e.g. the Kurile current). This effect is clearly brought out in the maps of temperature anomalies (Figs. 1 and 2, p. 12). These temperature anomalies are potent factors in maintaining the areas of permanent high and low pressures, and the circulation of winds round these in turn provide the driving force of the currents. The south-west winds, for example, on the south side of the Icelandic low, drive the warm waters of the Gulf Stream as the North Atlantic Drift towards the shores of Western Europe, banking up against the continental margin a depth of warm water such as is not equalled even in tropical zones and whose surface temperatures are far in excess of any other waters in similar latitudes. The analogous currents of the North Pacific, the Kuro Siwo and its continuation, the North Pacific Drift, although powerful and important, only give rise to a positive temperature anomaly of 20° in winter (cf. 40° in the North Atlantic). The greater effect of the North Atlantic Drift is explained as follows:

1. The form and position of north-east Brazil is such that much of the south equatorial current is deflected into the northern hemisphere, thereby increasing the Gulf Stream at the expense of the Brazilian current. The Philippines and New Guinea, which fulfil the same function in the Pacific, are less efficient because discontinuous.

2. The greater size of the North Pacific ocean involves greater dilution of the Kuro Siwo.

3. The S.E. trades, usually stronger and more constant than the N.E., blow beyond the Equator throughout much of the year, thus transferring much equatorial water from the southern into the northern hemisphere. This effect is more pronounced in the Atlantic than in the Pacific and further adds to the volume of the Gulf Stream.

Apart from occasional inversions due to differences in salinity, warm water is lighter than cold, and therefore floats on the surface where it is further warmed by insolation. It is clear, therefore, that where the prevailing wind, and therefore the surface drift of water, is on-shore, there the surface water is relatively warm and the warming influence

is carried inland. Where, on the other hand, the prevailing drift of
wind and water is off-shore, the warm surface water is carried away
and its place must be taken by water either moving along shore or,
more usually, rising from below and therefore much colder. As
examples we may quote: (1) the cold Benguela current off south-west
Africa, the result of the S.E. trades, and (2) the cold water which
rises off Cape Guardafui while the S.W. monsoon is blowing. The
latter, in parenthesis, exerts a cold influence and produces lowest
temperatures during the season which is really summer.

The East and West Shores of the North Atlantic Compared.
If, now, we consider the circulation of the North Atlantic it is clear
that there is a convergence of warm currents accompanied by on-shore
winds in low latitudes on the American shore, while the Eurafrican
shore, apart from the relatively small counter-current (the Guinea
current), is cooled by up-welling of cold water along the west coasts
of Spain and Africa. The western side of the Atlantic is therefore
hotter than the east, especially in winter, and has a smaller temperature
range. Compare the following pairs of stations in about the same
latitude and all at about sea-level:

Goree (Dakar)	.	.	Jan. 68·5	Mean Annual	74·8		Range	16·2
Vera Cruz	.	.	„ 71·4	„	„	77·4		„	11·0
Banana	.	.	July 72·5	„	„	77·0		„	9·0
Pernambuco	.	.	„ 75·2	„	„	79·0		„	6·3

The mean annual isotherms of 80° diverge on crossing the Atlantic
westward; they enclose only about 400 miles of coast in West Africa,
but on the American side the whole coast from 15°S. to 20°N. has a
mean annual temperature of over 80°. Proceeding northwards the
contrasts decrease until at about 30°N. there is little difference between
the two sides; the mean annual isotherm of 70° cuts both coasts in
about the same latitude. From this point northwards, however, there
is a striking difference in the opposite direction as the following figures
show:

Washington	.	.	Jan. 32·9	Mean Annual	54·7		Range	43·9
Lisbon	.	.	„ 49·3	„	„	59·5		„	20·9
			Difference 16·4	Difference	4·8		Difference	23·0

New York	.	.	Jan. 30·3	Mean Annual	51·8		Range	44·2
Oporto	.	.	„ 47·0	„	„	58·0		„	18·0
			Difference 16·7	Difference	6·2		Difference	26·2

{ St. John's . . . Jan. 24·2   Mean Annual 40·1   Range 36·0
{ Brest . . . „ 43·9   „   „ 53·6   „ 20·5

           Difference 19·7   Difference 13·5   Difference 15·5

{ Nain (Labrador) . Jan. — 7·1   Mean Annual 22·6   Range 54·0
{ Glasgow . . . „ 38·6   „   „ 47·3   „ 19·4

           Difference 45·7   Difference 24·7   Difference 34·6

The American coast suffers from the double disadvantage of continental conditions and the Labrador current, while the European is bathed in the warm waters of the North Atlantic Drift. The western side of the North Atlantic is thus heated by hot currents in the hot zone and cooled by cold currents in colder latitudes, while the east side is cooled by cold upwellings in the hot zone and warmed by a warm current in colder latitudes. The temperature gradient along the western margin of Eurafrica is therefore abnormally gradual, while that in eastern North America is the steepest in the world over such a great distance. Such a gradient has important economic and climatic results; it telescopes the temperature zones, giving a striking variety of produce, from the almost tropical cultures of Florida to the Arctic produce of Labrador, within a distance of 2,000 miles. Meteorologically it is an arrangement predisposing to remarkable and sudden changes of temperature as air may be imported from a locality not far distant but differing markedly in temperature. In eastern Asia there is a somewhat similar telescoping of temperature zones, but since the warm current (the Kuro Siwo) is less powerful, and the cold current (the Kurile current) has less ready access through the constriction of the Bering Strait, the gradient is less steep than in eastern America.

The Cold Currents of Western Margins in the Trade Wind Belt. In the southern hemisphere some excellent examples of the influence of currents on temperature can be traced round the coasts of South Africa and South America. The form of the mean annual isotherm of 70° is interesting. It is carried up the west coast of South America within 10° of the Equator by the cold Peruvian current, and down the east coast as far as 30°S. by the warm Brazilian current; across Africa it behaves in a similar manner. The current, rather than latitude, is here the control of temperature at coastal stations; there is less than 3° difference in the mean annual temperatures of Beira (76°) and Mombasa (78·5°) although they are 16° apart in latitude.

The cold currents which wash the western shores of continents are not merely offshoots of the Antarctic drift, they are reinforced by cold water welling up from greater depths as the surface water is skimmed off by the trade winds. Warmer water is therefore found, not equator-wards, but out to sea. The equivalents in the northern hemisphere, the Californian and Canaries currents, are not so strong nor so cold since wind circulation in the southern (water) hemisphere is stronger and less hampered by friction with land surfaces. Furthermore there is freer entry for Antarctic water than for Arctic, since the South Atlantic and South Pacific lie wide open to the south, while the North Pacific is constricted into the narrow Bering Strait and the North Atlantic waters are segregated by the shallow ridge of the Scoto-Icelandic rise. A further difference is to be found in the form of the coastal margins, which are more favourable for the free movement of water off the longitudinal coast of Chile and the plateau edge of south-west Africa than off the broken transverse coast-line of Eurafrica. In this last respect the Californian current resembles the Peru and Benguela currents and differs from the Canaries current.

Currents and Rainfall. It is quite in accordance with expectations that rainfall is heavy on coasts washed by a warm current and light where cold currents flow. Air passing over warm currents will be saturated at a high temperature and therefore a pregnant source of rain, while air which has passed over a cold current usually has its capacity increased by warming on coming into contact with land. Further, warm currents occur, with very few exceptions, where the wind is on-shore and cold currents where it is off-shore. The currents, therefore, only emphasize a distribution of rainfall which the prevailing wind determines. Thus warm currents increase the rainfall in British Columbia, the British Isles, in Japan (from the S.E. monsoon) and in Queensland (from the N.E. monsoon), etc., while cold currents contribute to the aridity of the Kalahari and Atacama deserts, Patagonia, etc. The influence of the cold Benguela current extends as far north as Banana at the mouth of the Congo, not far north of which it meets the warm Guinea current. Here are found some striking contrasts of rainfall. Banana, on the coast, has only 28 inches compared with over 60 inches in the interior, while Libreville, only 8° farther north along the coast, being under the influence of the warm Guinea current eddying round the Bight of Biafra, has nearly 100 inches. Cape Lopez divides the spheres of influence of the two currents and also the two rainfall types, but sometimes the warm waters of the Bight escape south of Cape Lopez and bring heavy falls of rain to the coast even south of Banana. A similar function is served by the bulge of the South American Pacific coast in Ecuador, which sharply divides the desert from the forest, the critical point being Cape San Lorenzo.

Influence of Lakes. To an extent roughly proportional to the size of the body of water lakes moderate the climate of their immediate vicinity. Lake-side stations on Lake Constance have a mean annual temperature 0·5° warmer than stations away from the lake, and the difference in autumn and winter is as much as 1°. The Great Lakes of North America produce more profound effects and the January isotherms are noticeably diverted from their course (see Fig. 16). The severity of cold waves is noticeably diminished here and the duration of the frostless season is noticeably longer. It is particularly on the eastern (lee) shore that these benefits are felt; the favourable conditions east of Lake Michigan are reflected in a fruit-growing area of some importance. Diurnal lake breezes are clearly recognizable round such bodies of water as the Caspian Sea and Victoria Nyanza, exactly

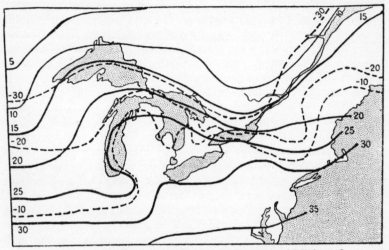

Fig. 16.—January Isotherms (full) and Mean Annual Minima of Temperature (dotted) over the Great Lakes

similar in character and origin to land and sea breezes. Humidity and precipitation are also increased, especially on the lee side, as, for example, on the western side of Victoria Nyanza from May to October while the south-east wind prevails.

Influence of Snow and Ice. Water bodies depend for their moderating influence on certain properties, some of which, such as the mixing of layers, disappear when the water freezes. The differences in behaviour of land and sea are therefore less pronounced in high latitudes, and lakes, when frozen, exert a diminished influence. The Gulf of Finland, before it freezes, brings to bear a beneficial influence on the climate of its shores, but in late winter and spring its influence declines. At Haparanda, for example, where the ice closes in in November and lasts until the end of May, it is 6° warmer in October than in April and in September than in May, but June, when the ice has disappeared, is only 2° cooler than August. In a similar manner a cover of snow delays the advent of spring by preventing the land from exerting its usual effects on warming up and by using up spring warmth in melting and evaporating the snow; a late spring is therefore characteristic of all snowy climates. Snow also increases the annual range of air temperature by increasing radiation during winter, but, on the other hand, being a bad conductor of heat, it preserves the warmth of the soil by preventing the escape of heat from the ground. For this last reason an early fall of snow may be the means of preserving autumn-sown cereals from winter frosts, as in Ontario where the snow comes before the severe frosts, and, in addition, supplies moisture, on melting, for the growing plant.

Influence of Physical Features. In shaping the smaller details of climate the minor relief features play an important role. The run of the feature lines has a not inconsiderable influence on the direction of winds, which tend to blow along rather than across them. Valley bottoms and low-lying land often suffer from inversions of temperature which may give rise to severe frosts, from which the valley sides, better drained of cold air, are exempt (see p. 276). The northern slopes of east-and-west mountain valleys in the northern hemisphere enjoy a more genial climate because of the greater intensity and duration of sunlight which they enjoy. Grain can be grown 1,000 feet higher on the southern slopes of the Alps than on the northern, while pasture, which extends to about 7,000 feet on the north side, occurs up to about 8,000 feet on the southern slopes. 'South aspect' is the main natural asset of many popular resorts such as Nice and Torquay. A backing of high ground to keep out northern influences is also of value, a function which the maritime Alps effectively perform for Nice, and the highlands of Devon less effectively for Torquay. Eastbourne is indebted to Beachy Head and the chalk downs for shelter from rough west winds and Bournemouth similarly shelters behind the Isle of Purbeck.

An interesting case of temperature control by relief is afforded by Siberia and the Amur basin, where cold stagnant air accumulates in winter in the basins ringed round by mountains. This intensely cold heavy air escapes over the lowest and easiest notches in the basin rim and brings intense cold to the towns situated in the gaps. Thus at Vladivostok, behind which is a gap at 600 feet, the January temperature is nearly zero, although only in latitude 44°N.; but farther north, where the mountain wall is more formidable, the temperature along the coast is milder. At Nicolaievsk, in the wider and lower gap of the Amur, the January temperature reaches 12° below zero, but still farther north the crescent of the Stanovoi shelters the shores of the Sea of Okhotsk and here is Ayan, 200 miles north of Nicolaievsk, enjoying a temperature of only 4° below zero.

Soils and Climate. Geological formation and the resultant soil type are also minor factors in determining climate. Dark coloured soils and surfaces absorb more of the sun's heat than lighter coloured ones, and are generally warmer by day, warming the air in contact and causing it to dance and form mirages as, for example, over tarmac roads. Dry soils, such as sands, have a low specific heat and respond rapidly to temperature changes, while wet soils, such as clays, retain moisture and are therefore conservative of heat and cold; the latter also predispose, by bad drainage, to fogs and mist. Further, by the influence which soils exert on the nature of the vegetation, they can powerfully modify the effects of climate, producing grassland (e.g. on chalk) where the climate is suited to forest, or desert (e.g. on loose sands) where there is enough rain to nourish grassland on any other soil, while limestone, in certain cases, gives rise to a type of scenery which is virtually desert, in spite of an adequate or even abundant rain.

Vegetation and Climate. Climate is the chief control of vegetation type (see p. 79) and the choice between forest and grassland is generally made by the rainfall, but there is no doubt that the vegetation in turn reacts strongly on the climate. The dense vegetation of the Selvas, for example, by its enormous transpiration, increases the humidity of the air and gives greater potential rainfall. This transpiration process has been utilized for the drainage of swamps, and there are many places in Italy and France (e.g. the Landes) which have been reclaimed by afforestation; the original grass and swamp vegetation, in these cases, had insufficient transpiration surface. Forests affect temperature, too, particularly the maximum temperatures, which they moderate by casting shade, by offering a large surface for radiation, by absorbing heat in the process of evaporation from the foliage and by the production of fog, mist and cloud which ward off the direct rays of the sun. They act as wind-breaks, affording shelter for crops and settlements, and by decreasing wind velocity they decrease evaporation and thus alleviate aridity; rows of trees are often planted for this purpose.

Microclimatology. The slight contrasts of climate that may result from small differences of aspect, slope, and form of the ground, the colour, moisture and texture of soils and surfaces, vegetation and plant cover, etc., constitute the study of microclimatology. It is a study of extremes rather than means, for temperature minima may result in killing frosts, dew formation and local fog, which are occasional occurrences rather than prevailing conditions; but the occasional extreme, by overstepping critical values, may often be significant and even vital for plants and crops, animals, insects or pests. Such contrasts in microclimatology have their best opportunity to develop during calm weather and anticyclonic conditions, for atmospheric turbulence rapidly equalizes temperature and humidity differences. As a means of heat transference turbulence is 10,000 times more effective than conductance; it is calculated that without turbulence and convection there would be no diurnal variation of air temperature more than six feet above the surface of the soil. Microclimatology is therefore essentially fine-weather climatology, or perhaps, since it is not concerned with means, it should be called micrometeorology.

SUGGESTIONS FOR FURTHER READING

Hann's *Handbook of Climatology*, vol. 1, is again the standard work on the subject-matter of this chapter, and contains copious references. Further information may be obtained from S. S. Visher, *Climatic Laws*, 1924; A. McAdie, 'Monsoons and Trades as Rain and Desert Makers', *Geog. Rev.*, 1922; R. C. Murphy, 'Oceanic and Climatic Phenomena along the West Coast of South America in 1925', *Geog. Rev.*, 1926; C. E. P. Brooks, 'The Influence of Forests on Rainfall and Run-off', *Q. J. Roy. Met. Soc.*, 1928.

The most compact treatment of micro-climatology is in R. Geiger's *The Climate near the Ground*, 1950. More severely meteorological is O. G. Sutton's *Micrometeorology*. W. G. V. Balchin and N. Pye have studied the Microclimatology of Bath, *Q. J. R. Met. Soc.*, 1947, and M. Parry of Reading, *Met. Mag.*, 1954, *Q. J. R. Met. Soc.*, 1956; *Weather*, 1950.

IV

AIR MASSES

Within recent years the whole subject of meteorology has undergone a revolutionary change in outlook and method, and undoubtedly the fuller understanding of atmospheric processes which the new outlook has placed at our disposal will, in due course, shed much light on the study of climates. The revolution began with the work of Professor Bjerknes in Norway during the war of 1914 to 1918, and has been powerfully accelerated by the demands for accurate weather forecasting for civil and service flying since then. Stated briefly the new outlook consists in the recognition of the fact that the general circulation of the atmosphere produces in certain places large masses of air with characteristic and well-defined physical conditions, especially of temperature and humidity. The strongly marked individual peculiarities of these air masses are acquired by sufficiently long residence in one place for the lower layers to acquire the physical characteristics of the surface on which they rest, and from these lower layers there takes place a steady and progressive transmission of characters to greater heights. The process may be completed in a few days, but the stay may often be longer than this. Eventually the air body becomes characterized by a considerable degree of uniformity of conditions horizontally and a clearly marked transition of characteristics vertically. The prerepuisite condition for the assumption of such characteristics is that of stagnant or outward spreading air, for under these circumstances there can be no disturbance by the introduction of alien elements by air from outside. Thus the sources of air masses are the great permanent or semi-permanent anticyclones of the earth's circulation. Conversely, areas of low pressure are regions of convergent air masses, which, if they are derived from source-regions with widely different temperatures or humidities may, on meeting, be strongly contrasted in these respects. Such boundary zones between air masses are often very narrow and precisely defined; the transition from one to the other is abrupt, and strong disturbances are likely to develop where they are in contact, giving rise to 'fronts'.

Reference to the mean sea-level pressure maps for summer and winter shows that the chief seats of the major anticyclones, and therefore the chief courses of uniform air masses are:

1. The polar regions.
2. The cold continental masses of Eurasia and North America in winter.

3. The 'Horse Latitude' high-pressure cells, especially over the oceans in summer, but to a less extent over such large land masses as North Africa and Australia in winter.

Air masses may therefore be described as:

1. 'Polar' or 'Tropical', which in general determines their temperature conditions, and through temperature, their capacity for moisture holding.

2. 'Continental' or 'Maritime', which in general determines the extent to which their moisture capacity is realized, i.e. their relative humidity.

The Life-history of Air Masses. When an air mass leaves its place of origin it begins to be affected by the physical conditions of the land or sea over which it travels and to undergo a progressive series of changes which affect the lower layers first, and are passed upwards, slowly in stable air but rapidly in unstable, until the whole mass has altered its character and ceases to be recognizable. The successive stages in this process may be described as 'modified', 'transitional', and 'neutralized'. This must not be taken to imply that a 'neutralized' mass is inert, in fact its energy and weather-making possibilities may increase the more it is modified, for example polar maritime air: it simply states that its properties are no longer those which characterized it at the outset. The weather phenomena that develop within a travelling air mass depend entirely on these changes, that is to say, on the life-history of the mass, and it is the average of these weather conditions, due to the average duration of experience of each air mass with its expectable characteristics, that constitutes the climate of a place. If, therefore, the genesis and movements of these masses are fairly regular, their behaviour might be expected to constitute the best basis for the understanding of climate. It is necessary, as a first step, to study firstly the properties of air masses in their various states of transition and secondly the interaction of air masses possessing different characters which follow from their difference of origin and history.

The Properties of Air Masses. The primary properties that characterize an air mass have been said to be its temperature and humidity, and the distribution of these elements vertically within the air mass. From these primary properties there are derived a number of secondary characteristics which vitally affect the weather and which are of value in identifying the mass; for example, the type of cloud and precipitation (hydrometeors) and the visibility. It is obvious that a simple statement in degrees is quite inadequate to describe the temperature conditions in a large air mass of considerable vertical and horizontal extent. In the first place if air moves up or down within the mass it is cooled or warmed adiabatically in the process. But the amount of this cooling or warming, provided that there is no condensation or evaporation, is fixed by the gas laws and is expressed by the adiabatic lapse rate of 5·4°F. per 1,000 feet. We may therefore standardize the height (which means pressure) and express the temperature as that which the air would have if raised or lowered adiabatically to a standard pressure of 1,000 mbs. This is the 'potential temperature'.

In the second place the air may gain or lose heat by condensation or evaporation, the gain or loss being equivalent to the latent heat of condensation of the water vapour condensed or the water evaporated. Here, then, is another reserve of heat which may be cashed, as it were, by condensation or expended in evaporation. But the amount of heat available in this way is determined by the humidity of the air and can be calculated. In assessing the temperature of an air mass it is assumed that all this reserve of heat is cashed; the temperature thus obtained is the 'equivalent temperature' of the mass. Combining equivalent and potential temperature we have the 'equivalent potential temperature' which is the temperature that a given mass would have if lifted until all its moisture was condensed and then returned adiabatically to a pressure of 1,000 mbs.

This value is constant vertically and is a 'conservative' character of the air mass concerned, independent of changes due to expansion and compression, or to condensation and evaporation.

We now need an expression for the moisture content of the air which shall also be constant vertically and 'conservative'. Relative or absolute humidity would not do, as these change with height (i.e. with pressure), and so we use an expression defined in terms of weight, which does not change with vertical movement. This is the 'specific humidity', defined as the mass of water vapour (grams) present in unit mass (kilogram) of air.

These concepts are necessary for any precise work on air masses, and especially for weather forecasting, but expression in more general terms will usually be sufficient for climatic description and explanation.

Stability. A clear understanding of what is meant by stability and instability is essential for a proper understanding of weather and climate, for on this depend most meteorological phenomena. A punch-ball suspended between floor and ceiling by elastic is in stable equilibrium because if displaced in any direction it will return eventually to its original position. A pole balanced on its end is unstable because, when it is disturbed from its precarious equilibrium, forces come into play which disturb it still further and cause it to fall flat.

Now if a mass of dry air is raised it expands and cools at the dry adiabatic rate (γd) of 5·4°F. per 1,000 feet of ascent. If this rate of cooling makes it colder than the surrounding air it will be heavier than it and will sink back; it is stable. If, on the other hand, the adiabatic cooling rate is less than the rate of fall of temperature (lapse rate) of the

surrounding air the lifted mass will be warmer and lighter than its surroundings, and will have the urge to rise still farther; it is 'absolutely
unstable'. Thus whether a mass of dry air is stable or unstable depends
on whether its lapse rate (γ) is less steep or steeper than the dry adiabatic lapse rate (γd). Lapse rates that give 'absolute instability' are

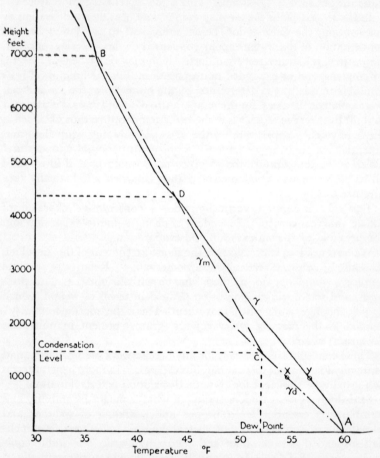

Fig. 17.—Temperature-Height diagram, conditional instability

exceptional, but there is another factor to be considered which alters
the adiabatic lapse rate, and that factor is condensation. When this
occurs latent heat is given out which delays the rate of cooling, so that
when condensation is occurring the ascending air cools at the 'saturated
adiabatic' (γm), or 'moist adiabatic' rate which is the dry adiabatic
minus the latent heat. Since the amount of latent heat made available
depends on the amount of water vapour condensed, the moist adiabatic

will vary with the amount of water vapour present in saturated air, which, in turn, depends on the temperature of the air. At very low temperatures (say −30°F.) the water vapour present is so small that γm is practically the same as γd, but at mid-latitude temperatures of about 50°F. it is about 3° per 1,000 feet, which is about the average lapse rate (γ), and at high temperatures (say 85°F.) it is less than 2° per 1,000 feet. It will be seen that in cold climates the air is likely to be stable even when saturated, but in hot climates saturated air is likely to be unstable up to a considerable height. It is a very common condition that the lapse rate is such that the air is stable as long as it is unsaturated, but becomes unstable when condensation occurs. This is known as 'conditional instability' and is represented graphically in Fig. 17.

Suppose the temperature of the air at sea-level is 60° and that the lapse rate is represented by the full line AB. Some of this air at a temperature of 60° is now forced upwards from sea-level; its rate of cooling is given by the line AC (γd). At 1,000 feet (point x) it will be at 54·6° (60°−5·4°). It is colder than the surrounding air at that level whose temperature is at the point y. Suppose next that the dew-point of this air is 52°, this temperature will be reached at 1,481 feet (point C). The rate of cooling from now on is slowed down to the moist adiabatic rate shown by the curve CDB. Note that the ascending air is still heavier than the surrounding air and continues to be so until the point D is reached. Thereafter it is warmer and therefore unstable until the point B is reached, after which the rising air is again colder and heavier than its surroundings, and has returned to stability.

We see that air may be stable when unsaturated but unstable when condensation is occurring (conditional instability), and this is a very common state of affairs in nature. The nearer such air is to its dew-point the more likely it is to reach the instability state and to produce rain, which, when once started, feeds the action and prolongs and intensifies the precipitation. This 'trigger' action may be supplied by vigorous turbulence or by orographic forced ascent.

Diurnal Variation of Stability. The earth receives heat from the sun by day, but loses that heat by radiation at night; the temperature changes thus brought about at the earth's surface are communicated to the lower layers of the air, thus affecting the lapse rate. The nocturnal cooling diminishes the lapse rate, and if there is little wind to mix the layers an actual inversion of temperature may be produced (curve A over land in Fig. 18). The diurnal warming of the earth, communicated to the surface air produces a temperature maximum here in the early afternoon, and the lapse rate is steepened and may exceed the adiabatic (curve B over land in Fig. 18). In this case instability arises and convection currents are produced. Air over land, then, has its stability increased by night and reduced by day.

Over the sea conditions differ, for the sea surface neither absorbs nor radiates heat readily, so that its temperature shows little or no diurnal variation; diurnal variations of stability are correspondingly small, but such as occur are in the opposite direction, for fairly active radiation goes on from the upper surface of cloud or from layers of air with high moisture content, and here the temperature falls by night, so that the temperature lapse between sea surface and this level is steepened and a nocturnal tendency towards instability is developed.

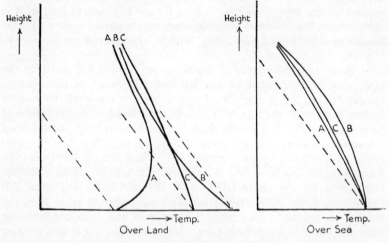

Fig. 18.—Diurnal Changes of Lapse Rate
A—early morning   B—midday   c—evening
– – – Dry adiabats

For the reasons given above all instability phenomena tend to reach a nocturnal maximum over the sea and a diurnal maximum over the land, and this tendency qualifies the inherent stability or instability of air masses described previously. Thus cold air masses, with an inherent tendency to instability, passing on to land in summer will produce cumulo-nimbus cloud and rain, perhaps with thunder, during the afternoon, as occurs in temperate continental climates, but may be made stable at night by the nocturnal radiation. Passing out to sea cold masses, despite their inherent tendency to instability, may be rendered stable by day, but will have their instability reinforced by night.

The phenomena just described explain the development of weather within air masses; we are now in a position to consider the stability and other characteristics of warm and cold masses, which, in general terms, show a high degree of uniformity over large areas. Although an air mass, as a unit, is identified with a pressure type, the same pressure type does not necessarily indicate the same air mass. It is a common experience from weather forecasting in our own climates that two high-pressure systems, even in the same locality, may be accompanied by widely different weather conditions, depending on their different histories. The fundamental concept is that of the air mass with all its concomitant weather characteristics; the pressure type is only one of the grounds for suspecting its existence. By adopting the air mass as a unit the great advantage is obtained of relating together all the meteorological characteristics, instead of, as in orthodox climatology, splitting them up into the description and statistical statement of the component elements. The method represents a considerable step towards fulfilling one of the first desiderata of any system of climatic description and classification, that of making it truly genetic.

Warm and Cold Air Masses. It should be clearly recognized that these are not necessarily warm and cold in an absolute sense, but only relatively. Thus a warm mass is one which is warmer than the surface over which it is passing and is therefore being cooled from below. In the process it acquires certain characteristics, especially an increased stability, which determine the weather within it. Conversely cold masses, passing over warmer surfaces, acquire increased instability. It has been suggested that confusion might be avoided by adopting terms which emphasize these dominant characters and referring to warm masses as 'stable', and to cold masses as 'unstable' or 'labile'.

Cold Air Masses. These have their main sources in polar or arctic regions, but owing to the extension of the continental anticyclones in winter they may spread far down the interior plains of the U.S.A., and as far south as the Himalayan mountain barrier in Asia. In their place of origin they are characterized by:

1. Low temperature due to loss of heat by radiation.
2. Low specific humidity due to the low moisture capacity of air at low temperatures.
3. Stable stratification due to intense cooling of lower layers and a consequent small lapse rate.

On leaving its place of origin a cold air mass begins to be warmed in its lower layers by the warmer sea or land; the results are:

1. The lapse rate is steepened, sometimes to such an extent that instability results and strong ascending currents arise.
2. The humidity increases, especially over a warm ocean.
3. The combination of these two factors produces cumulo-nimbus clouds and precipitation in the form of short, sharp showers with turbulence, gusts and squalls; bright intervals occur between showers.
4. Visibility is good between showers because of the initial purity of the air and the dilution of impurities by the strong turbulence.

The strong inherent tendency towards instability in cold air masses encourages mixing not only by turbulence but also by convection, which is much more effective. Its properties thus become modified more rapidly than in tropical masses. Furthermore the properties of one part of a cold air mass over hot land, for example, may be greatly changed, while those of other parts, over cold water for example, may not. Thus discontinuities may arise within the air mass itself and active fronts may develop.

As examples of cold air masses may we quote the Canadian air reaching the sea along the southern and eastern coasts of the U.S.A. in winter, and the winter monsoon of China, so cold and dry over the continent, but so unstable and rain-bearing where it passes out into the Pacific, towards and beyond Japan.

Tropical Maritime. As has already been said the chief sources of these air masses are the sub-tropical anticyclones which persist throughout the year over the oceans about latitude 30° N. and S. In its place of origin the tropical maritime air is characterized by:

1. High temperatures derived from the warm sea over which they lie.
2. High humidity in the lower layers, since abundant moisture is available and the capacity of the air for moisture is high at high temperatures.
3. A fairly stable stratification.

On leaving its place of origin such tropical maritime air as moves to lower latitudes usually becomes cooled in its lower layers by contact with the relatively colder sea or land over which it passes; the results are:

1. The lapse rate is diminished and even inverted, stability is thus greatly increased, convection is made impossible and turbulence is greatly reduced.
2. The cooling of the lower layers with their high specific humidity results in a great increase in relative humidity, and dew-point is soon reached.
3. The combination of these two factors produces fog, or, if turbulence is strong enough, stratus cloud. Cloud cover is continuous and drizzle or steady rain is produced, especially if the air is lifted, for example orographically.
4. Visibility is low because of the high moisture content and because impurities are trapped beneath the inversion or retained in the lower layers in the absence of convection.

The inherent tendency to stability in tropical maritime air practically precludes convection as a mechanism for the modification of properties. The process of modification is carried out mainly by turbulence, which is not only markedly less effective, but affects only the lower layers of air.

If tropical maritime air invades a hot continent in summer, or moves towards yet warmer seas, it becomes, in effect, a cold air mass since it is travelling into regions hotter than its source. It therefore shows, in a marked degree, the instability characteristic of cold air masses, and since the moisture charge is high, the instability rain so produced is heavy. Most east coasts in mid-latitudes derive their summer maximum of rain from these sources, e.g. the Gulf Atlantic States of the U.S.A., Natal, New South Wales, etc.

Tropical Continental. North Africa alone in the northern hemisphere produces a tropical continental mass in winter (the Harmattan), but in summer the arid western regions of the U.S.A. also become a source of tropical air. In the southern hemisphere Australia and, to a less extent, South Africa contribute air of this type. In its source region tropical continental air is dry and, in summer, very hot and unstable. The low humidity results in a high condensation level and no precipitation, despite instability. Leaving its source region it acquires moisture in its lower layers which results in convective instability and North African air, for example, contributes summer rain to south-east Europe. In winter it is dry and warm, but stably stratified because occurring in anticyclonic regions of descending air. Travel to cooler regions increases its stability, but its high temperature provides energy where it meets cold air masses, as, for example, along the Mediterranean front.

Convergence of Air Masses, Fronts. The migration of air masses from their source must result eventually in their meeting and interference, and this will occur in regions of low pressure which, by their very nature, must be regions of convergence. Experience shows that although there is inevitably some marginal mixing and incorporation, the air masses tend to retain their individual characteristics fairly clearly defined, especially in the upper air, which is less disturbed by the turbulence that affects the lower strata in contact with the earth's surface. The convergence results in the displacement from the surface of the lighter and warmer by the denser and colder air, and this ascent produces condensation and precipitation. The severity of the disturbance that accompanies this convergence is proportional to the contrasts of temperature and humidity of the air masses involved, for the temperature contrasts provide the main source of energy for the generation of storms. Storminess is greatest where polar air meets tropical, as it does along the polar fronts, and much less where tropical meets tropical along the equatorial front. Even here, however, considerable contrasts may occur. The Doldrum belt in the Atlantic is stormiest near the African coast because the convergent trade winds here have different life-histories and therefore different properties. Storminess is much less and the belt of disturbance is narrower in the mid- and western Atlantic because here the trade winds have a similar length of trajectory over warm seas of approximately the same temperature. But nearer to the North American continent storminess increases again because North American continental air masses, of very different nature, introduce real 'frontal' conditions.

Form and Inclination of Frontal Surfaces. On a stationary earth warm and cold air masses would be in equilibrium with the

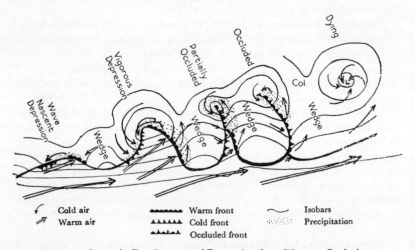

Fig. 19.—Stages in Development of Depression from Wave to Occlusion

warmer air resting on the colder and separated from it by a horizontal surface, but on a rotating earth they are in equilibrium when the separating surface is inclined to the horizontal at a small angle, which depends on the density discontinuity, the wind force and the latitude; normally, this angle is about 1 in 100 along the polar front.

In plan the shape and direction of a frontal surface may take any form depending on the extent and direction of flow of the two masses concerned; it is, in fact, like the front between two opposed armies. At this stage the equilibrium is stable and the front is stationary, the air streaming parallel to the front; but this state can hardly persist. The differential movements of air set up wind shear which produces waves, as does wind blowing over water; the frontal surface bulges up and down, and the line of intersection of the front with the earth's surface oscillates horizontally. The advance of a portion of one mass, making a salient in the line, sets up instability, and at the tip of the salient a small low pressure will form because warm air has displaced the cold. This is the birth of a depression which grows rapidly.

The Development of Depressions. Since the wave is unstable, its amplitude increases, and within twenty-four hours it has become a fully developed depression. The tongue of warm air, generally of tropical origin, thrusts deeper northwards towards the developing centre of low pressure; it is known as the 'warm sector', and its advancing edge is known as the 'warm front'. Here the warm air is rising up

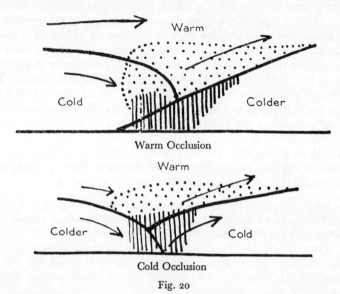

Warm Occlusion

Cold Occlusion

Fig. 20

the slope (about 1 in 100) of the cold air, and is consequently cooling by expansion, cloud and precipitation being formed in the process. The cold air swings round the salient of warm air and attacks it in the rear, thrusting underneath and displacing the warm air upwards forming a 'cold front'.

The slope of the discontinuity surface here in the rear is steeper (perhaps 1 in 50), cooling is more rapid and precipitation is heavier. The system as a whole travels generally eastwards, the best guide to its direction and speed being given by the wind in the warm sector, not the surface winds which are retarded in speed and 'backed' in direction by friction, but the geostrophic wind at about 2,000 feet. This is the state of affairs in a vigorous depression; the weather associated with it is described below.

Occlusion. The warm front loses part of its forward motion by having to rise over the colder air and so is gradually overtaken by the cold front which travels about 5 m.p.h. faster. The warm sector is thereby lifted off the ground and an occluded front results. The cold air in rear has now come into contact with the cold air in front, but these two streams may not have the same temperature and density. If the cold air in rear is colder than that in front it thrusts underneath and begins to lift it as at a normal cold front; this is known as a cold occlusion. If the opposite is the case the cold air in rear rides over the colder air in front, as at a normal warm front, and this is known as a warm occlusion.

The process of occlusion is gradual, beginning at the tip of the warm sector which is gradually squeezed out, the movement of the warm and cold fronts being rather like that of the closing blades of a pair of secateurs. The process takes two or three days to complete, which may be compared with the twelve to twenty-four hours for the formation of the fully developed depression from the initial wave disturbance. Thus a cyclone during most of its life-history is in a partially occluded state and is undergoing progressive occlusion. After this the occluded front gradually dissolves and the cyclone becomes a large whirl of more or less homogeneous air.

Weather Along the Polar Front. According to its location in relation to the position, at the time, of the polar front, a region in its proximity may have any of the following types of weather:

1. Polar air (depressions to the south or east): cold, dry, unstable, good visibility, cumulus and perhaps cumulo-nimbus cloud, local showers, perhaps thunder if air is very cold above.

2. Tropical air (depressions to the north or west), warm, moist, stable, visibility poor, stratiform cloud, sea and coast fog. During summer cloud clears by day, during winter cloud persists through the day.

3. Disturbed westerly type. Polar front.

The weather associated with depressions is described in all textbooks of meteorology, and will only be treated scantily here by means of the diagrams and sections of Fig. 21.

The following will serve as a sample of weather during the passage of a depression:

	Surface Wind	Temp.	Dew-Point	Cloud	Visibility	Barometer
Pre-Warm Front	S. to S.E.	45°	30° (outside rain) 43° (in rain belt)	Ci.→ As.→ St.→ Ns.	Good	Rapid fall
Warm Sector	S.W. (veer)	60°	58°	St. Sc.	Moderate to poor	Steady
Post-Cold Front	N.W. (veer)	48°	32°	Cu. Cb.	Good	Rapid rise

These weather types must be regarded as travelling across sea and land, undergoing, meanwhile, changes due to: (1) the progressive development of the disturbance from wave to cyclone to occlusion; (2) the development of local weather in the systems due to relief, etc.

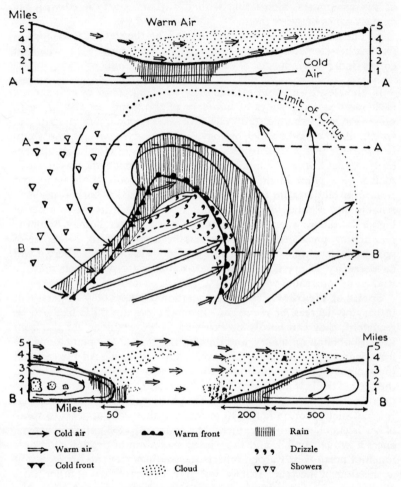

Fig. 21.—Depression Structure and Weather

Air Masses and Climate. From what has been said the application of air mass study to climatology may be expected to apply, in generalized terms, in the following ways:

1. Regions, such as those of air mass sources, which are under the influence of a single air mass throughout the year, or the season,

will have reliable climates whose characters are the sum total of the characters of the air mass concerned.

2. Regions of conflict between air masses will have changeable weather made up of samples of weather types associated with each of the air masses, alternating with disturbances associated with the contact zones between them.

3. The climate of any place, as defined by the statistical average of its weather-elements, reflects the relative frequency with which it comes under the influence of each air mass and front.

If, therefore, we could satisfactorily show the areas of influence of each mass and the zones of influence of the fronts we should be in possession of a valuable comprehensive scheme of climatic description, which, like the modern meteorology, would be dynamic. It would also be synthetic in that it considers all the elements of climate simultaneously, and as a whole. Climatology has already gone some way towards such a view-point in the consideration of monsoon, trade wind and equatorial climates, for these are already recognized as fairly consistent, reliable and regular thermo-dynamic systems. It is in the weather of higher latitudes, where non-periodic elements enter largely, that the practice of piecemeal description has been resorted to in despair of recognizing system. If the dynamic method is to be applied here it will be necessary to crystallize out quasi-stationary types which may be studied as integrated wholes.

Spells of Weather. These quasi-stationary types may be persistent in varying degrees, or recurrent. Frontal types are more likely to be recurrent; they can hardly be expected to be persistent, for the front itself is not often stationary, and the weather type at any point along the front is in a state of change as waves move along it. All the same, if the general position of the frontal zone moves over only a narrow width, the area of the belt affected will be subjected to a constant procession of depressions, each with a fairly constant pattern of changeability, so that it is not such a contradiction in terms as at first sight appears to speak of persistent changeable weather. The region affected naturally experiences a heavy rainfall, and the winter maximum of precipitation over much of north-west Europe reflects the frequency of this arrangement at this time of the year. It may, however, be established at other times of the year, as for example in the wet summer of 1903 when the rainfall of the British Isles as a whole was 146 per cent of the normal from May to October.

Weather types dictated by a constant type of air, either at its source, or at a certain stage in its transformation, are more likely to be constant from day to day as long as they remain established and to have a higher degree of persistency. The westward extension of polar continental air over the Low Countries, and even the eastern parts of the

British Isles, is a fairly common occurrence in the early months of the year, and often results in quite prolonged drought accompanied by cold east winds. The frequent occurrence of these dry spells is recognizable in the mean rainfall figures, and accounts for the February-March minimum of precipitation over a wide area (cf. p. 211).

Certain popularly recognized spells of characteristic weather, such as the 'Indian summer', the 'Ice Saints', etc., are probably due to a tendency for certain types of air flow to be established at fairly regular times of the year. Statistical examination shows, however, that their recurrence is not nearly so regular as popular opinion makes out.

At the other extreme the reliable climates of the equatorial zone, with its monotonous regularity of daily weather sequence, provides an example of practically permanent influence of a single air mass. Tropical and monsoon climates enjoy a high degree of reliability in each of the seasons of alternating air mass domination, but even here the control may break down temporarily and spells of abnormal weather may occur, such as 'breaks in the monsoon'. Ultimately it will be possible to achieve the aim of air mass climatology by defining all climates in terms of the seasonal frequency, on a statistical basis of spells of broadly uniform weather type.

Generalized Circulation Scheme. The principal effect of the contrast of land and sea surfaces is to modify or reverse the influence of the planetary circulation in the direction of producing winter highs over continents and winter lows over oceans, summer lows over continents and, less markedly, summer highs over oceans. These pressure systems may be regarded as monsoonal in the broad sense. Thus a number of semi-independent cells or wheels of circulation develop, which cog into each other along peripheral or frontal zones. Such a system, simplified and idealized from the autumn circulation of the North Atlantic and its neighbouring lands is shown in Fig. 22.

Wherever the air masses travelling outwards from the centres of dispersal are contrasted in temperature, there will arise great temperature discontinuities and a concentration of isotherms; in fact, the general conditions are ripe for frontogenesis.[1] The principal areas of frontogenesis, characterized by frequent and vigorous cyclonic activity are indicated in Fig. 22, their actual positions will be described later.

The system of circulation wheels, replacing the misleading concept of circulation belts, shows that the width of the areas of polar or of westerly winds is very variable, and that in places these wind zones narrow down and almost disappear. The polar circulation represents an attempt, dictated by thermal gradient, at a return current of air towards the great low-pressure area of vertical removal at the Equator.

[1] For a discussion on the conditions required for frontogenesis reference should be made to Petterssen's *Weather Analysis and Forecasting*, pp. 239–73, or to the original papers quoted in his bibliography.

But the attempt is normally frustrated by the intervening westerlies. The obstacle, however, is weak in places, and it sometimes happens that powerful outbreaks of polar air burst right through to join up with the trade winds and surge powerfully equatorwards. When this occurs cold waves will be experienced along the path of the polar air. Such bulges of the polar front will clearly have their best chances of success where mountain barriers impede the westerly winds, i.e. east of the meridional mountain ranges, such as the Rockies and the Urals, for these paths are situated to windward of the mountains from the point of view of the polar winds, but to leeward from the point of view of the westerlies.

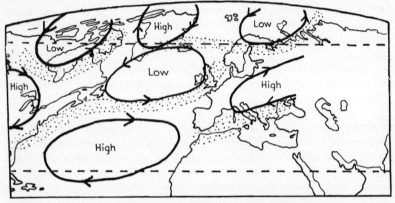

Fig. 22.—Generalized Circulation Wheels and Frontogenetic Regions in North Atlantic and Bordering Lands, Autumn

Air of polar origin, thus irrupting into the hot zone becomes pronouncedly unstable and produces unusually heavy convectional rain; could it reach the equatorial front it would make conditions there ripe for active frontogenesis.

Conversely the western slopes of the meridional mountain barriers present the most favourable corridors for westerly winds to penetrate far to the north, for the mountains give protection from the generally shallow polar continental air. In this way mild conditions are carried far northwards into the north-east Atlantic and Pacific oceans and seas. Fig. 22 shows only the surface winds and the horizontal circulation wheels, but the picture would be very incomplete if we ignored the vertical circulation which may also be compared with a number of inter-cogged wheels, though the upper half of the wheels, which must remain largely hypothetical, is omitted in Fig. 23 for this reason.

Combining the vertical and horizontal circulations we obtain the concept of the air masses as rolling balls or pillows of air of simple form but very complex movement. It is, however, with the horizontal circulation that the geographer and climatologist are primarily concerned, provided that the vertical structure is borne in mind when considering frontal zones.

Omitting and neglecting all the details of the distribution of land and sea we might expect to identify a succession of weather zones, in accordance with the general system of circulation and wind belts, from North Pole to Equator as shown opposite.

But such a succession is rendered practically unrecognizable by the profound changes in the character of the air impressed upon it by the

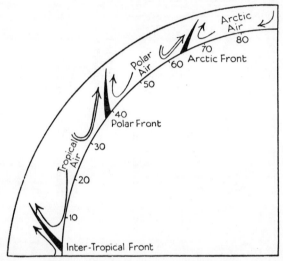

Fig. 23.—Generalized Vertical Circulation

variations of temperature and humidity over the continents and oceans. The most serious divergences from the ideal scheme are concerned mainly with wind directions, which are deflected by monsoonal influences, in the broad sense, and with the latitude of the boundaries between zones. Thus the polar frontal zone does not by any means follow a parallel of latitude, but in the North Atlantic follows a line whose mean position extends from the tropics off Florida to the Arctic Circle off Norway, and is clearly closely related to the trends of the shores of the North Atlantic and its fringing continents. Thus the hypothetical zonal arrangement has no reality and must be replaced in practice by a geographical arrangement that takes cognizance of the facts of land and sea distribution.

January Circulation. Fig. 24 shows the average conditions during the mid-winter months of the Northern hemisphere.

We are at once impressed by the complication of the North Atlantic and North Pacific regions and the adjacent land areas. By comparison

6

Approx. N. Lat.	Pressure	Temperature	Wind	Air	Stability	Cloud	Hydrometeor
90–70	High	V. Cold	N.	Arctic and Polar Source	Stable	St.	Ice Crystal Fog
70–55	Lower	Cold	N.	Polar Transitional	Unstable	Cu.	Showers
55–50	Low	Variable	Variable	Polar Front	Unstable	Ns.	Rain
50–45	Higher	Mild	S.W.	Tropical Transitional	Stable	St.	Drizzle
45–40	High	Warm	W. (Light)	Tropical	Moderately Unstable	Cu.	—
40–20	V. High	V. Warm	Calm	Tropical Source	Descending Stable	—	—
20–10	Lower	Hot	N.E. (Trade)	Tropical Transitional	Decreasingly Stable	Cu.	Showers
10–0	Low	Hot	E. (Light to Calm)	Equatorial	Unstable	Cu. & Cb.	Rain

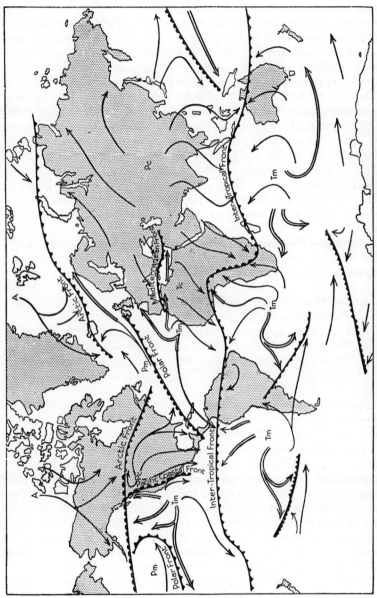

Fig. 24.—Air Masses and Fronts, January and February (mainly after Bergeton)

the Asiatic and Indian Ocean region is straightforward, as is also the Southern hemisphere. Even after making allowance for the more limited knowledge available for these areas it is obvious that the degree of complication must be much less because of the less intimate intermingling of land and sea. In consequence the system of source regions in the southern hemisphere is simple and the arrangement more nearly zonal. As there is no large land mass in the latitude belt 40° to 60° S. there are no source regions for polar continental air. The Antarctic continent supplies a source of Antarctic air, but the land mass is central and symmetrical to the hemisphere; polar maritime air develops around it. There is no tropical continental air in January (summer), and the dominant air mass is the tropical maritime that has its source in the three cells of high pressure in the South Pacific, South Atlantic, and South Indian Oceans. Tonguing into these from the north are the low-pressure cells of South America, East Africa, and North Australia, which are in effect extensions of the doldrum belt, the equatorial front.

Northern Hemisphere Winter. Fig. 25 shows a polar view of the principal air masses and frontal zones. The principal source regions are clearly marked, comprising:

1. The North Polar region of Arctic air.
2. The Siberian region of Polar continental air.
3. The Canadian region of Polar continental air.
4. The Tropical North Pacific region of Tropical maritime air.
5. The Tropical North Atlantic region of Tropical maritime air.
6. The North African region of Tropical continental air.

Spreading outwards from their source regions these masses undergo gradual transitions as previously described and eventually meet and interfere. Thus the Canadian polar air is modified on crossing the east coast of Canada and the U.S.A., and eventually meets the tropical maritime air along the Atlantic polar front oscillating in position between the West Indies and Great Lakes, but most frequently to be found extending from Florida to Iceland. Arctic air from Greenland and the Pole meets modified Canadian or modified Atlantic air along the Atlantic arctic front from Cape Farewell to Novaya Zemlya, where Siberian air also enters the field and may give rise to a front against Arctic air. Similarly frontal conditions may develop across Canada where Canadian and Arctic air are in juxtaposition.

Siberian air, modified on crossing the coast of Manchuria and China, generally forces its way over Japan before encountering Pacific tropical air and forming the Pacific polar front which, in mid-winter, is generally to be found about 500 miles out in the ocean, north-eastwards from the Philippines. Pacific tropical air comes in contact with

Canadian polar air along the west coast of Canada and the U.S.A. forming another Pacific polar front continued north-westwards as the Pacific arctic front where Arctic air meets Pacific air either of tropical or modified polar origin.

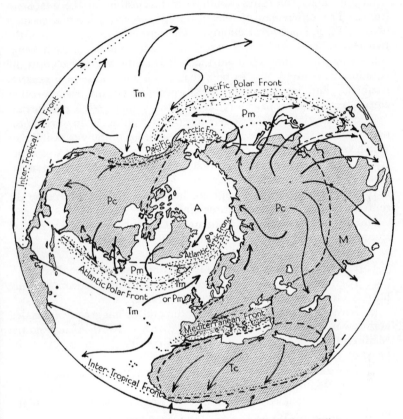

Fig. 25.—Air Masses and Fronts, Northern Hemisphere, Winter

Pc	Polar Continental	Tm	Tropical Maritime
Pm	Polar Maritime	A	Arctic
Tc	Tropical Continental	M	Monsoon

∴ Frontal Zones

Finally, Eurasian polar continental air, drawn into the Mediterranean low-pressure area, meets tropical air of Atlantic or African origin and generates an active front which is quite distinct from the Atlantic polar front. Depressions travelling along it may branch north-east-wards into south Russia, or south-eastwards into Persia and north-west India.

It is necessary to emphasize that the positions of the fronts quoted above can only be considered as their mean position in the mid-winter months. They are subject to very considerable oscillation and displacements according to the temperature distribution and the strength of the air flows concerned.

At first sight the convergence of air masses of tropical origin in the equatorial belt (doldrums) would seem favourable for the genesis of fronts. The differences of temperature, however, seldom exceeding 2° or 3°F., are not great enough. The temperature contrasts of the two hemispheres are at their greatest in late summer of each hemisphere, at which time the doldrum belt is farthest removed from the equator. It is interesting to note that this is the season of maximum frequency and intensity of tropical revolving storms, though other causes, explained on page 123, are probably more often responsible.

The equatorial front of convergence is driven far to the south (10° over the sea, 15° over the land) by the vigorous outflow of Asiatic air which constitutes the winter monsoon. This air undergoes a gradual transition from polar continental characters in its source region in the heart of Asia to equatorial characters at the equatorial front. There is therefore no sub-tropical high in the Indian Ocean to become the source region of tropical maritime air.

Northern Hemisphere Summer. Comparing the air masses of summer (Fig. 26) with those of winter (Fig. 25), we notice:

1. The source region of arctic air remains practically unaltered.

2. The source regions of polar continental air have shrunk to quite narrow bands in northern Canada and Siberia.

3. The source region of tropical continental air has greatly expanded and now includes the heart of Asia together with North Africa and the Mediterranean lands. Another tropical continental mass covers the heart of Western North America.

4. The sources of tropical maritime air in the sub-tropical highs have increased in size and moved northwards.

5. Southern hemisphere tropical air extends some way into the northern hemisphere.

6. The monsoon lands are under monsoon air which is really tropical maritime or equatorial air. It is separated from tropical continental by the great mountain ranges.

7. The eastern half of North America is invaded by tropical maritime air, thus having conditions that closely resemble the monsoon lands of south-eastern Asia, though less hot, humid and persistent.

The chief frontal areas now surround the Arctic air, occurring chiefly where Arctic air meets air from the warm continents. These temperature contrasts are slight in mid-latitudes and only become great enough for vigorous frontogenesis in high latitudes. Even there the available

energy is much less than in winter and cyclonic activity is correspondingly feebler. Wind force is also lighter in summer, and this also
contributes to a weakening of cyclonic energy.

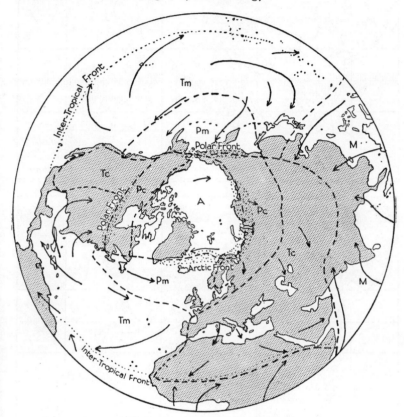

Fig. 26.—Air Masses and Fronts, Northern Hemisphere, Summer

Pc Polar Continental	Tm Tropical Maritime
Pm Polar Maritime	A Arctic
Tc Tropical Continental	M Monsoon

∴ Frontal Zones

Air Mass Climatology. It is worth while to repeat, for it provides
the most rational and embracing explanation of climatic phenomena,
that the two most important elements in climate are governed by the
nature and the source of the air mass. In the first place the temperature
of the day is most reliably forecast by predicting the type of air which
will be in residence in or in passage over the region involved. Similarly,
in retrospect, the temperature of the month or of the year is seen to
have been governed by the relative frequency of occupation or duration
in passage of air masses affecting the region. In the second place the

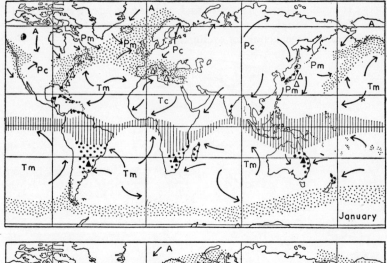

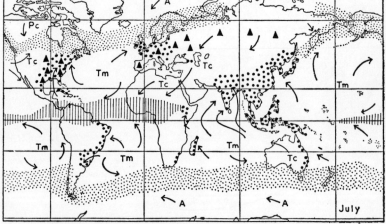

▦ Frontal and orographic rain from Polar and Arctic Fronts.
▥ Frontal and convectional rain from Equatorial Front.
⦿ Orographic and monsoonal rain from moist onshore masses.
▲ Summer convectional instability over land.
△ Winter convectional instability over sea.

Fig. 27.—The Principal Causes of Rainfall

air mass characters, combined with the local relief, provide the key
to precipitation at any point of time, while over the month, or season
or year the outstanding features of the distribution of rainfall receive
a rational explanation in the seasonal frequency of dominance of the
principal air masses. On this basis the régime of each rainfall region
is explained in Fig. 27.

SUGGESTIONS FOR FURTHER READING

The concept of Air Masses finds a place in the newer text-books on meteorology, and the following are recommended: S. Petterssen, *Introduction to Meteorology*; Petterssen, *Weather Analysis and Forecasting*; H. R. Byers, *Synoptic and Aeronautical Meteorology*; D. Brunt, *Physical and Dynamical Meteorology*. Suggestions for its application to climatology are made by T. Bergeron in 'Ritchtlinien einer dynamischen Klimatologie', *Met. Zeitschr.*, 1930 and by A. Austin Miller in 'Air Mass Climatology', *Geography*, 1953. Air Masses are the fundamental units used in *Everyday Meteorology* by A. Austin Miller and M. Parry, 1958. J. Namias writes on *An Introduction to the Study of Air Mass and Isentropic Analysis*, and J. Gentilli on 'Air Masses of the Southern Hemisphere', *Weather*, 1949. The literature has grown immensely in recent years; a list of the more important contributions is given in Petterssen's *Weather Analysis and Forecasting*.

V

THE CLASSIFICATION OF CLIMATES

Need for Classification. The almost unlimited combinations of climatic factors acting on an almost infinite variety of topography produce a bewildering number of geographical climates, and it is clear that any system of classification adopted can recognize only the broadest types unless it is to become unwieldy. But in spite of the seeming complexity it becomes clear on closer examination that certain combinations of climatic elements repeat themselves with some degree of regularity in different parts of the world, and it is convenient to recognize each type and to give it a name.

Climatic Provinces. Supan, in 1896, suggested a division of the world into 35 provinces, each characterized by a certain combination of climatic elements. These provinces are reproduced in Bartholomew's *Atlas of Meteorology* (the 35th, not mentioned there, is the Antarctic province). This is simply a list, not a classification, but these provinces might be arranged in groups together (e.g. the desert provinces, 12 Saharan, 14 Kalahari, 17 Inner Australian, and 31 Peruvian) to form a classification. The question now arises as to what principles should be adopted in sorting and co-ordinating geographical climates, and clearly this will depend on the purposes for which the classification is required.

The Objects of Classification. Classifications are of two kinds, on the one hand those intended to show genetic relationship and on the other 'Classifications of Convenience' based on certain similar effects, by the co-ordination of which easier memorization is effected. One might, for example, group together all the desert climates irrespective of their cause of aridity; that would be a classification of convenience without genetic basis. Or one might group together all the monsoon climates irrespective of their temperatures and rainfalls; that would be a classification for which there would be a true genetic basis, namely, the fundamental control of the seasonal wind reversal, springing from the same ultimate cause, but it would bring together climatic types inducing such enormously different biological responses as those in the Ganges delta and in Manchuria. The ideal classification is one which combines the advantages of both, and thus groups together provinces with a similarity of elements resulting from a similarity of causes.

Pure and Applied Climatology. If climatology is to be considered as a science in its own right it will eventually be compelled, as other sciences have felt compelled, to adopt a genetic classification, probably

78

with the air mass as the fundamental unit, because, in the light of present knowledge, the air mass appears to be the parental stock from which climatic types evolve by environmental influence. This gradual evolution of finally distinct types has been treated in the previous chapter.

More often, however, climatology is not a 'pure' but an 'applied' branch of knowledge, studied as an environmental influence in relation to agriculture, health, air conditioning, geography, etc. Each of these branches of knowledge makes its own demands and is interested in some aspects of climate and not in others; eventually each will define climates in the manner most closely applicable to its own concern and will use methods of assessment best suited to its needs. Thus medical climatology is well on the way to defining 'environmental warmth' in terms of the rates of heat loss from the body; agricultural climatology is devising elaborate assessments of the moisture factor affecting the needs and the use of water by growing crops. It is fairly safe to assume that these devices will find their place in the delimitation and description of medical and agricultural climates. But the geographer, for whom this book is primarily intended, is concerned with climate as a regional influence and as one of the means of defining geographical regions.

Geographical Climatology. Geographical regions are of many orders of magnitude, major, minor and minute. The smallest sub-regions are generally defined in terms of geology, soil, population density, economic preoccupations or some such factor varying rapidly from place to place. Climate is never used on this scale, for within such small distances it varies only slightly and that in respect of micro-climates only. Regions of the next order of magnitude are most frequently defined in terms of relief units. It is true that sudden changes of altitude or relief often bring about marked changes of climate; the Great Valley of California, for example, is clearly marked off by its heat and aridity from the mountains that overlook it. But the relief is the real determinant of geographical character and this is recognized in the name by which it is known in geographical literature. It is only over very large distances that climatic contrasts are powerful enough to be the real determining factor of geographical individuality. The geographer, therefore, finds climate useful, in fact indispensable, in the recognition of major regions on a sub-continental scale, and on such a scale the finer and subtler points of differentiation are not only irrelevant, but are impossible to set down on a map.

The Importance of Biological Responses. As a geographical science climatology is vitally concerned with the causal relationships between certain distributions, and regards as essential criteria the influences which climatic conditions exert on the sum total of human activities. Perhaps the truest indicator of these is afforded by vegetation

type which, in turn, invokes certain clearly defined physical and
economic responses. It is unfortunately true that vegetation reflects
not only the sum total of climatic influences, but also of soil conditions,
ground-water supply, human interference and a host of other controls,
yet in broad outline it cannot be doubted that climate is the predominant control of major vegetational regions. Other factors may
considerably modify the margins of the type, but climatic provinces,
on a large scale, may fairly safely be judged by the characteristic
physiognomy of their natural vegetation. Any classification which
develops and stresses such relationships is of value to the geographer,
but to choose vegetational environment as a basis of classification is
only a first step. The data of climatology are not the facts of vegetational
distribution but the statistical values of the climatic elements, and our
next step is to discover those elements which most potently influence
vegetation types, for we may then accept these as the true criteria. It
may be said at once that temperature and rainfall are the elements
required; either by actual amount or seasonal distribution their control
is paramount.

Classifications Based on Temperature. The temperature control
of climatic zones was recognized by the Greeks, and although the
classical division of the earth into torrid, temperate and frigid zones
was delimited by the mathematically and astronomically defined lines
of the tropics and the polar circles, there was, as the names imply, a
recognition that temperature was all-important. But such a scheme fell
short in ignoring the effects of land and sea. These zones are, however,
still recognized as the fundamental divisions, although under new
names and delimited in a different manner, namely, by actual temperature values which reflect marine and continental influences as well
as insolation values.

Although the interval of 10°C. seems to have been an attempt to
regiment climate into the decimal system Supan chose the mean annual
of 68°F. (20°C.) as the limit of hot climates. It conveniently encloses the
trade wind belt and practically coincides with the limit of palms and is, in
most respects, a very convenient boundary, although its advantages are
coincidental rather than causal. The polar zones he proposed to define
by the isotherm of 50° (10°C.) for the warmest month and in this he
appeared to have a line of real significance, for it is summer warmth
that matters in cold climates. A single month, however, is not enough
and trees need at least 3 months of adequate temperature to get through
their reproductive cycle, which begins at about 43°F. or 6°C., as
Schimper showed. They are indifferent to a really cold winter which
they avoid by a resting period. Supan's limit seemed at the time to provide a useful transition from the Arctic forest to the treeless tundra.

Köppen's Temperature Zones. These apparently significant
figures (68° and 50°) were adopted by Köppen in a more complete

classification, based on the duration of each station above, between or below these critical values as follows:

1. Tropical belt. 12 months above 68°.
2. Sub-tropical belts. 4–11 months above 68°; 1–8 months between 50° and 68°.
3. Temperate belts. 4–12 months between 50° and 68°.
4. Cold belts. 1–4 months between 50° and 68°; 8–11 months below 50°.
5. Polar belts. 12 months below 50°.

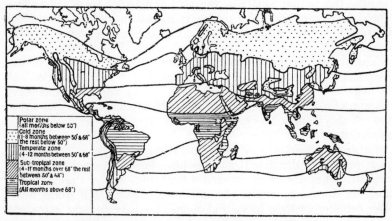

Fig. 28.—Köppen's Temperature Zones (simplified)

A much simplified map showing the distribution of these zones is given in Fig. 28, but it should be realized that the temperature values used are actual values not reduced to sea-level; the 'isotherms' are therefore not isotherms in the usual meaning of the term. On a map of such a small scale there is inevitably much confusion in mountain districts where the zones are crowded together and the map has been considerably simplified for the sake of clearness.

This classification at least has the merit of symmetry and simplicity, but it is hardly to be expected that climatic types will conform to so rigid a scheme. For example, the isotherm of 50° for the coldest month becomes the poleward boundary of the sub-tropical climates (at sea-level). But this line runs through Sardinia, the toe of Italy and Morea, thus excluding the whole of the northern Mediterranean and the greater part of the peninsulas of Italy and Greece from the sub-tropical climates. The Californian Mediterranean climate also becomes 'temperate', being grouped with the cool temperate climate of British Columbia. Cape Town and Swanland, too, are excluded from the sub-tropical climates.

The Algebraic Classifications. The most widely used classification today is a later one, also devised by W. Köppen who devoted a long, prolific life to the search by trial and error for a series of formulæ which would define vegetation regions (called climatic regions) in terms of various values of temperature and rainfall. Each climatic type is designated by a shorthand notation replacing the well known, but imprecise and rather lengthy terms in general use. For example the Mediterranean climate of Cape Town becomes *Csb;* *C* (Mesothermal) because the coldest month has a temperature between 18°C. and −3°C., and the warmest month exceeds 10°C.; *s* (Summer-dry) because the wettest month of the cold season has more than three times the rain of the driest month of the warm season; *b* (equable) because the warmest month is below 22°C., but at least four months exceed 10°C.

The system has been applied to most parts of the world in the *Handbuch der Klimatologie,* edited by Köppen and has been tried out by many authors in other regions, both small and large. It is, therefore, necessary for all students of climatology to be familiar with its principles, though they cannot be expected to remember its many complicated definitions of types.

Other classifications which resemble Köppen's in being quantitative and in using a shorthand notation have been suggested by Thornthwaite. They are based on careful field study in the United States, and the equations, empirically discovered, are applied world wide. The defining formulæ in such classifications have become extremely elaborate and appear to give a highly precise definition. But those who have tried them out generally find errors that make it necessary to suggest alterations and refinements in order to obtain a closer correspondence in the area of their study. So before accepting such classifications, or any one of them, let us see whether there is good reason for acknowledging, or hope of discovering, such a precise relationship between vegetation and the value of climatic elements as is implied to exist. It must be remembered that:

1. All vegetational (and climatic) boundaries are of a transitional nature; there is never a sharp change point.

2. Though vegetation is mainly controlled by climate, it is also influenced by edaphic, biotic and microclimatic factors that displace the boundaries locally.

3. 'Mean Values' of climatic elements are unreal; extremes are often significant limiting values.

4. Temperature and rainfall form the basis of the classifications; other qualities of the climate, e.g. wind or sunshine, influence vegetation but are ignored. The complexity and the variable nature of these interacting factors defy expression in a single formula of manageable proportions.

5. Neither climates nor vegetation belts are static. So soon after a glacial period fluctuations of climate are still occurring, for which abundant evidence is available; if vegetation belts are limited by climate, as is assumed, they may be expected to follow in the wake of changing climates, with a lag that has not been determined.

The Climatic Year. The values used in such classification are, of

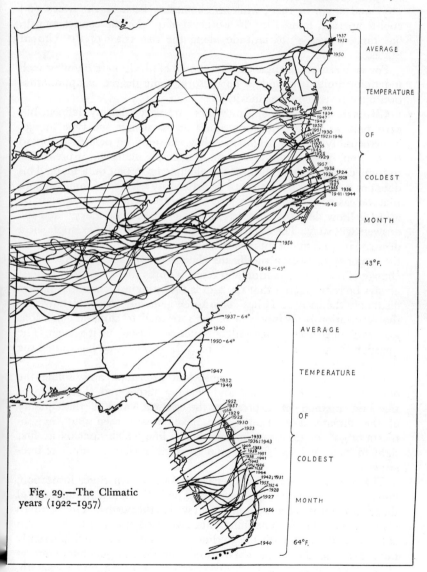

Fig. 29.—The Climatic years (1922–1957)

course, mean values over the period of 35 years, but in any individual year the figures are departed from quite considerably. This does not affect the choice of crops selected in anticipation of the year, nor is it considered in the choice of long-term, e.g. tree crops.

The northern limit of the tropical climates, defined by the isotherm of 64 for the coolest month, has extended as far north as Savannah in 1937 and again in 1940, while the boundary between the warm and cool-temperate, limited by the isotherm of 43° for the coolest month, has ranged over 4° of latitude along the east coast of the United States.

For such reasons as these the attainment of precision in defining such indefinite limits is bound to be illusory: some degree of approximation must be accepted as inevitable.

Climatic Boundaries. Since these are never sharply defined, but exist only as broad transition zones across which one type merges imperceptibly into its neighbour, they can only be defined in broad terms, and, other things being equal, a simple definition is preferable to some complex value difficult to calculate or to plot on a map. This does not mean that the limit can be arbitrarily selected as some isopleth that, on casual inspection, seems to fit the case. The first question to be asked is 'how does the climate change across the boundary?' and the answers will vary. The essence of the transition may, for example, be a decrease in quantity of summer warmth or the onset of a dry season. In the majority of cases a significant change of climate is reflected in a change of natural vegetation and often of cultivated crops, but it must always be remembered that the aim is to distinguish climates that are distinct in themselves. It has come to be generally accepted, however, that vegetation is an indicator of the sum total of climatic conditions and it will be in accordance with accepted practice to use it as such. The question then becomes 'what is there in the climate that has caused the vegetation to change?'

As a general principle we must first determine and isolate the operative climatic influence and, having discovered that, seek for the best method of expressing the critical value of that quality of the climate. For instance, we have already seen that the isotherm of 50° for the warmest month is a line which appears at first sight to have real significance and to circumscribe the limits of tree-growth.

The Timber-line, a Temperature Limit. On closer inspection, however, it fails to do so everywhere. The isotherm lies to the south of the timber-line in marine climates, but well to the north of it in Northern Canada and Siberia. The fuller data now becoming available in high latitudes tend to discredit the 50° line, which always seemed inherently improbable, for a tree does not live and reproduce by July temperatures alone. Some measure of the duration and intensity of summer warmth, adequate for growth, is needed. Considering duration first, a growing season of three months above 43° F. gives a closer correspondence, but has the same failing, though to a less degree, that the line runs too far north in maritime climates and too far south in continental areas. The higher temperature of the midsummer months in the latter seems to be vital. If this is estimated as 'Accumulated temperature' by adding up the degrees above 43° for each month we may express this value in 'month-degrees'. A very close correspondence appears to exist between the line of 18 month-degrees and the timber-line in all but the most maritime climates of the Northern Hemisphere. With another empirical approach, Thornthwaite disclosed a close correspondence with a 'Temperature Efficiency Index' of 16, obtained by summing 12 monthly $T-E$ indices, each calculated from the formula $i = \dfrac{T-32}{4}$. In these ways a nearer approach to perfection is achieved, but only at the cost of increased complication. Because it is easy to recognize by inspection of the monthly means of temperature, the criterion of three months above 43° will be adopted here as the limit of cold climates, despite its minor inaccuracies.

Aridity Index. The definition of aridity is equally elusive. The simplest arbitrary limit of the desert might be taken as the 10-inch annual isohyet, and a dry month might be defined as one with less than 1 inch of rain. Aridity, however, is not solely a matter of rainfall, it is increased by high evaporation which, in turn, is accelerated by high temperature, low atmospheric humidity, strong wind and low pressure. Any formula which professes to define aridity must incorporate these variables, but most of them only take cognizance of temperature. Many authors regard 20 inches as the desert limit in hot climates and 10 inches in cold climates. De Martonne defined a dry climate as one in which the mean annual rainfall (measured in cm.) is less than double the mean annual temperature (measured in degrees centigrade). Later he proposed a modified 'Index of Aridity' $I = \dfrac{P}{T+10}$. Köppen proposed a rather elaborate formula which incorporated provision for different types of seasonal incidence. Thornthwaite, using empirical methods, fixed the limit of desert climates by a Precipitation Effectiveness Index of 16 obtained by summing twelve monthly P/E ratios calculated from the formula $P/E = 11 \cdot 5 \left(\dfrac{P}{T-10} \right)^{\frac{10}{9}}$. Other tentative estimates of Saturation Deficit, Precipitation/Temperature ratios, etc., have been proposed, varying in their complexity, but if the conclusions are set down and compared as nearly as is possible, bearing in mind their different forms, considerable discrepancies are revealed in the delimitation of aridity, as shown by the following table:

7

Annual Rainfall defining the desert limit according to	Mean Annual Temperature					
	°C. °F.	°C. °F.	°C. °F.	°C. °F.	°C. °F.	°C. °F.
	25 77	20 68	15 59	10 50	5 41	0 32
	cm. in.	cm. in.	cm. in.	cm. in.	cm. in.	cm. in.
Köppen (1918) .	32 13	29 11	26 10	23 9	20 8	16 6
Köppen (1923) .	41 16	36 14	31 12	26 10	21 8	16 6
Thornthwaite (1931)	37 15	30 12	25 10	20 8	15 6	— —
De Martonne (1) .	50 20	40 16	30 12	20 8	10 4	— —
De Martonne (2) .	35 14	30 12	25 10	20 8	15 6	10 4
Miller (1950) . .	40 15	35 14	30 12	25 10	20 8	15 6

It is clear that no unanimity exists as to the limit of aridity, or at least that the values of the climatic elements are incompetent to define it. Refinements of accuracy, for what they are worth, are only attained at the sacrifice of simplicity, and the boundary, arrived at by most of the above methods, is a value difficult and tedious to work out and set down.

Any of the above formulæ may be adopted as the boundary of the desert climate. De Martonne's (2) has the merit of simplicity; if the data are in metric units one can fairly quickly determine whether the climate of a place lies beyond the limit of aridity (i.e. if it is a desert). But a still simpler formula can be deduced from Fig. 30, in which the index letter of the vegetation is placed at the point defined by its mean annual temperature and rainfall. The line whose formula is $R = \dfrac{T}{5}$ appears to mark off the deserts fairly satisfactorily, except, perhaps, on the border of the Mediterranean climates, where the concentration of rain in the cold season results in diminished loss by evaporation. So when using Fahrenheit degrees and inches of rain, as is the practice in Britain and America, it is convenient to notice that if the annual rainfall is less than a fifth of the mean annual temperature the vegetation will probably be 'desert', with more it will be scrub or steppe or savanna, according to the temperature.

From the examples quoted above the conclusion is reached that a single climatic element can seldom satisfactorily define any climatic region, and where such a limit is used it must be interpreted liberally and regarded only as the best available.

Climate and Vegetation. It has already been laid down as a general principle that a satisfactory classification of climates must reflect the climatic control of vegetation; that climatic provinces must coincide as closely as possible with the major vegetational regions of the globe. It seems likely, therefore, that the means of classification may best be discovered by a brief examination of these controls.

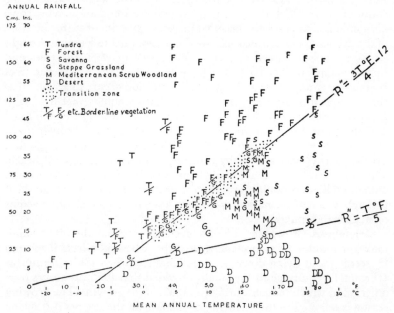

Fig. 30.—Climatic Requirements of some Major Vegetational Formations

Broadly speaking, vegetation type (i.e. forest, grassland or desert) is conditioned by humidity, either in the form of rainfall or, more rarely, as moisture of the air; the flora (i.e. selva or taiga, savanna or steppe) by temperature. There are exceptions to this general statement; the poleward limit of forest growth, for example, is set by temperature, and furthermore soil conditions (edaphic factors) may overrule climate, causing grassland to grow where the climate would indicate forest (e.g. limestone soils) or forest where the climate would indicate grassland (e.g. water-holding clays), but in the main it expresses the true relationship between climate and vegetation.

Forest Climates. The essential requirement for tree-growth is a constant supply of moisture available for the roots. A season of prolonged drought is a condition adverse to but not prohibitive of forests, for the soil may store up sufficient moisture to supply the tree through the dry season. Monsoon forests, for example, survive a drought of four to five months with the aid of the heavy rainfall which precedes it. Still longer drought can be safely endured when adaptations exist for the economical utilization of water supply. Various checks on transpiration, such as reduced leaves, thickened cuticle, woolly or hairy leaf coverings serve this purpose; other trees store up water in trunks or leaves, others dispense with leaves during the dry season. But under such adverse conditions tree-growth seldom attains the status of forest which is typically hygrophilous; woodland having these xerophilous characteristics becomes increasingly evident as the duration of the dry season increases; such woodland, adapted to seasonal drought (tropophilous) is known as scrub. The space between the trees and bushes opens and becomes filled in with herbaceous plants, the trees become rarer and rarer and a gradual transition occurs to grassland as the drought grows longer. In climates that combine great heat with great drought in a prolonged dry season, even grasses are inhibited and the transition from savanna to desert takes place through thorn forest and thorn bush to open scrub, with bare earth between the dry, brittle, thorny bushes or fleshy cactuses, as in the Caatingas of Brazil. The degeneration of sclerophyllous woodland in Mediterranean climates follows a similar pattern, a transition to desert being attained through a scrub formation, variously known as chaparral, macquis or mallee and mulga.

Grassland Climates. The life cycle of grass, on the other hand, makes it indifferent to prolonged drought, whose ill-effects it escapes by seeding and dying down, to start life anew when the rain returns. Its rainfall requirements are small, provided that the rain occurs over a period in which temperatures are sufficient for growth. The growing season is adapted to the incidence of rainfall; on the steppes it is spring and early summer, in the savanna summer, while in desert regions grasses and other plants seize the opportunity presented by any adequate fall of rain at any season to begin rapid growth. The summer rains of continental interiors are well suited to grassland, but the prolonged winter drought makes these areas unfavourable for trees which, generally speaking, prefer maritime or sub-continental climates with their more regular moisture supply.

Clearly the length of the dry season rather than the actual amount of rain is the chief factor in determining vegetation type; forest, which is hygrophilous, passing into scrub, which is tropophilous, and that into grassland and finally desert, which is xerophilous, as the dry season becomes longer.

Types of Forest. The principal varieties of forest type are:

1. The broad-leaved evergreen.
2. The broad-leaved deciduous.
3. The coniferous.

In the first of these growth is continuous throughout the year and such forests can exist only where temperature and rainfall conditions are favourable for continuous growth all the year round; i.e. where the mean temperature never falls below 43° and where a well-distributed rainfall gives an unfailing water supply. These include the equatorial and tropical rain-forests with little or no seasonal variation of temperature and the sub-tropical rain-forests such as are found in extra-tropical latitudes on the eastern sides of continents up to about 40°, as in South Africa from Knysna to Natal, in New South Wales, South China and Japan and the south-eastern states of the U.S.A. On western margins in similar latitudes the drought of summer demands economy of water; the trees and shrubs are of a xerophytic type and rarely attain the status of true forest.

In the second type there is a resting period, at the beginning of which the tree sheds its leaves to grow a new leaf cover as soon as conditions again become favourable. This resting period is imposed in hot climates by drought but in cold climates mainly by temperature,[1] i.e. there is a cold season with temperatures below 43° which checks active growth. The isopleth of six months with 43° (6° C.) thus marks approximately the boundary between the evergreen and the broad-leaved deciduous forest.[2] The position of this line is shown in Fig. 31.

The third type also undergoes a long period of enforced inactivity imposed by the physiological drought (=cold) of winter. But by the survival of the leaves through this resting period the coniferous tree is ready to begin assimilating without delay as soon as temperatures become favourable, whereas the deciduous tree must spend some time in growing its assimilating organs. Further, the coniferous type of fructification has this advantage, that it is pollinated one year and dispersed the next, whereas the deciduous tree has to pass through the whole process in one year. Coniferous forest is clearly able to exist with a much shorter growing season than the deciduous type. Length of growing season, i.e. the number of months with temperatures above the basal temperature for growth, is clearly the important control of forest type. On the map (Fig. 31) are inserted figures for numerous stations showing the number of months at each with temperatures above 43° F., i.e. the duration of the growing season. A strong coincidence is revealed between the southern limit of the taiga and a six-month growing season. In other words, the deciduous forest requires

[1] It is probable, also, that the darkness of winter in higher latitudes has some influence on the adoption of the deciduous habit, for the photosynthetic function of the leaves must be much reduced in activity at this season.

[2] See A. F. Schimper, *Plant Geography*, p. 417. Note that the figure 42° has often been used as the zero of plant-growth, and is adopted by the Meteorological Office for the calculation of accumulated temperatures (*Book of Normals*, Section II, 5.C). Nevertheless, Schimper's figure of 43° seems to give better results in the case of tree-growth and will be adopted here.

at least six months with temperatures adequate for growth, and if this is not available the deciduous forest gives way to the coniferous. The six-month isopleth runs from Bergen, through Oslo, Gefle, Helsinki and Leningrad to Kazan on the Volga. In North America, owing to the extraneous factor of post-glacial dispersal, the deciduous forest is not so clearly developed, but both the Pacific coniferous forest and the mixed temperate forests of the Atlantic margin contain deciduous species and correspond to the temperate deciduous forests of western Europe. The sub-arctic forest, which corresponds to the taiga, shows the same broad relation to the six-month isopleth. It will be noticed, however, that there are minor exceptions in the heart of North America, in the extreme east of European Russia and in Manchuria (in fact, in the more extreme continental climates); deciduous forest thrives here with a growing season as short as five months. This is doubtless to be accounted for by the combination during the summer months of abnormally high temperatures and adequate rain. At Blagovestchensk, for example, although the temperature is above 43° from May to September only, the figures for these five months are 50, 63, 70, 66, 54. This may be compared with Riga where, though May to October have temperatures above 43°, the figures for these six months are 51, 60, 64, 63, 55, 44. Adding

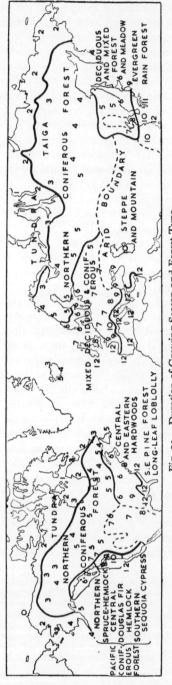

Fig. 31.—Duration of Growing Season and Forest Type

Figures indicate the number of months with mean temperatures above 43°

together the excess over 43° for each month, these might be stated as 88 and 79 month-degrees respectively (cf. p. 10 and p. 33).

Deciduous trees are essentially plants with two distinct and strongly marked habits of life, hygrophilous and xerophilous, and require a long transitional period in which to pass through their rather long and elaborate stages of transformation. Such prolonged spring and autumn are found only in marine climates, the transition from winter to summer and back again in continental climates being sudden and complete, and consequently better suited to the coniferous evergreen habit.

Further, the coniferous tree can exist on a much smaller rainfall (about 15 inches) than the deciduous. East of Kazan, for example, the deciduous forest dies out altogether, not because of temperature, but because of deficient rainfall. Eastward from this point, therefore, the southern boundary of the taiga is set by aridity, approximately by the $\frac{T}{R}=5$ isopleth which divides it from the steppe. Kazan is therefore a meeting-point of three vegetation types, the deciduous forest, the coniferous forest and the steppe. There is, then, a double reason for the replacement of deciduous by coniferous forest on passing from western margin cool-temperate climates towards the continental interiors, namely, the degeneration of the seasons of spring and autumn and the increasing aridity.

Fig. 32 shows graphically the duration of the growing seasons at eight stations in forest climates of different types. The upper curve is the curve of monthly temperature and the lower line is the 43° line. The area enclosed between the two lines is the growing season and is shaded except where, as at Athens, absence of rain sets a check to natural growth. Akassa illustrates the 'selva' type of evergreen forest with an abundant margin of temperature in all months and an adequate rainfall, well distributed. Rangoon illustrates the deciduous monsoon forest type, growing despite a dry season of four or five months. Galveston illustrates the sub-tropical evergreen type of warm-temperate rain forest, while Athens illustrates the sclerophyllous woodland (evergreen) with adequate temperatures in all months but a summer season of inactivity during drought. Paris illustrates the typical temperate deciduous forest with a growing season of eight months and transitional seasons with temperatures just about the zero of growth. Harbin illustrates the rather abnormal deciduous forest of Manchuria with a growing season only about five months long but with a large margin of temperature in the midsummer months. Trondheim illustrates the maritime type of coniferous forest climate with plenty of rain, but little surplus warmth in a growing season five months long, while Yakutsk illustrates the continental 'taiga' type with a very short growing season and no transitional period; note the steep rise of the temperature curve and its rapid passage from temperatures below to temperatures

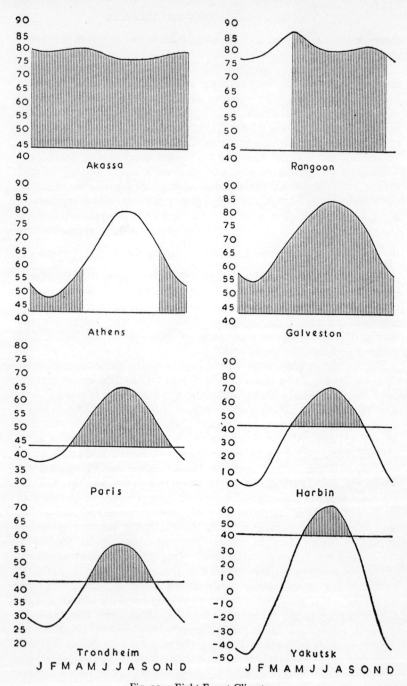

Fig. 32.—Eight Forest Climates

above the basal temperature for growth in May and back again in September.

This short discursion into plant geography brings to light the vital part played by two very important considerations:

1. The seasonal distribution of rainfall, and particularly the length of the dry season, if any.

2. The seasonal distribution of temperature and particularly the length of the cold season, if any.

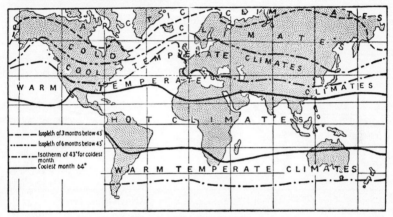

Fig. 33.—Temperature Zones

The former is more important in low latitudes and will be used to subdivide the hot climates, the latter is more important in high latitudes and will be used to subdivide the temperate and cold climates.

Homologous Climates. A classification based on these two considerations would have as its foundation the variation of two climatic elements, but these result from the interplay between the two fundamental factors of the planetary circulation and the distribution of land and sea, for the prevailing wind, acting on the continental blocks, will tend to produce corresponding climatic types in corresponding latitudes and in corresponding positions with relation to the coast-line. The trade winds, for example, blowing on-shore on eastern margins in, say, lat. 20°S. give constant rains and high temperatures; but blowing off-shore on western margins in the same latitude, they give rise to deserts, whose temperatures are moderated along the littoral by the upwelling of cold water which the winds induce. Similar combinations of elements, in fact, will result from similar causes, and it should be possible to devise a classification based on a similarity of factors and having, therefore, a similarity of elements, thus possessing the twofold advantages of convenience and a foundation in first causes.

SUBDIVISION OF THE ZONES

Fig. 33, showing the duration of the temperature seasons, and Fig. 34, showing the seasonal distribution of rain, should now be compared with maps showing the distribution of pressure and the prevailing wind directions for January and July.

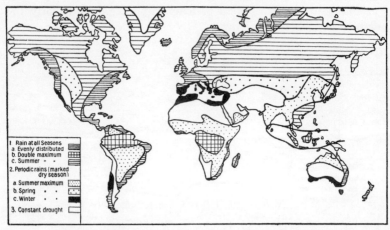

Fig. 34.—Seasonal Distribution of Rain (Highly Generalized)

Subdivision of the Hot Zone. Supan had delimited the hot climates by the mean annual isotherm of 68° (20° C.); actually 70° gives a better limit in many respects. Köppen, in defining the Tropical belt as having '12 months above 68°' really adopted the isotherm of 68° for the coldest month as the boundary. Later he preferred the isotherm of 64° (18° C.) which in fact runs very close to the mean annual isotherm of 70°. The essential quality of the hot climates is that they never experience temperatures too low for plant growth; even the 'winter' is hot, and crops requiring considerable heat for ripening can be grown, moisture permitting, at any time of the year. There is therefore much to be said for delimiting climates which are 'continually hot' by means of a 'coldest month' isotherm, and it will be used here. Defined in this way the hot climates, except where influenced by altitude, never experience temperatures too low for active plant growth; rainfall availability will therefore be the deciding influence on vegetation type. Within the zone there are four types of rainfall régime:

1. Rain at all seasons, with double maximum.
2. Rain at all seasons, evenly distributed.
3. Periodic rains with marked summer maximum.
4. Constant drought.

Turning to the pressure and wind map these are seen to correspond to:

1. Regions which are visited twice a year by the belt of calms and rains as it follows the sun, but which are never far enough from it to suffer drought.

2. Regions where the trade winds are on-shore and therefore bring relief rains, except when, locally and temporarily, the low-pressure belt of calms brings convectional rains.

3. Interior regions visited by the doldrum belt of calms and rains once a year, or, if twice, at intervals so close together as to merge into one wet season. For the rest of the year they are visited by the trade winds bringing a dry season. Beyond the limit of the swing of the rainfall belt these regions receive summer rains from the trades, drawn farther inland by a monsoonal effect.

4. Interior regions perennially under the influence of the trade winds blowing over land.

These correspond to four types of climate known as:

1. Equatorial.
2. Tropical (Marine) or Trade Wind Coast.
3. Tropical (Continental).
4. Desert.

The first three types require further subdivision, for in addition to the normal type they include distinct varieties characterized by a monsoonal régime.

The Temperate Zone. Subdivision of the hot zone has been made on the basis of seasonal régime of rainfall, but in the temperate zone the consideration of greater significance is the seasonal régime of temperature. Within the tropical zone the apparent movement of the sun results in only small changes of temperature but in considerable changes of rainfall; the seasons here are the 'wet' and the 'dry' seasons. In the extra-tropical zones, on the other hand, seasonal changes of a planetary nature produce changes of temperature which are often very pronounced, and although there may also be important changes in rainfall, temperature is the real determinant of season. Towards the equatorward limit of the temperate zone the winter is only a 'cool season', not severe enough to prevent plant growth completely, but as we proceed polewards the winters become progressively more severe. Thus we may distinguish between: (1) Temperate climates without a cold season (Warm Temperate or Sub-tropical) and (2) Temperate climates with a cold season (Cool Temperate), separated by a temperature of 43° for the coldest month, a value which, as we saw on p. 89, divides the evergreen from the deciduous forest (see Fig. 31). The second type requires further subdivision according to the

duration of the cold season. We have seen that the restriction of the
season of growth sways the balance of favour from the deciduous to the
coniferous habit and that the transition occurs approximately when less
than six months become available for growth. Further, the increasing
length of the cold season polewards restricts agriculture and gradually
forces on man other means of living such as fishing, hunting and stock-
keeping. We may therefore subdivide the cool temperate zone into:
(1) Climates with a short cold season (1–5 months below 43°), and
(2) Climates with a long cold season (6–9 months below 43°).

This last is by no means 'Temperate' although included in Supan's
temperate zone; it would be better described as a 'cold climate', in
which case Supan's Cold Caps may be promoted to 'Arctic Climates'.

We have now recognized three subdivisions of the temperate zone
according to the length of the cold season, but each of these embraces
at least two different types according to distance from marine influence,
which, in these latitudes, comes chiefly from the West. This marine
influence makes itself felt in: (1) a restriction of the annual and diurnal
range of temperature and (2) a modification of the seasonal régime of
rainfall, which in marine climates has a uniform distribution with a
tendency to an autumn and winter maximum (orographic and cyclonic),
but in continental climates has a spring and summer maximum (convectional). As an approximate guide a difference of 30°F. between
the mean temperatures of the hottest and coldest months may be
accepted; both in Europe and in western North America it runs
fairly close to the line dividing predominant winter rain from a summer
maximum.

The Warm-Temperate Zone contains the following rainfall
régimes:

1. Rain at all seasons, evenly distributed.
2. Periodic rains, spring and early summer maximum.
3. Periodic rains, winter maximum.
4. Constant drought.

Turning to the pressure and wind maps these are seen to correspond
to:

1. A region with a restricted development on eastern margins
alternately under the influence of the trades which, coming off the
sea, bring rain, and the westerlies which, though coming overland,
bring a certain amount of rain from cyclonic storms. The former are
more effective as rain-bearers than the latter and there is a tendency
to a summer maximum, which is exaggerated wherever there are
monsoons (e.g. China).

2. A region bordering the interior deserts on their poleward margins
and receiving cyclonic and convectional rains in spring and early
summer. Much of this is virtually desert, bordered by steppe.

3. A region with a restricted development on western margins alternately under the influence of the calms of the high-pressure belt (or the trades, both of which are desiccating agents) in summer and the cyclonic westerlies in winter, and therefore enjoying winter rains and summer drought.

4. A region in continental interiors, isolated from the sea and receiving only scanty convectional rains.

These correspond to four types of climate known as:

1. Eastern Margin Sub-tropical.
2. Steppe.
3. Western Margin Sub-tropical (Mediterranean).
4. Desert.

The eastern margin sub-tropical is sometimes called 'Chinese', but since the essence of the Chinese climate is its monsoonal character, the name 'Chinese' is unfortunate as applied to the normal *facies*. It is better to consider the monsoonal variety as a separate type of which South China may be used as the example.

The Cool-Temperate Zone and the **Cold Zone** each contain the following régimes:

1. Rain at all seasons, evenly distributed; tendency to winter maximum.
2. Rain at all seasons, summer maximum.

As has been already pointed out these correspond to: (1) Marine; and (2) Continental varieties.

The Arctic Climates. Two types of climate are here included, 'snow and ice' climates and 'tundra'. Supan marked them off from the 'temperate' climates by the isotherm of 50° for the warmest month which virtually coincides with the poleward limit of trees. Reasons have been given for preferring a definition based on the availability of warmth throughout the growing season. This is best expressed as an accumulated temperature, but for the sake of simplicity the duration of the season will be adopted here without reference to the intensity of the warmth. The last of the trees grow with three months above 43°, beyond that line, generally speaking, is tundra (see Fig. 31). Beyond the tundra, where temperatures do not rise above freezing-point for long enough to melt the snows at 'midsummer' is perpetual snow and ice. Gunnar Anderson distinguished High Arctic climates, in which vegetation is absent, from Low Arctic climates 'with bushes and peat bogs' (tundra) and separated them by the isotherm of 6° C. (43° F.) for the warmest month. This is conveniently consistent with the use of 43° hitherto and it will be adopted here. Both climates have maritime and continental varieties, but they differ unimportantly in the biological responses they invoke and there is no purpose in subdividing them.

Two types of climates, although occurring in all of the foregoing zones, are sufficiently distinct to receive separate recognition and to be considered separately. These are:

1. **Mountain Climates**, i.e. those whose climatic and weather conditions are determined chiefly by their relief and altitude and in which the influences of latitude and proximity to sea are strongly modified thereby.

2. **Desert Climates**, i.e. those in which the dominant characteristic is aridity. This character, with its concomitant effects on temperature range, etc., gives a unity to deserts all the world over in whatever zone they are found. As a convenient approximate limit the formula proposed on p. 86, R (in.) $= \dfrac{T}{5}$ (°F.) may be adopted.

The desert climates may be subdivided into: (a) Hot deserts, with no cold season, i.e. no month below 43°; (b) Cold deserts, with cold season, i.e. one or more months below 43°. Since the presence or absence of a cold season is the real distinction between the two types, these may be conveniently divided by the isotherm of 43° for the coldest month, the line already adopted as the limit of the warm temperate climates. It is true that this line has not the same significance here, since vegetation is scanty or absent, but it is probable that deserts with such winter temperatures will experience considerable night frosts in winter.

We are now in a position to set down in tabular form the classification of climates which will be followed. The figures given are intended as approximate guides only and are not to be accepted too literally, or applied too rigidly.

A. Hot climates. Always hot. No month below 64°.
    1. Equatorial. Double maximum of rain.
    1m. Equatorial. Monsoon variety.
    2. Tropical, marine. No real dry season.
    2m. Tropical, marine. Monsoon variety.
    3. Tropical, continental. Summer rain.
    3m. Tropical, continental. Monsoon variety.

B. Warm temperate or sub-tropical. No cold season, i.e. no month below 43°.
    1. Western margin (Mediterranean). Winter rain.
    2. Eastern margin. Uniform rain.
    2m. Eastern margin. Monsoon variety. Summer maximum of rain.

C. Cool temperate.[1] With cold season, i.e. 1–5 months below 43°.
　1. Marine. Uniform rain or winter maximum.
　2. Continental. Summer maximum of rain.
　2m. Continental. Monsoon variety. Strong summer maximum.

D. Cold climates. Long cold season, i.e. 6–9 months below 43°.
　1. Marine. Uniform rain or winter maximum.
　2. Continental. Summer maximum of rain.
　2m. Continental. Monsoon variety. Strong summer maximum.

E. Arctic climates. Very brief warm season, i.e. less than 3 months above 43°.
Ice climates. Always cold. No month above 43°.

F. Desert climates. Rainfall (in.) less than $\frac{1}{5}$ of temperature (°F.)
　1. Hot deserts. No cold season, i.e. no months below 43°.
　2. Cold deserts. With cold season, i.e. one or more months below 43°.

G. Mountain climates.

The generalized distribution of these types is given in Fig. 34 and the schematic distribution on an ideal continent in Fig. 35.

SUGGESTIONS FOR FURTHER READING

Köppen's classifications appeared in *Pettermanns Mitteilungen*, 1918, and was reviewed and commented upon by R. de C. Ward in the *Geog. Rev.*, 1919, and by James in the *Monthly Weather Review*, 1922. His revised classification is defined and the types mapped in his *Klimakarte der Erde*, Justus Perthes, Gotha, 1928. Thornthwaite's classification is described in the *Geog. Rev.*, 1931. The 1931 paper supplies references to numerous publications on cognate subjects. Moisture efficiency, evaporation and indices of aridity are discussed by De Martonne in *Compt. Rend. Acad. Sci.*, 1926; by Thornthwaite in *Geog. Rev.*, 1948, and by Penman in *Proc. Roy. Soc.* (A), 1948. A commentary on Thornthwaite's classification by P. R. Crowe will be found in *Geographical Studies*, 1954. C. H. Merriam drew attention to the importance of accumulated temperatures in the *Nat. Geog. Mag.*, 1895. The calculation of day-degrees on this basis is described in Air Ministry Form 3300. See also A. Austin Miller, *Advancement of Science*, 1950, *Proc. Inst. Brit. Geog.*, 1952 and *Report of I.G.U. Congress*, Washington 1952. *Wetter und Leben* first published in 1952 provides a vehicle for articles on biogeography, as do the *Reports of The International Society of Bioclimatology and Biometeorology*, from its foundation in 1958.

[1] The extreme annual range of temperature experienced in these climates might be said to make the adjective 'temperate' a misnomer. The term, however, is too well established to be summarily dismissed from a classificatory system which uses in the main the traditionally descriptive terms.

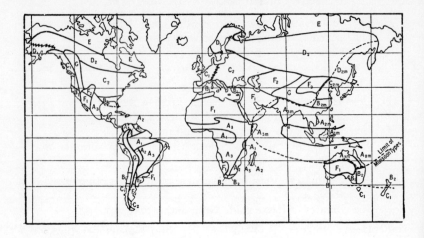

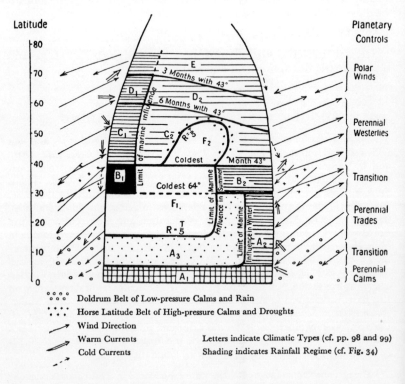

Fig. 35.—Distribution of Climatic Types and their Schematic Arrangement on an Ideal Continent

VI

EQUATORIAL CLIMATES

Hot Climates in General. As defined by the isotherm of 64° for the coldest month, the Hot Climates occupy a broad belt extending well beyond the tropics in each hemisphere and occupying more than half the surface of the globe. Notwithstanding the fact that three-quarters of this huge area is water, such an enormous extension entitles these climates to full and careful attention. Furthermore, this zone is one as yet very under-developed and one with a great agricultural future whose realization will be closely bound up with the study of its climate. Unfortunately our knowledge of these climates is based on observations at comparatively few stations, huge areas still having less than one station to 4,000 square miles, and much yet remains to be done in the way of intensive study; but in this respect it is fortunately one of the characteristics of the zone that climatic phenomena possess a high degree of reliability and regularity, and perhaps require less minute analysis than the more intricate phenomena of temperate latitudes. The great extension of the sea helps to preserve the uniformity of the zone, much of which is necessarily marine or littoral, but two large continental blocks, Africa and South America, project across the zone and offer good opportunity for study of the continental type with the minimum modification by marine influences. In the Indian and West Pacific oceans there is a monsoonal régime which introduces peculiarities calling for special mention, and these will be dealt with later. The symmetry of Africa about the Equator, coupled with the fact that it is better known than South America, makes it peculiarly suitable for illustration, and for this reason examples will be taken chiefly from that continent.

The hot climates lie within the sphere of the N.E. and S.E. trade winds which meet in the equatorial belt of calms. The flanking high pressures of the horse latitudes, the sources of the trade winds, effectively exclude external influence from colder latitudes, making for a stability of meteorological conditions which is a marked characteristic of the zone. Both the seasonal and the diurnal régime are controlled almost exclusively by the annual and diurnal march of the sun, so that it is sometimes said that there is no weather here, only climate.

The annual migration of the sun between the tropics brings, generally speaking, two maxima of insolation to the equatorial zone and a single maximum to the tropical zone (see p. 34), this single maximum being reached at about 12°N. and S. The march of temperature, subject to certain modifications, follows the march of the sun, and the rainfall belt, subject to a lag which is often considerable, follows the sun also. Thus, broadly speaking, double maxima of temperature and rainfall are characteristic of the equatorial, single maxima of the tropical zone.

THE EQUATORIAL CLIMATES

Temperature. Throughout this climatic type temperatures are uniformly high and extremely constant, the excessive summer temperatures which occur at the tropics are never recorded, nor are their low winter temperatures, but the thermometer hovers about the 80° mark throughout the year. Maximum temperatures do not necessarily correspond to maxima of insolation, since cloudiness and rainfall considerably moderate temperature; and in this zone, where the variation of insolation is never great, their effects are particularly noticeable. Even at the tropics the length of the day varies only between $10\frac{1}{2}$ and $13\frac{1}{2}$ hours and the minimum midday elevation of the sun is 43°, while within the limits of equatorial climates the variations of insolation intensity and duration are considerably less even than this. To this the constant humidity and cloudiness add their stabilizing influence and the annual range of temperature is consequently very small indeed.

Annual Range. The mean annual range near the Equator is, on an average, about 5°, at Pará (1° 27′ S.) it is less than 3°, at Akassa (4° 15′ N.) it is 4°. Over the oceans it is less still and many island stations have a mean annual range of less than a degree (Jaluit, Marshall Is., 0·8°). Extreme temperatures show the same constancy and the same small range as mean temperatures. The thermometer rarely rises above 100° in the equatorial zone, or falls as low as 60°, i.e. the extreme range seldom exceeds 40°.

Diurnal Range. By contrast with the extremely small annual range the diurnal range seems considerable, though it may not be more than 15°. There is sometimes a nocturnal fall of as much as 25° and the night temperatures (although, perhaps, well above 60°) may appear uncomfortably cold, especially to a native population inadequately provided with clothing and protection; wherefore it is often said that 'night is the winter of the tropics'. The native peoples are very sensitive to the slightest fall of temperature and will often light fires if the temperature falls below 70°; Europeans, too, lose their power of resistance to cold in these climates of uniform heat.

Relation between Temperature and Rainfall. At Bolobo (see Fig. 36) the mean temperature varies only 2·2° throughout the year and there are two maxima, reached at the equinoxes. Rainfall and humidity, being remarkably uniformly distributed over the year, do not affect temperature which follows closely the vertical sun. But at

Lagos (Fig. 36) the curve of temperature shows a marked drop in May and June when the rains begin and actually reaches a minimum in August, although insolation is then at a maximum. The drop in temperature is clearly related to the cooling effect of the rains and of the clouds which accompany them.

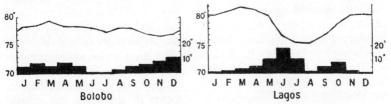

Fig. 36.—Annual March of Temperature and Rainfall in Equatorial Climates

Pressure and Winds. Characteristically the equatorial belt is a region of calms or very light winds; the evil reputation for ill-health and discomfort which it has earned is to be attributed to the hot, moist, stagnant air. The reason for this light air movement is not far to seek; on the monthly isothermal maps the 70° isotherm never invades the tropical climates and the belt with mean temperatures exceeding 70° is very wide. Within this broad expanse of uniformly hot land small areas only exceed 90° and these are almost entirely the local products of the deserts. Here, then, is the most uniform temperature distribution to be found anywhere in the world; such feeble temperature gradients beget a feeble pressure gradient and air movement is consequently slight. The pressure gradient decreases towards the centre of the zone and finally ceases in a region of calms, where escape is found in upward ascent of an enormous body of air with resultant expansion and cooling and consequent heavy precipitation. Towards the tropics the barometric gradient is steeper and vigorous air currents are directed equatorwards. These are the trade winds, blowing with great regularity and fulfilling an important function by importing vast quantities of moisture from the warm oceans over which they have passed to be ultimately condensed by the ascension at the belt of calms. The swing of the wind belts (see p. 25) brings the beneficent influence of the trades to different parts at different seasons, but wherever and whenever they blow they are a welcome feature for the stimulation which they give. Although they are warm winds their effect is pleasantly cooling, for they fan and ventilate the body, encouraging evaporation from the skin. Exposure to the trade winds is a desirable asset in a house in these latitudes and sites are chosen with this object in view.

Land and Sea Breezes. As the seasonal temperature changes cause a seasonal swing of the wind belts, so the diurnal changes bring about a diurnal change of wind direction, the land and sea breezes (see p. 43). The mechanism, however, is not the same in the two cases, for while the seasonal winds depend on seasonal variations of insolation (i.e. the solar régime), the diurnal winds depend on the differences in behaviour of land and sea towards insolation (i.e. on physical influences). It is especially in the belt of calms, where the pressure gradient is small, that the diurnal changes have the best opportunity of producing effects, and where, consequently, land and sea breezes are most reliable. In the trade-wind belt the diurnal changes only modify the prevailing winds, in the doldrum belt they produce an actual reversal of direction. Although restricted to a narrow coastal zone, the sea breeze has an importance out of proportion to its limited distribution, for the majority of European settlements in this zone are coastal trading settlements, and the residential districts in these are usually those best favoured by the sea breeze, whose daily arrival is impatiently awaited for its wholesome and cooling influence.

Rainfall and Seasons. In the absence of any marked temperature changes in this zone, rainfall becomes the determinant of season. The principal rains are those of the convectional type and reach a maximum shortly after the passage of the zenithal sun, about April and November on the Equator. But this simple scheme is subject to considerable modifications for local reasons. It frequently happens that one of the maxima is more pronounced than the other (the 'greater rains' and the 'smalls' of the Gold Coast) and often one is suppressed (see Pará). In most cases the spring maximum following the northward passage of the sun is considerably greater than the autumn maximum following its southward return. Entebbe has 25 inches during the three months of the first maximum (March, April, May) and only 13 inches in the second (October, November, December); Lagos has 39 inches in May, June and July and only 16 inches in September, October and November. The graphs for Djole and Mwanza (Fig. 37, p. 105) provide more examples. Although not by any means universal, this marked asymmetry of the curve seems to indicate a real difference in the influences at work in the two rainy seasons. It is very clearly marked in Africa where the contrasts in relative proportions of land and sea in the two hemispheres are most pronounced, and it appears to be the greater area of water in the southern hemisphere which causes the effect, since during its passage over the southern hemisphere the sun has evaporated a greater quantity of water which is available for condensation in the zone of ascension.

The equatorial rainfall type does not extend far on either side of the Equator; within a few degrees there enter two dry seasons following the solstices and becoming gradually more and more pronounced. At about 5°N. and S. these dry seasons are more or less equal in duration and intensity, but polewards one becomes gradually longer and the other gradually shorter. In the coast zone of British Guiana (7°N.), for example, four seasons are recognized:

1. The long wet season. Mid-April to mid-August.
2. The long dry season. Mid-August to mid-November.
3. The short wet season. Mid-November to end of January.
4. The short dry season. End of January to mid-April.

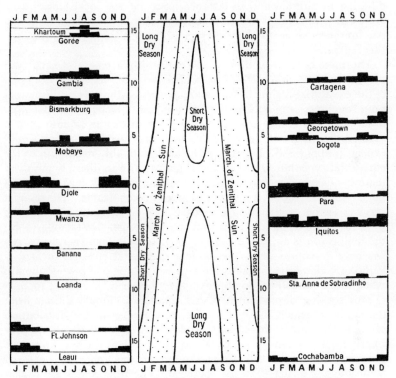

Fig. 37.—March of Rainfall in Equatorial and Tropical Climates

And of these the long dry season alone can be relied upon to remain true to type. Further polewards still, the short dry season becomes only a break in the rains and is eventually eliminated; the long dry season has assumed the proportions of a drought and the equatorial climate has passed into the tropical.

Fig. 37 reproduces de Martonne's diagram showing the ideal relationship of the rainy seasons to the zenithal sun. The region with no dry season is seen to extend only about 2°N. and S. of the Equator; from this point to nearly 15°N. and S. there should be two dry seasons, after which the short dry season dies away and there should be a single dry season and a single wet season in which there may be two maxima of rain.

On either side of this ideal arrangement are placed some actual rainfall graphs for certain stations, African stations on the left, South American on the right, the base of each graph being placed approximately on its own latitude. A warning should be issued against a too literal acceptance of these graphs since they are not strictly comparable, differing in altitude and in distance from the sea; local factors, too, are often potent influences in modifying the planetary régime. They do, however, serve to show certain important departures from the theoretical.

The most obvious of these is the entire displacement of each type some degrees to the north of its proper latitude. The best examples of the equatorial régime in Africa are given by Mobaye and Bismarckburg roughly 4° and 8°N. of the Equator respectively; and by Georgetown (6°N.) in South America. Djole, almost on the Equator, has a marked dry season in July and August, a régime such as one would expect in about 8°S. Bogotá (5°N.) has the double maximum of the equatorial type, but Quito on the Equator has a southern hemisphere régime. There is a double rainfall maximum at Mwanza (2° 31′ S.) but by Tabora (5° 1′ S.) this is scarcely recognizable. However it survives farther south along the coasts, e.g. at Banana (6°S.) on the west coast and at Lindi (10°S.) on the east coast. By Loanda (9°S.) the dry season is already 6 months long, but Gambia (nearly 11°N.) has only 4 months dry. The typical equatorial climate, in fact, is found, not along the Equator, but between 2° and 8°N., while the full swing of the rainfall belt is approximately between 18°N. and 10°S.

This striking displacement of the equatorial climate springs from the contrasts which the two hemispheres present in the distribution of land and sea. In these low latitudes, where temperatures are constantly high, the land is nearly always hotter than the sea, and since there is more land north of the Equator in this zone, the northern hemisphere is hotter than the southern. The S.E. trades are cooled in passing over seas, while the N.E. trades, coming chiefly over land, are hotter and therefore ascend sooner; the point of meeting of the two currents and the upward escape is thus north of the Equator. A further factor, though one of little importance, is the longer sojourn of the sun in the northern hemisphere (in aphelion) than in the southern (in perihelion) by about 8 days. This displacement of the heat equator (see Fig. 27) has other important results which will be dealt with later.

Nature of Equatorial Rain. Rainfall throughout the equatorial zone is heavy and reliable: 50 inches to 80 inches is usual, 100 inches is not rare and 200 inches is exceeded in places. Grey skies are the rule during the rainy seasons; the morning may dawn clear and sunny, except for morning mists which quickly melt away, but as the day wears on dark clouds roll up and violent storms occur, usually with an accompaniment of lightning and thunder. The daily incidence of rain is strikingly regular at any given place, often so regular that appointments are made for 'after the rain' much as one might make arrangements for 'after tea'. The time of this daily maximum varies from place to place according to local conditions, but on land it is nearly always between midday and midnight and usually about 3 or 4 p.m., i.e. following the greatest heat and convectional action. In spite of the violence of these storms and the strong convection which characterizes them, there is no violent vortical motion such as occurs in convectional storms of higher latitudes, since, by reason of proximity to the Equator, the deflection by the earth's rotation is extremely feeble.

Humidity and Cloudiness are greatest during the wet season and serve to make the heat more trying; for although day temperatures in the equatorial climates seldom exceed 100° and are usually between 80° and 90°, the conditions are the most oppressive, unhealthy and enervating that can be experienced. Where humidity is highest there is little relief even at night, the body perspires freely, but there is little evaporation to cool it, and in the sultry, hot-house atmosphere all strength of body and will seem to ebb away. On the coast the arrival of the sea breeze brings some relief, but when it dies away the sweltering heat seems to settle all the heavier.

High Altitude Equatorial Climates. The most noticeable influence of altitude is a lowering of temperature, and for this reason, if for no other, the climates of high altitudes in equatorial zones would merit separate treatment. But there are also changes in many other climatic elements which confirm the type as quite distinct. At altitudes above 10,000 feet even on the Equator the conditions are those of mountain climates and as such will be treated in a later chapter, but at altitudes of 5,000 to 10,000 feet the climatic conditions are more truly described as 'modified equatorial'. The temperatures at these altitudes are those of the temperate zone, it is the 'tierra fria' of Latin America (see p. 129).

Since each 300 or 400 feet makes a difference of about 1° in the temperature, highlands offer refuges from the stifling heat of the lowlands. Parts of South America are particularly fortunate in the close proximity of highland to coast, for while commercial activity is concentrated in the hot coastal zone, administration can be centred quite close at hand in a climate with great advantages of comfort and health. Caracas (4,000 feet) is only 6 miles distant (25 miles by rail) from its port, La Guaira, but has a mean temperature nearly 10° lower, while the difference of humidity makes the contrast seem even greater.

The mean annual range of temperature is typically small, less even than at low levels: Quito varies only 0·7° between the hottest and coldest

months, Bogotá 1·6°. The diurnal range, however, is considerable
(30° or 40°) and frost at night is a common occurrence at altitudes of
8,000 and 10,000 feet, while precipitation frequently takes the form
of sleet and snow. But the air temperatures give an inadequate idea
of the physiological sensations; for the clear and rarefied air promotes
rapid radiation at night and interferes little with the intense insolation
by day. The contrast between conditions in sunshine and shadow is
a revelation. These contrasts are exaggerated at still higher levels and
will be dealt with more fully under mountain climates.

The rainfall régime is true to type and the two equinoctial maxima
are usually clearly recognizable, but above the zone of maximum
precipitation the amount of rain is substantially decreased; Quito has
42 inches compared with over 80 inches in the lowlands which lie to
east and west. There is the same regular diurnal incidence of rainfall
and a regular diurnal variation of cloudiness, for the rainfall is of the
same type (convectional) and obeys the same controls. Mornings are
usually fine and clear but clouds gather in the late morning and early
afternoon for the regular afternoon storm, often a thunderstorm with
hail.

Vegetation and Cultivation. The constant heat and humidity
of the equatorial climate supports a characteristic vegetation type, the
'equatorial rain forest' or 'selvas', with its astounding profusion and
vigour of growth, especially of foliage. Tall tree-tops form an irregular
canopy beneath which grows a layer of shorter trees and finally a
tangle of creeper and undergrowth so dense as to be almost impenetrable, while parasitic and epiphytic plants, often with gorgeous
flowers, add to the display of vegetational extravagance. A marked
absence of seasonal rhythm results from the monotony of the climate,
flowering, fruiting, seeding, growing and dying occurring side by side.
There is no resting period, no check to the rapid growth; no clear
annual rings appear in the wood of the trees.

There is little rhythm, either, in agricultural practice, but the dry
season (or the periods of minimum rainfall) become the harvest seasons
for most crops. Rubber is gathered in the dry season, which in the
northern half of the Amazon basin is August to February and in the
southern half is May to October.

Economically the wealth of vegetation is somewhat misleading, for
the profusion is chiefly of vegetative parts, but palm-oil and rubber
are important bases of industry and wealth. There is, moreover, an
unfortunate tendency to a sporadic occurrence of species which much
reduces their value; rubber trees, for example, are scattered through
the Amazon forests in such a manner that the collection of wild rubber,
for long Brazil's chief source of wealth, was unable to compete with the
concentrated cultivation of the plantation product. Mahogany, green-
heart and other timber trees suffer from the same disadvantage.

The exuberance of growth of the forest makes clearance for cultivation difficult to effect and expensive to maintain, especially since there is no real dry season when the aid of fire can be enlisted. Plantation cultivation, however, is rapidly spreading over the zone and replacing the old wasteful exploitation; it does much to avoid the leaching by heavy rain in areas cleared of their vegetation. This is chiefly a response to the demand from temperate countries for tropical produce both for food and for raw material, a comparatively new departure which is accelerating development, and with which is associated the growth of settlements, often of considerable size, for trading purposes. The difficulty of penetration retards development; roads and railways are difficult and expensive in construction and upkeep (the 225 miles of the Madeira–Mamoré railway cost over £5,000,000 in cash and incalculable loss of health and life). Airways now carry passengers between towns, but penetration is chiefly by rivers, which provide the only clear paths, and settlements are scattered along their banks.

Vegetation and Altitude. At higher levels the vegetation type undergoes a steady change, but so much depends on aspect in relation to sun and winds, on soil and other factors that it is difficult to generalize. Up the Andean slopes the dense forests of the Amazon give way, at about 5,000 feet, to forest of a sub-tropical nature with tree ferns and cinchona; this degenerates into bush and scrub with bamboo and fuchsia at about 8,000 feet and eventually into grassland at about 10,000 feet. In East Africa, where the rainfall is less, tropical grassland begins at 3,000 feet, but is followed by sub-tropical forest above and this by temperate grasslands; but in Kamerun, reached by saturated air currents from the south-west, dense forest extends up to 6,000 feet.

REGIONAL TYPES

The Amazon Basin opens to the east, and offering free access to the trade winds, has the widest expanse of uniformly heavy rainfall to be found anywhere in the world, Pará has 87 inches, Manaos 66 inches, and more than 70 inches probably falls over an area of more than two million square miles; but recording stations are unfortunately few. Westwards, as the Andes are approached, the rainfall is heavier still since the easterly winds are forced to ascend; Iquitos has over 100 inches annually. This terrific precipitation (nearly 3,000 cubic miles a year) ensures an abundant supply of water, and the discharge of the Amazon far surpasses that of any other river. The double rainfall maximum which one expects to find is not recognizable and, as far as can be ascertained from the scanty records available, a summer rainfall is general. Most of the basin lies south of the Equator and the important southern tributaries have a southern hemisphere régime with a marked

dry season in August and September when the level of the main river
falls. In March and April these southern tributaries flood and the
level of the Amazon may rise as much as 40 feet, converting the whole
basin into a vast swamp.

Separated from the Amazon basin by the Andean divide is another
area of heavy rain, the west coast of Colombia, washed by the warm
waters of the equatorial counter-current, which receives more than
100 inches annually, and, in places, as much as 300 inches occurs;
Buenaventura has 281 inches.

The Congo Basin. Much of the Congo basin has an equatorial
climate, but the rainfall is considerably less than in the Amazon basin,
for the entry of the easterly winds into the Congo Basin is obstructed
by the plateau of East Africa and they have lost much of their moisture
before their ascent in the zone of calms. However, 50 inches occur over
a very large area (cf. 70 inches or 80 inches in the Amazon) and provides an adequate supply for the river's discharge, which is the second
largest in the world (Congo average annual discharge 419 cubic miles,
Amazon 528 cubic miles). Thanks to the symmetrical arrangement of
the Congo basin on either side of the meteorological Equator, the river
floods twice a year, in May and December, these being the times when
the northern and southern tributaries respectively have their maximum
precipitations. At no time does the level fall so far as to cause serious
inconvenience to navigation.

The Guinea Lands, also, although between 5° and 10° N., have a
climate which is virtually equatorial, but in which a certain monsoonal
influence is an essential constituent. The key to this climatic province
lies in the existence of the great land mass of West Africa, always subjected to intense insolation. Even in winter the southern Sahara and
the Guinea lands are hotter than the sea, consequently the doldrum
belt never moves south of the Equator but swings between the coast in
January and 10° or 15°N. in July. In July the S.E. trades are drawn
across the Equator as a S.W. monsoon, and in January the belt of calms
lies over the coast, though even then the prevailing wind is south-west.
From June to November 75 per cent of winds on the Gold Coast are
from the south-west, and from December to May 64 per cent. Thus the
coastal lands rarely experience the benefits of the N.E. trade wind to
which they would be entitled by virtue of their latitude, for these hardly
ever penetrate south of Freetown. From here northwards the coast is
moderately healthy, but the south coast formerly earned the reputation
of the 'White man's grave'.

Both winter and summer conditions bring rain to the coast lands,
Akassa (144 inches) and Lagos (72 inches) have the double maximum,
but in Liberia and Sierra Leone, where the coast, backed by the Futa
Jalon Highlands, is at right angles to the S.W. wind, the régime is
distinctly monsoonal. Freetown (175 inches) and Konakry (170 inches)

have single maxima in July or August, corresponding to the greatest vigour of the S.W. monsoon. The most important rain here is, in fact, orographic rather than convectional, and the same is true in Kamerun where again the coast is at right angles to the S.W. monsoon and is backed by high land. Kamerun Peak rises above 13,000 feet and on its western (i.e. windward) slope more than 400 inches of rain fall in a year.

There is a narrow strip of littoral east of Cape Three Points where the rainfall suffers a startling diminution. In the midst of a region with 80 inches or more this area has less than 40 inches, and in places as little as 20 inches. The rainfall along the coast from west to east is as follows: Axim 81 inches, Sekondi 40 inches, Cape Coast Castle 33 inches, Accra 27 inches, Christiansborg 21 inches, Kwitta 22 inches. But these peculiarly low figures hold only for a narrow strip; Aburi, 30 miles behind Accra, has 45 inches, and Kumasi, 100 miles farther inland, has 58 inches. The cause of this phenomenon is obscure, but it is probably to be explained by the pull of the Guinea current which causes the up-welling of cold water in lee of Cape Three Points, thus decreasing the moisture capacity of the winds passing on-shore over it. At the same time the change in direction of the coast makes the winds more oblique to the coast-line. The lower humidity reacts on temperature and Kwitta, because drier, is 5° hotter than Axim during the heavy rains.

The equatorial climates do not extend completely across Africa, the eastern plateau having a continental type of climate much modified by two influences, namely, its altitude and the proximity of the Asiatic and Abyssinian monsoons. Much reduced though the rainfall is, it still retains the typical double maximum over much of this area. The resultant vegetation is grassland (altitude savanna) and it forms a natural continuation of the tropical savannas with which it is better described (see p. 129).

There is a return to equatorial conditions in the coastal lowlands behind Pemba and Zanzibar, where, however, the effect of the monsoons is clearly felt (cf. p. 154). Zanzibar has a double rainfall maximum, but with monsoon winds; Tanga shows a tendency to a third maximum of rain in July between the normal equatorial maxima which occur in April and November.

MONSOONAL VARIETY

Situated between the two great monsoon foci of Asia and Australia, the East Indian islands present a variety of the equatorial climate into which modifications are introduced by the seasonal wind changes.

Pressure and Winds (Fig. 38). In January there is a difference of pressure of 3 mm. between 10°N. and 10°S. and the so-called 'West Monsoon' is at its height. By April the isobars are more or less

symmetrically arranged about a low-pressure trough just south of the
Equator in South Celebes and north New Guinea; light breezes and
calms are now the rule. By July the pressure difference is about
3 mm. in the opposite direction and the so-called 'East Monsoon'
is at its height. By October or November there is a return to the
symmetrical arrangement about a low-pressure trough with light
and variable winds focusing on Celebes and New Guinea. Such, in
broad outline, is the scheme of winds—two periods of steady air flow
(the monsoons) alternating with two periods of indefinite circulation at

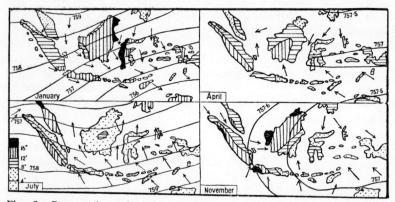

Fig. 38.—Pressure (in mm.) winds and rainfall (in inches) in the Malay Archipelago

the changes-over. But in actual practice there is no such simplicity, for
in the first place the change-over of the monsoons is not simultaneous
throughout the archipelago, but follows the sun; for example, the south-
east monsoon prevails south of the Equator in April, but is not established north of it until May. In the second place, so complex is the
intermingling of land and water and so varied is the relief, that local
influences profoundly modify the circulation and even in places render
it completely unrecognizable.

The velocity of the monsoon is greatest and its direction most constant in the west, over the open Indian Ocean, where friction and interference are least; but even at its height, land and sea breezes and mountain and valley winds are nearly always able to make themselves felt in
spite of the prevailing wind. The direction and force of the wind
depends largely on the direction and strength of the relief; where the
air flow is impeded by mountainous islands set across its path (e.g. Java
during the east monsoon), the wind is 'stowed' and concentrated in
gaps and on the high plateaux where the velocity is greatly increased.
Warmed by descent on the leeward side, these winds have a foehn-like
nature, their dry heat doing considerable damage to the more sensitive
crops, especially tobacco. Such a wind is the 'Koembang' of Java, and

the 'Bohorok' which visits the plains of Deli, Sumatra, during the east monsoon.

On the whole, winds in the East Indian Archipelago, though extremely complicated, are regular and reliable at any given locality, even to their diurnal variations. Once the local peculiarities, numerous though they are, are thoroughly mastered, the general absence of strong winds favours shipping on the seas and agriculture on the islands. High wind velocities are very rare and only the most northerly islands are visited by typhoons. There are, however, local disturbances peculiar to limited areas, such as the 'Sumatras' of the Malacca Straits, sudden squalls with violent thunder and lightning and heavy rain which occur, always at night, during the south-west monsoon (the East Monsoon of the East Indies).

Rainfall. Several factors combine to make the East Indian islands one of the largest areas of uniformly heavy rain in the world. They are most of them mountainous, they rise from a very warm sea, they lie in the equatorial belt of ascending air currents and they stand in the path of a double monsoon. Less than 40 inches is an extreme rarity, 80 inches is more the rule and 150 inches is by no means unusual, while 268 inches is recorded from Craggan in the mountains of Java. The air is so near the saturation-point that practically any ascensional movement gives rise to precipitation at once. The ascent is brought about in one of two ways:

1. Local heating effects which give rise to local showers and thunderstorms, especially during the calms which accompany the changes of the monsoon. At this time of the year there is generally to be recognized a regular diurnal migration of the belt of clouds. In the morning it lies in the low valleys where the temperature is inverted, but the mists quickly vanish before the early sun and the mornings are clear and fine. The moisture is carried upwards by the ascending currents of daytime and condenses as cumulus cloud with a flat base coinciding with the level of the saturation temperature. Rising higher, the cumulus becomes cumulo-nimbus and thunder showers follow in the afternoon; Buitenzorg is the most thundery place on earth.

2. Forced ascent giving orographic rain, especially characteristic of the monsoon months, distributed, naturally, on exposed slopes and increasing with altitude up to about 3,000 feet. These rains are more continuous than the other type, they give the greatest daily amounts and account for most of the floods. They do not, of course, possess the same regularity of diurnal incidence.

Seasonal Distribution. Thus we have a combination of equatorial and monsoon régimes. The former would tend to give a typical double maximum, but such is not often recognizable owing to interference by

TEMPERATURE

Station	Lat.	Long.	Alt. (ft.)	J	F	M	A	M	J	J	A	S	O	N
OCEAN ISLAND	1°S.	170°E.	177	81	81	81	81	81	81	81	81	81	81	81
MALDEN ISLAND	4°S.	155°W.	21	82	82	82	82	82	82	82	82	83	82	82
SEYCHELLES	5°S.	55°E.	15	80	80	81	82	81	80	78	78	79	79	80
JALUIT	6°N.	170°E.	Coast	81	81	81	80	80	80	80	80	80	81	81
AKASSA	4°N.	6°E.	20	78	79	80	80	79	77	76	76	76	77	78
CAPE COAST CASTLE	5°N.	1°W.	Coast	80	79	81	81	80	78	77	76	77	79	80
DUALA	4°N.	10°E.	26	80	80	80	79	79	77	75	75	76	76	78
YAUNDÉ	4°N.	12°E.	2,461	74	74	74	72	72	71	70	71	71	71	72
LULUABURG	6°S.	22°E.	2,034	76	76	76	77	77	76	77	76	76	76	77
NEW ANTWERP	2°N.	19°E.	1,230	79	80	79	78	79	78	77	76	77	77	78
MOMBASA	4°S.	40°E.	50	80	80	82	81	78	77	75	76	77	78	79
ZANZIBAR	6°S.	39°E.	56	83	83	83	81	79	78	77	77	78	79	81
PARÁ	1°S.	49°W.	42	78	77	78	78	79	79	78	79	79	79	80
MANAOS	3°S.	60°W.	144	80	80	80	80	80	80	81	82	83	83	82
IQUITOS	4°S.	73°W.	328	78	78	76	77	76	74	74	76	76	77	78
GEORGETOWN	7°N.	58°W.	6	79	79	80	81	81	80	81	82	83	83	82
NAIROBI	1°S.	37°E.	5,495	64	65	66	65	63	61	59	60	63	66	64
ENTEBBE	0°	32°E.	3,842	71	71	71	70	70	69	69	69	69	70	70
QUITO	0°	79°W.	9,350	55	55	55	55	55	55	55	55	55	55	54
BOGOTÁ	5°N.	74°W.	8,730	58	58	59	59	59	58	57	57	57	58	58
COLOMBO	7°N.	80°E.	24	80	80	82	83	83	82	81	81	81	81	80
SINGAPORE	1°N.	104°E.	10	80	80	81	82	82	81	81	81	81	81	81
JAKARTA	6°S.	107°E.	23	78	78	79	80	80	79	79	79	80	80	79
FORT DE KOCK	0°	101°E.	3,500	69	69	70	70	71	70	69	69	69	70	69
PADANG	1°S.	100°E.	3	79	80	79	80	80	79	79	79	79	79	79
BANDOENG	7°S.	108°E.	2,366	72	72	72	72	72	72	72	72	72	73	72
SOERABAJA	7°S.	113°E.	16	79	79	79	80	80	79	79	79	80	80	80
KUTA RAJA	6°N.	95°E.	Coast	—	—	—	—	—	—	—	—	—	—	—
PONTIANAK	0°	109°E.	10	78	79	79	79	80	80	80	79	79	79	78
MENADO	1°N.	125°E.	28	77	77	78	78	79	79	79	80	80	79	79

RAINFALL

D	Yr.	Ra.	J	F	M	A	M	J	J	A	S	O	N	D	Total
81	81	0·3	11·5	8·9	8·6	8·1	5·6	5·1	6·8	3·9	5·2	5·6	5·7	8·9	83·9
82	82	1·1	4·1	2·0	4·7	4·6	4·2	1·9	2·0	1·6	0·9	1·0	0·8	0·8	28·6
80	80	3·8	16·9	12·5	9·1	7·0	6·0	4·5	2·6	2·3	5·7	5·0	9·4	13·5	94·5
81	81	0·8	11·5	11·9	17·9	14·0	19·8	15·8	15·5	13·8	13·6	11·4	14·0	17·3	176·5
79	78	4·0	2·6	6·5	10·0	8·6	17·0	18·6	10·1	9·3	19·3	24·7	10·6	6·5	143·8
80	79	5·7	0·5	1·3	2·4	3·5	7·2	10·2	2·2	0·9	0·9	2·4	2·8	0·8	35·1
79	78	4·7	1·9	3·7	8·0	8·9	12·0	21·5	29·3	27·2	20·7	16·9	6·3	2·6	159·0
73	72	3·7	1·6	2·7	5·9	9·1	8·1	4·5	2·6	3·3	7·6	8·9	5·9	2·0	62·2
77	77	1·5	7·2	5·4	7·9	6·1	3·1	0·2	0·1	2·5	6·5	6·6	9·1	6·6	60·8
78	78	3·8	4·1	3·5	4·1	5·6	6·2	6·1	6·3	6·3	6·3	6·6	2·6	9·3	66·9
80	79	6·5	0·8	0·9	2·3	7·8	13·7	3·6	3·5	2·2	1·9	3·4	5·0	2·2	47·3
83	80	6·1	2·8	2·2	6·1	14·1	10·7	2·0	2·4	1·7	2·1	3·6	7·5	5·4	60·2
79	78	2·7	12·5	14·1	14·1	12·6	10·2	6·7	5·9	4·5	3·5	3·4	2·6	6·1	96·2
81	81	3·0	9·2	9·0	9·6	8·5	7·0	3·6	2·2	1·4	2·0	4·1	5·5	7·7	69·8
78	77	4·0	10·2	9·8	12·2	6·5	10·0	7·4	6·6	4·6	8·7	7·2	8·4	11·5	103·1
81	81	4·0	7·9	4·6	7·2	6·0	11·1	11·7	9·9	6·5	3·1	2·9	6·7	11·1	88·7
63	63	6·4	1·9	4·2	3·7	8·3	5·2	2·0	0·8	0·9	0·9	2·0	5·8	3·5	39·2
70	70	2·7	2·6	3·6	5·8	9·8	8·5	5·1	3·0	3·0	3·1	3·5	4·9	5·1	58·0
55	55	0·7	3·2	3·9	4·8	7·0	4·6	1·5	1·1	2·2	2·6	3·9	4·0	3·6	42·3
58	58	1·6	3·7	3·5	4·5	9·6	6·5	3·2	2·6	3·3	2·9	8·4	9·6	5·6	63·4
80	81	3·2	3·3	1·9	4·3	9·7	10·9	7·3	4·4	3·2	4·8	13·4	11·8	5·1	80·1
80	81	2·3	9·9	6·6	7·4	7·6	6·7	6·8	6·8	7·9	6·8	8·1	9·9	10·6	95·1
79	79	1·8	13·0	12·8	7·8	5·1	4·0	3·7	2·6	1·7	2·9	4·5	5·5	8·5	72·1
69	69	1·8	9·2	7·0	9·2	10·2	7·6	5·6	3·8	6·4	6·4	9·0	9·0	10·4	94·0
79	79	1·0	13·5	9·9	11·9	14·0	12·6	13·0	11·8	13·7	16·1	20·0	20·7	19·4	177·6
72	72	1·0	7·6	7·1	9·6	9·0	5·2	3·6	2·6	2·3	3·6	6·7	8·9	8·5	74·7
80	79	2·0													
—	—	—	5·9	3·5	3·5	4·2	6·1	3·6	4·3	4·4	6·8	7·2	7·2	8·3	65·0
78	79	2·2	10·8	7·9	9·8	10·8	10·7	8·7	6·3	8·9	8·4	14·8	15·7	13·2	125·9
78	79	2·3	18·6	14·4	10·3	8·0	6·6	6·5	4·9	3·8	3·4	4·8	8·6	14·7	104·6

the latter. The double maximum may, however, be identified fairly
clearly at several stations in the immediate vicinity of the Equator,
e.g. Pontianak and Padang. At southerly and easterly stations in this
zone there is a tendency for the maximum at the earlier change of the
monsoon (April) to exceed that at the later change (October) e.g.
Manokwari (see Fig. 39), Tondano and Bandoeng, while at northerly
and westerly stations the opposite obtains, e.g. Penang (see Fig. 39),
Kuta Raja, Pontianak. The autumn maximum, in fact, tends to predominate over the spring maximum in each hemisphere. Any departures from this formula, and they are the rule rather than the exception,

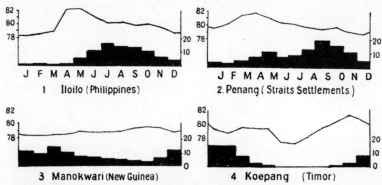

Fig. 39.—Temperature and Rainfall Régimes in the East Indies. (For explanation,
see text)

are accountable by exposure to one or other monsoons, and numerous
interesting examples of compound rainfall régimes occur. Still farther
from the Equator a single maximum becomes the rule, a season of
distinctly lower rainfall appears (e.g. Batavia), becoming progressively
more and more marked until there is an actual dry season, e.g. Koepang, Timor (see Fig. 39). It is to be noted, however, that the dry
season is July to September (Australian type) even north of the Equator
(e.g. Menado, Celebes), the Asiatic type with the dry season from
January to March being practically limited to the north of Sumatra and
the Chinese Sea. In other words, the E. monsoon is the dry season,
the rains coming more particularly from the W. monsoon. This is
easily understood when it is realized that these islands are much nearer
to the Australian than the Asiatic monsoon centre; the E. monsoon is
thus a descending wind, passing overland from the Australian high
pressure, and is therefore doubly dry, while the W. monsoon is an
ascending wind passing almost entirely over a warm sea towards the
Australian low and is fully saturated. The importance of a passage
over water as a source of moisture is well illustrated by a comparison of
the figures of Koepang, Timor, whose monsoon travels over Northern

Territory, with Toeal in the Kei Islands, whose monsoon comes over the Gulf of Carpentaria; the former has five months with less than 1 inch of rain, the latter none.

Summarizing, we may say that there are four rainfall régimes represented, the last two of which are more properly tropical:

1. Double maximum with the earlier one predominating (in the south and east).

2. Double maximum with the later one preponderating (in the north and west).

3. Single maximum with dry season July to September (Australian type).

4. Single maximum with dry season January to March (Asiatic type).

Temperature. As would be expected in a thoroughly marine equatorial climate the sea-level temperatures are monotonously high (about 80° throughout the year) and humidity is constantly excessive. The highest monthly averages of wet-bulb temperature at Batavia are in the neighbourhood of 76° and the mean maximum about 79°. These do not appear excessive when compared with, say, Hanoi, whose June mean is 80° and which reaches 83° in the heat of the day, but there is no relief throughout the year. The conditions at Batavia, then, though always enervating and unpleasant, are never unbearable, and moreover they can be avoided by retreat to near-by hill stations. The temperature decreases steadily with altitude and it is the hill station which made the energetic Dutch colonization possible. Night frosts are experienced on the high plateaux, especially in the dry season when nocturnal radiation is excessive and inversions occur, but not on the mountain tops where air movement is freer. The annual variations of temperature are so small as to be of more meteorological interest than geographical importance, but three main types may be recognized corresponding to:

1. A northern hemisphere régime with a maximum in the northern summer, e.g. Iloilo, Philippines (see Fig. 39). This is actually tropical, not equatorial.

2. A central equatorial régime with two maxima approximately corresponding to the zenithal sun, when the heat combined with the high humidity and calm air is most oppressive, e.g. Batavia.

3. A southern hemisphere régime with maxima in the southern summer, e.g. Koepang (see Fig. 39) and Port Moresby. This also is actually a tropical climate.

As was the case with the rainfall in the region of the double maximum, one maximum usually preponderates over the other and we may subdivide (2) into:

9

(*a*) An area in the north and west in which the first (April and May) maximum is greater, e.g. Penang (see Fig. 39), Fort de Kock and Padang.

(*b*) An area in the south and east in which the second (September and October) maximum is greater, e.g. Manokwari (see Fig. 39), Soerabaja and Bandoeng.

A comparison of the temperature and rainfall curves in Fig. 42 emphasizes the cooling effect of rain, the greater maximum of temperature coinciding with the lesser maximum of rain, and vice versa. Close examination shows, too, that in all cases the maxima do not occur at the time of greatest insolation but at the end of the dry season when the cloudless sky permits free inward radiation, but when the humidity, daily increasing as the rains approach, appreciably impedes loss by outward radiation. The greatest heat is recorded when the rains are late and their cooling influence is delayed.

For notes on reading, see end of next chapter.

VII

TROPICAL CLIMATES

Beyond the limits of the swing of the equatorial rainfall belt, in the latitudes where the trade winds blow throughout the year, are the greatest deserts of the world. Between these and the equatorial climates lies a belt, which, for part of the year, is under the influence of the trades, but for the rest of the year experiences an invasion by the belt of convectional rains; here are found the tropical climates with alternate trade-wind and doldrum influences. In continental interiors and up to the sea-board on western margins the season of trade-wind influence corresponds to the dry season; but the trades, though generally drying winds, are rain-bearing where they blow off the sea; consequently there are certain areas on eastern margins which do not suffer drought from their sojourn in the trade-wind belt.

Here, then, are two fundamental types of tropical climate, continental and marine, the one with, the other without, a pronounced dry season. The latter are restricted to a narrow strip on eastern margin, but extend some distance beyond the swing of the equatorial rains, as far as the trade winds are found (see Fig. 35). In lee of these areas which lie beyond the poleward extensions of the equatorial rains, the rainfall decreases steadily westward (away from the marine influence) and in this direction the savanna passes gradually into desert. But this transition area of gradually decreasing rainfall experiences a régime almost identical with the tropical continental climates and can be included with them, although its causes and the mechanism of its rainfall are different. In South Africa, for example, the equatorial rainfall belt reaches only about as far south as the Zambesi and most of the rain south of this comes from the trades. But the seasonal régime of this rainfall is controlled by a form of monsoonal action, for in winter the pressures over the land are high and the trades cannot penetrate, while in summer the continent develops a low pressure which reinforces the trades and draws the rain-bearing currents far inland. Thus there is a marked summer maximum and a winter drought which is practically as complete as that of the normal tropical type (e.g. Kimberley).

MARINE TYPE

Rainfall. Near the poleward margin of this type rain is derived almost entirely from the trades, but stations nearer the Equator receive orographic rain from the trades for part of the year and convectional rain from the low-pressure belt of calms for the remainder. In having no dry season the rainfall régime agrees with the equatorial, although from different causes, and it produces similar reactions. Much of the east coasts of Brazil and Africa, most of Central America and the windward slopes of islands in the trade-wind belt enjoy a rainfall of over 50 inches and are heavily forested. Havana, on the latitude of the driest part of the Sahara, has over 50 inches and the driest month claims nearly 2 inches.

Naturally the aspect of a station with regard to these rain-bearing winds plays a great part in determining its rainfall. The extraordinary contrast between the windward and leeward coasts of Hawaii has

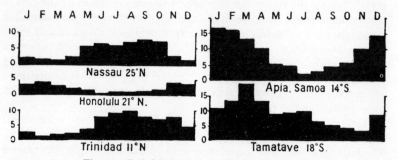

Fig. 40.—Rainfall Régimes on Trade-wind Coasts

already been quoted (p. 38), but nearly every island shows the same character in a more or less marked degree according to its relief. The Blue Mountains of Jamaica rise above 7,000 feet, Port Antonio on the windward shore has 140 inches, Kingston on the leeward shore has less than 40 inches; Colon, at the Atlantic end of the Panama Canal, records 130 inches, Balboa, at the Pacific end only 70 inches. The lower islands have a lower and less reliable rainfall since the trades pass over without forced ascent and therefore without much precipitation (the Bahamas have only about 50 inches and in some years less than 30 inches), which is rather unfortunate as the lower and flatter islands are usually better suited by their relief to agriculture, but often suffer from drought. Being chiefly orographic in nature, these rains do not show the same regularity of diurnal incidence as the convectional rains of the doldrum belt, sometimes they are heavy at night when the land surface is colder, sometimes during the day because the sea breeze strengthens the prevailing trade winds on the eastern side.

The seasonal incidence sometimes shows a winter maximum when the trades are strongest (e.g. Honolulu, Fig. 40), but more usually an autumn maximum while sea temperatures are high and land temperatures are falling (e.g. Nassau and Tamatave, Fig. 40). On the equatorward edge of the belt these are usually overshadowed by a heavy summer rainfall due to invasion by the doldrum belt (e.g. Trinidad and Samoa, Fig. 40). Furthermore, the rainfall is not entirely orographic; the land surface sets up convection currents which bring rain and this action is naturally strongest during the period of strongest insolation, a factor which further conduces to a summer maximum. In winter this is inoperative and lee coasts, especially, are fairly dry.

Temperature. In temperature, as in rainfall, the tropical marine type resembles the equatorial. Temperatures are uniformly high (Kingston and Mozambique each have a mean annual of 79°) and the yearly range is small (Kingston 6°, Mozambique 9°). But the uniformly high temperatures are less trying than in equatorial climates because of the steady fanning of the trade winds for most of the year, and only bring discomfort during the season of calms combined with high humidity which comes about the solstice.

Winds. The prevailing wind is the constant trade wind, modified by land and sea breezes. The exposure of east coasts to the prevailing wind has the result of silting up the harbour mouths and the surf makes them difficult of approach; the best ports are therefore to be found on the south and west coasts, e.g. Kingston (Jamaica), Castries (St. Lucia), etc.

Hurricanes. It is during the season of calms that there are apt to occur the violent storms known as hurricanes, typhoons or tropical cyclones, a feature of the trade-wind belt, which, though irregular and infrequent, is very important, resulting in considerable loss of property and life.

	J	F	M	A	M	J	J	A	S	O	N	D
Northern Hemisphere .	1	1	1	1	2	4	16	25	25	15	6	3
Southern Hemisphere .	26	22	23	9	4	0	0	0	0	1	4	11
Bay of Bengal and Arabian Sea	3	0	1	12	19	19	2	3	5	17	13	6

AVERAGE ANNUAL FREQUENCY OF TROPICAL CYCLONES. PERCENTAGE OF TOTAL

Although liable to occur at any time of the year, there is a danger period which, as the accompanying table shows, follows after the farthest poleward migration of the belt of calms, i.e. in autumn. In the Indian Ocean, the necessary conditions for these storms are fulfilled during the period of calms which come at the changes of the monsoon, i.e. there are two annual maxima. Fig. 41 shows the distribution of these storms and their most frequent tracks. It will be seen that there are six principal centres in which about 90 per cent of tropical hurricanes originate and these lie in the belt of calms at its farthest limit. Here is found the necessary combination of conditions, namely:

1. Calm air which allows intense heating of the lower layer, thus giving rise to a condition of instability followed by rapid inversion.

2. Distance from the Equator sufficient to produce the necessary vortical movement.

In each ocean it will be noticed that the storm area is on the western side, since here the currents have piled up the warm waters of the

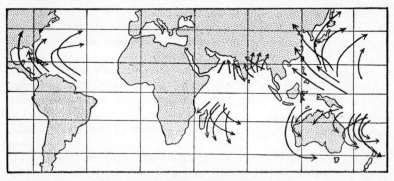

Fig. 41.—Distribution of Tropical Cyclones

equatorial belt and thus provide a generous supply of water vapour which nourishes the storms.

It will be noticed, also, that there is no storm area in the South Atlantic off Brazil, where it might be expected, as the belt of calms never passes into the South Atlantic (cf. p. 45). The track of the storm in each case is a simple parabolic curve, first westward, then poleward, then eastward. By reference to the isobaric map this curve will be seen to round the western ends of the permanent oceanic 'highs'; the storms make use of weak points in the high-pressure belts, occasioned by the seaward extension of the continental summer 'lows', to break through into the westerly circulation where they can often be traced as normal cyclonic storms. The West Indies and Florida, Bourbon and Mauritius, the Philippines and Formosa suffer severely from the visitations of these storms. Lives are lost (50,000 in the Swatow typhoon of August, 1922), plantations are ruined and extensive damage to property occasioned by them. The growing of certain delicate crops, e.g. cacao, is out of the question in areas frequently visited, on account of the high wind velocities attained. The torrential rain which these hurricanes bring contributes to a second maximum of rainfall which can often be recognized in the rainfall graph.

CONTINENTAL TYPE

In continental interiors and on leeward coasts a very different type of climate is experienced. Here the trades are dry winds and the season of their dominance is the dry season. Normally there is a single rainy season and a single dry season of greater or less duration; this marked seasonal incidence, judged by the vegetational responses it invokes, is the criterion of the climatic type.

The Dry Season. After the rains the soil is saturated, the air is moist and the nocturnal fall in temperature produces a copious dew, but under the drying influence of the trades the effect of the past rains is quickly obliterated, grass and shrubs begin to wither and die down, the soil becomes dust, woodwork cracks, the rivers dwindle and the level of water falls in ponds and lakes. Relative humidity is low (60 to 70 per cent) and may fall as low as 10 per cent when strong winds blow from the deserts, such as the Harmattan of West Africa. This wind has a humidity of less than 30 per cent and it often penetrates into the humid coastal zone, where it is much appreciated in spite of its heat; for its dryness is invigorating and it cools by promoting evaporation. Here it is known as 'the doctor' because of the improved health and comfort which it brings, but the dryness which suits man is inimical to plants, and trees and crops may be dried up and ruined. Inland, where it blows almost constantly in the dry season, the heat and dust associated with it are trying in the extreme, the skin dries and cracks, and chills and colds are rife because, in the drier air, the greatly increased diurnal range of temperature places a severe strain on the constitution.

The natives take advantage of this dry season to burn the bush and to prepare for cultivation; other bush fires are less intentional, but the smoke of these fires, together with the dust which the hot dry wind often brings, fills the air and turns a bright cloudless sky into a dark haze through which the sun shows red and dull.

Temperature. Under cloudless skies and in the dry air much higher temperatures are recorded than in the equatorial zone. Mean temperatures of over 90° for the warmest month are frequent, while daily maxima at the height of the hot season frequently exceed 110°. At night the fall of temperature is rapid, and before dawn 50° may be recorded, while night frosts are by no means unknown in the 'cold' season; such extreme ranges, although stimulating, are somewhat trying. Temperatures soar higher and higher as the season advances and as the sun approaches the zenith, but the rains are approaching and their arrival brings some alleviation of the extreme heat. At Mongalla the mean for March, before the rains, is 83° and for July only 76°.

The maximum temperature at Timbuktu (Fig. 42) is reached in May, after which the arrival of the rains drives the temperature down nearly

10° by August, in spite of increasing insolation. When the rains cease there is a recovery to a second maximum in September, after which the falling off of insolation causes the temperature to fall to a January minimum. At Cuyaba the maximum is reached in October, after which temperatures are remarkably uniform while the rains last; at Zomba the rise of temperature is checked by the rains in October and falls more than 6° to a second minimum in December, although insolation is strongest then.

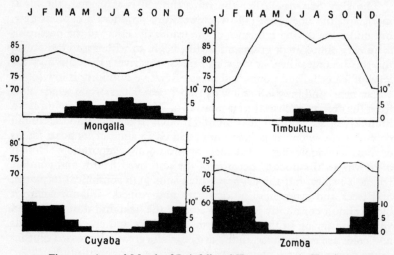

Fig. 42.—Annual March of Rainfall and Temperature in Tropical Continental Climates

This fall in temperature with the arrival of the rains is probably not appreciated, for the increased humidity makes the heat harder to bear, i.e. 'sensible temperature' does not decrease although actual air temperature does. The season just preceding the rains is the most unpleasant of the year, for humidity is steadily increasing and as yet the cooling and purifying influence of the early showers is not felt.

The Tornado Season. The arrival of the rains is announced by violent storms known in West Africa as *tornadoes*. In Sierra Leone the first of these occur in April, three months before the rains, and they increase in frequency until June when the real rains begin. They are rare during the rains but recur when these end in September, becoming less and less frequent until December. They bring practically the only rain which falls outside the rainy season. They are extraordinarily sudden and violent and they usually occur at night, the darkness helping to make them more impressive, but they last only a short time and affect a relatively small area. Although usually preceded by flickering lightning, they are sometimes entirely unannounced; the calm air suddenly becomes a rushing wind, the silence is rent with peals of thunder, and rain falls in sheets. In half an hour it has passed and all is quiet again. Elsewhere the season of violent storms is shorter, but everywhere thunderstorms precede the rains.

The Wet Season. The effect of the rains, when they come, is magical; the dust becomes mud, the dry water-courses become torrents, the withered vegetation bursts into life and in three weeks a dusty grey landscape has clothed itself in the fresh green of grass and leaves. The conditions during the wet season may be described in almost the same words as the equatorial climates; there is the same high humidity, the same heat with the same small diurnal range of temperature, but there are usually spells of fine weather when there is a temporary return to something like dry season conditions.

The duration of the rains decreases steadily polewards, so also does the amount of rain and its reliability; 50 inches on the equatorial margin becomes 10 inches on the fringe of the desert; 20 inches to 40 inches is a usual amount, but a station recording a mean annual rainfall of, say, 30 inches may have more than double this in one year and half this amount the next. This unreliability is unfortunate since the tropical climates, although at present largely pastoral, are otherwise well suited to agriculture.

The fact that the rain all occurs in summer when temperatures are high makes growth at this season very vigorous, but it also results in a high rate of evaporation and a considerable loss by run-off. Over the greater part of the Union of South Africa the evaporation is more than 90 inches, while the rainfall averages about 30 inches, i.e. evaporation is three times the rainfall. This disadvantage is increased by the intensity of the falls; 4 inches often fall in 24 hours and 2 inches in an hour. At many stations in Rhodesia the average fall per rainy day is half an inch, while more than half the rain in the Transvaal falls at this rate. These points must always be borne in mind in the selection of crops for cultivation and it must be realized that only a small proportion of the recorded rainfall is effective; 30 inches is frequently inadequate for agriculture.

The departure of the rains, like their arrival, is marked by violent storms, gradually decreasing in intensity. Temperatures, which had fallen with the advent of the rains, now probably rise again, though not so high as before, since the angle of the sun is declining; diurnal maxima rise higher and nocturnal minima fall lower. With the withdrawal of the sun to the other tropic a minimum is reached soon after the solstice. The mean temperatures of the coldest months are usually in the neighbourhood of 70°, except on high plateaux, and it is only relatively a 'cold season'. Conditions at this time are the most pleasant of the year, for the freshening effects of the wet season are still to be seen and felt, while the heat and dust of the height of the dry season have not yet brought discomfort.

Thus it is usually possible to make a threefold division of the year comparable with that recognized in India, where, however, it is much more clearly defined, into a hot season, a wet season and a cool season. Although astronomically winter, the sun being in the other hemisphere, the dry season is generally spoken of as summer since it is bright and sunny and, in the absence of cloud and rain, frequently hotter than the wet season. Throughout Latin America it is 'verano' and the rainy season is 'iverno', in which we may recognize the familiarity of an exotic people with the summer drought and winter rain of their native Mediterranean régime.

High Altitude Tropical Climates. As in equatorial climates the effect of altitude is to change the value but not the distribution of climatic elements. The single maximum of temperature and of rainfall at or near the solstice still holds good, but temperatures are lower and rainfall, above a certain level, is less. Mexico City (7,500 feet) has a mean annual temperature of 60° and 23 inches of rain, Cochabamba (8,000 feet) of 63° and 18 inches of rain. The lower rainfall gives a tendency towards aridity which is increased by the clearness of the air, the low atmospheric pressure and the tendency of strong winds, all factors which promote evaporation. Further, the rainfall has a high periodicity and falls in the form of heavy showers, thus increasing run-off and diminishing its efficacy; irrigation is therefore generally necessary for successful cultivation.

The mitigation of temperatures by altitude has important results on cultivation and settlements. Europeans can grow sub-tropical and even temperate crops in tropical latitudes, hill stations provide relief and renewed health for those compelled to live for the rest of the year at lower levels. The Spaniards found it possible to colonize Mexico and Peru, displacing the indigenous civilizations of the Aztecs and Incas; the Dutch settled on the South African plateau, and Europeans are now firmly established on the East African plateau even on the Equator.

Temperatures here are practically those of sea-level at the Cape, in fact over the whole of the plateau south of the Equator there is a remarkable uniformity of temperature, decreasing altitude compensating for increasing latitude, as the following maximum and minimum figures show:

					Max.			Min.		
Nairobi .	. Lat.	1° 25′S.	5,450 feet.		(Mar.)	65		(July)	58
Salisbury	. ,,	17° 54′	4,880 ,,		,,	(Jan.)	69	,,	,,	56
Bulawayo	. ,,	20° 2′	4,470 ,,		,,	,,	72	,,	,,	57
Johannesburg	,,	26° 11′	5,925 ,,		,,	,,	67	,,	,,	51
Graaff Reinet.	,,	32° 16′	2,500 ,,		,,	,,	72	,,	,,	56
Cape Town .	,,	33° 56′	40 ,,		,,	,,	70	,,	,,	55

The great diurnal range makes frost a real menace towards the margin of the zone and at higher altitudes, imposing a limit on more delicate crops. Sugar-cane, whose upper limit can conveniently be taken as the dividing line between the 'tierra templada' and 'tierra fria' (see p. 129), is grown up to 6,000 or 7,000 feet in equatorial latitudes in Peru, but at Jujuy (lat. 23° 53′ S.) it is not seen above 3,000 feet.

Vegetation of Tropical Climates. The heavy and regularly distributed rainfall and the constant high temperatures of the marine type support a tropical rain forest practically indistinguishable from the equatorial, but as we go inland from the trade-wind coasts or polewards

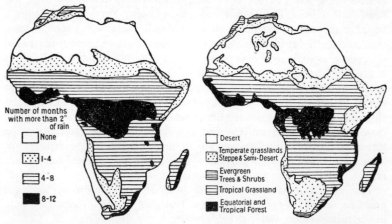

Number of months
with more than 2″
of rain

☐ None

▨ 1-4

▤ 4-8

■ 8-12

☐ Desert

▨ Temperate grasslands / Steppe & Semi-Desert

▤ Evergreen Trees & Shrubs

▥ Tropical Grassland

■ Equatorial and Tropical Forest

Fig. 43.—Duration of Rainy Season in Africa

Fig. 44.—Vegetation of Africa

from the Equator in continental interiors forest growth becomes impossible as the drought lengthens, and the selvas gradually pass into woodland and scrub with scattered trees in a dominant setting of tall grass (the savanna) (Figs. 43 and 44). By contrast with the uniformity of the equatorial régime the key-note of tropical continental climates is the marked seasonal rhythm; and by contrast with the perennial activity of the selvas the savanna vegetation grows with astonishing rapidity in the wet season and remains practically dormant through the dry. The trees of the savanna are not forest trees, they are markedly xerophilous, hard, gnarled and thorny, with reduced leaves (acacias) or with devices for the storage of water (bottle trees and baobabs); most of them are deciduous, shedding their leaves at the commencement of the dry season and passing a xerophytic existence until the rains return. The heat and drought of the Harmattan often splits the trunks of trees, and from these splits a gum is exuded; the 'gum arabic' exuded by certain acacias is an important article of commerce in the Sudan. But it is the grass which is characteristic, grass which grows strongly to a height of six feet or more during the rains, but dies down to the roots in the dry season.

The indigenous economy of tropical grasslands is pastoral and, since pasture is poor in the dry season, partly nomadic. The native tribes of the African savanna are dependent on their cattle; extensive cattle ranching is practised on the campos of Brazil and the llanos of Venezuela, while the monsoon grasslands of Australia, somewhat similar in nature, are the home of a cattle industry on organized lines. There are attempts by native tribes at agriculture on a small scale, maize or millets being sown at the beginning of the rains, but as yet little has been done to bring these lands under the plough. Yet with an adequate water supply they will yield good crops of millet, maize, tobacco, cotton, etc., and their future development is assured along these lines; rapid progress is being made everywhere. The summer incidence of rain is a great advantage, but its unreliability is a drawback, which can, however, often be overcome by irrigation. It should be noticed that the incidence of rain in tropical climates is similar to that in monsoon climates, but that the amount is generally much less; thus they grow cotton, but not jute; millets, but not rice; coffee, but not tea. The torrential nature of the rain tends to an early exhaustion and impoverishment of the soil, so that careful conservation and replenishment are necessary for continuous cultivation.

The rhythmic climate imposes a rigid rhythm on agricultural practice; water for growing plants is available, without irrigation, only in the rainy season, while the dry season is the time of harvest. Coffee, in Brazil, is harvested from April to September, the dry season, and in Venezuela from September to May. Sugar-cane in Rio de Janeiro, where there are summer rains, is cut from March to October; in Pernambuco, where there are winter rains, from October to March.

With its strong seasonal changes the climate is much more stimulating than the dull monotony of the equatorial zone, and the savanna tribes are vastly superior, mentally and physically, to the forest dwellers. Poor pastoralists though they are, they cherish their independence and do not readily accept subordination.

Vegetation and Altitude in Tropical Climates. In the tropics of Latin America, where the highlands rise from the dense forests of a marine tropical climate, it is usual to recognize three zones of climate and vegetation. The limits of the zones differ from place to place, but the average altitudes are as follows:

1. The *Tierra Caliente* extends from sea-level to 3,000 feet and embraces the hot steamy coasts of Mexico and Brazil with their dense forests producing rubber and cacao.

2. The *Tierra Templada*, between 3,000 and 6,000 feet, has mean annual temperatures between 65° and 75° and is characterized by an unusually small annual range of temperature (often only 4° or 5°). Its altitude corresponds to a zone of cloud which forms at the saturation level when the air rises up the slopes by day and a zone of mist at night when the clouds condense. The atmosphere of damp warmth keeps temperatures uniform and supports a forest vegetation of extraordinary density with tree ferns, lianas, etc. Under cultivation the zone produces maize, coffee, tea, tobacco and cotton.

3. The *Tierra Fria*, between 6,000 and 10,000 feet, has mean annual temperatures of 54° to 65°. The lower temperatures and decreased rainfall bring about an impoverishment of the forest which passes into scrub and grassland and the crops are those of the temperate zone, maize, wheat, lucerne, etc. At the same time conditions have become more favourable for settlement and especially European settlement; the density of population in the 'tierra fria' of Mexico is four times that of the 'tierra caliente'.

A somewhat similar succession of zones is found in tropical Peru, Bolivia, Ecuador and Colombia, where the trades, after crossing the interior savannas, are again forced upwards by the Andes and give rise to a strip of 'Montana' forest similar in nature and origin to the forests of the marine tropical type.

But in East Africa the highlands, e.g. Kilimanjaro, rise from the xerophilous 'altitude savanna' of a continental climate. This savanna is followed by a wetter zone with temperate or sub-tropical rain forest, clothing the slopes between 3,000 and 9,000 feet, above which is steppe with tussocky grass.

In a zone in which the trade winds play such an important part it is easily understood that aspect with respect to these will profoundly modify climate and vegetation. The rivers of Colombia, the Magdalena and the Cauca, lie open to the north-east winds and send arms of dense forest up into the grassland and woodland of the plateau. The upland ridges of the East African plateau receive an adequate rain, but the north-south depressions, sheltered from the rain-bearing winds, are often so dry that they are basins of inland drainage, while some of them are actually drying up and soda is obtained from the disappearing lakes, e.g. Lake Magadi and Natron Lake.

REGIONAL TYPES

Africa. North and South Africa present a striking contrast in the latitudinal extension of their tropical climates; north of the Equator the strip of savanna between forest and desert is only about 600 miles wide, but south of the Equator it stretches through more than double this width and, on the eastern side of the continent, almost down to its southern extremity. In North Africa the extent of the tropical climates is the extent of the swing of the equatorial rainfall belt, for the trade winds have come overland all the way and bring no rain; the margin of the desert is therefore sharply defined and runs almost due east and west. In the southern half of the continent, on the other hand, the trades come over the sea, the continent is narrow and offers a high barrier to the passage of the winds, the eastern side is therefore well watered and a belt of forested coast-line backed by savanna extends nearly the length of the continent.

Nigeria and the Guinea Lands. Behind the coastal forest belt the S.W. monsoon of the wet season (see p. 110) alternates with the N.E. trades of the dry season, the latter often strengthened and blowing out from the desert as the hot dry *harmattan*, sometimes temporarily bringing Saharan conditions, with humidities as low as 20 per cent, right down to the coast. A somewhat similar climate extends right across Africa as far as the headwaters of the Nile and it is from the summer rains that the White Nile gets some of its water. But the regularity of the discharge of the White Nile, which is so important for Egyptian agriculture, follows from the drainage of the lakes of the great plateau where the régime is equatorial.

East Africa. There is no real dry season here but two maxima occur (see Nairobi and Entebbe), one in February to May, known as 'the maize rains', the other in October to December, known as 'the millet rains', but except locally round Lake Victoria the rain is below normal (Nairobi 40 inches). Farther east the rainfall is still less (Makindu 20 inches) and near the coast there is virtually desert (Kismayu 15 inches), which becomes more and more complete to the north-east of Somaliland.

The whole region is affected by the proximity of the two monsoon centres of Abyssinia and India, which receive much of the rain which would otherwise fall here. In the northern hemisphere winter the winds are north-east, reinforced trades blowing parallel to the coast. They would not normally bring much rain, but in actual fact they bring practically none, for the African continent develops a high pressure of its own which deflects these winds and gives them a direction which is actually off-shore; not until the coast changes direction at Mombasa is 20 inches recorded in winter. In summer there is a strong air flow to the focus of the Indian monsoon and winds are south-west, again parallel to the shore, and the only rain which falls is from the branch of this current which is drawn off to the Abyssinian focus. It is during the change-over in May and June that this coast gets its rain (see Mogadiscio); Kismayu has more than half its rain in these two months.

The East African plateau is a highly favoured region with an adequate but not excessive rain well distributed through the year, a vegetation easily cleared for cultivation, and temperatures modified by altitude to an extent that makes it a suitable field for European settlement. The mean temperature of the hottest month at Nairobi is 65°, not much hotter than London, and of the coldest 58°, the equivalent to an English June, and although the diurnal range is rather large, it is not excessive, and there is no danger from frost. The altitude and the bright light are stimulating, perhaps too stimulating after a time, and Europeans find the climate physiologically congenial as well as agriculturally productive.

Continuity of the African Savannas. By causing a break in the forest belt the East African plateau completes the connection between the northern and southern savannas, thus facilitating free movement throughout the length of the continent; a fact which has been of great significance in the spread of species of animals and plants and which has also affected the migrations of human groups.

Discontinuity of South American Savannas. The Venezuelan and Brazilian savannas are isolated, since the orographic rains along the Andean slopes support a dense forest, the Montana, which terminates the savannas on the west. This fact may account for the poverty of the mammalian faunas of the South American tropical grasslands as compared with the African. For whatever may have caused this depopulation, and this is clearly of quite recent date, their repopulation would be a slow process in the absence of immigration from other regions.

The Tropical Climates of South America are full of anomalies, imperfectly understood because of insufficient data. North of the Equator the tropical climates are represented in wide stretches of the Guiana Highlands and in the llanos of the Orinoco. South of the Equator there is a much fuller development in the campos of Brazil (e.g. Cuyaba) as far south as Asuncion in Paraguay. A feature of the climate here, as in the middle Amazon, is a winter anticyclone bringing cold waves known as 'friagems' or 'surazos', when the temperature may fall below 50° and cause considerable discomfort.

The savanna type of grassland passes eastward into a more xerophytic type, the caatinga or thorn scrub of the projecting nose of Ceará and Pernambuco. Here is a region, centred round the middle course of the São Francisco, with 30 inches or less, but rendered still more inhospitable by the unreliability of the rains. The dry season is normally six months long, but it often happens that the rains fail to arrive to break the drought and famine results, with loss of most of the cattle and depopulation of the area. This aridity is as yet not satisfactorily explained, although it is generally attributed to the presence of high land to the east; this high land, however, does not receive excessive rain to compensate, as this explanation would lead one to expect.

Along the east coast, from Bahia southwards, summer is the rainy season, but between here and Cape San Rocque there is an anomalous winter rainfall maximum. At this season the S.E. trades are directly on-shore, while in summer, which is relatively dry, the prevailing wind is north-east, i.e. parallel to the coast.

Central America and Mexico. In the 'American Mediterranean' the prevailing wind throughout the year is the N.E. trade, and the islands and coasts are wettest where most exposed to this. But the simple circulation is modified by monsoonal indraught which brings summer rains to the whole of the area. In winter the warm waters of the Caribbean and the Gulf of Mexico produce a low-pressure centre round which there is a counter-clockwise circulation, more or less parallel to the coast-line; rainfall is heaviest where the lie of the coast is across this wind swirl, e.g. in Honduras. Winds along the Mexican coast are northerly and sometimes strong (Nortes), a continuation of the American 'Norther', and they are occasionally felt high up on the plateau (Papagayos). These northerly winds bring cold which is intense for the latitudes and the fall of temperature is often sudden and unpleasant.

SUGGESTIONS FOR FURTHER READING

Besides A. Knox's *Climate of Africa*, 1911, the available volumes of the 'Handbuch der Klimatologie' and the appropriate chapters in the *Oxford Survey of the British Empire*, the following articles, etc., on the Climatology of Equatorial and Tropical Africa will be found useful. R. Miller, 'The Climate of Nigeria', *Geography*, 1952; H. G. Lyons, 'Meteorology and Climatology of German East Africa', *Q. J. Roy. Met. Soc.*, 1917; J. J. Craig, *Rainfall of the Nile Basin*, Cairo, 1913; C. E. P. Brooks, 'The Distribution of Rainfall over Uganda with a Note on Kenya', *Q. J. Roy. Met. Soc.*, 1924; C. E. P. Brooks, 'Rainfall of Nyassaland', *Q. J. Roy. Met. Soc.*, 1919; C. E. P. Brooks, 'Rainfall of Nigeria and Gold Coast', *Q. J. Roy. Met. Soc.*, 1916; N. P. Chamney, *Climatology of the Gold Coast*, Bull. 15, Dept. of Agri., Accra, 1928; G. G. Auchinleck, *Rainfall off the Gold Coast*, Bull. 2, Dept. of Agri.; H. Schmidt, *Der Jahrliche Gang der Niederschläge in Africa*, Deutsche Seewarthe, 1928. There is a good collection of maps and charts in M.O. 492, *Weather of the West Coast of Tropical Africa*, H.M.S.O., 1943. B. J. Garnier has begun a series of *Climatic Observations at University College Ibadan* including measurements of evapotranspiration.

For South America the following may be consulted: C. de Carvalho, *Météorologie du Brésil*, 1917; G. G. Chisholm, 'Meteorology and Climatology of Brazil' (a review of the above), *Scot. Geog. Mag.*, 1917; 'Climate in North-East Brazil', *Geog. Teacher*, 1926 (a summary and review of 'Causas provareis das seccas do nordesti Brasileiro', J. de S. Ferraz); 'Climatological Data for Northern and Western Tropical South America', Supplement to *Monthly Weather Review*, March 1928; B. Franze, *Der Neiderschlagsverhältnisse in Sud Amerika*, Petermanns Mitteilungen, Suppl. No. 193, 1927.

For Central America and the West Indies: 'Climatological Data for Central America', *Monthly Weather Review*, 1923; 'Rainfall Maps of Cuba', *Monthly*

Weather Review, 1928; 'Climatological Data for West Indian Islands', *Monthly Weather Review*, 1926; 'West Indian Hurricanes and other Tropical Storms of the North Atlantic Ocean', Supplement to *Monthly Weather Review*, 1924; C. F. Brooks, 'Notes on the Climate of Panama', *Geog. Rev.*, 1920; *The Oxford Survey of the British Empire*, and Government Handbooks.

For the East Indian Archipelago: *Het Klimaat van Nederlandsch Indië*, by Dr. C. Braak, in three volumes, each with summaries in English; C. E. P. Brooks, 'The Climate of the Fiji Islands', *Q. J. Roy. Met. Soc.*, 1920; and the 'Atlas van Tropisch Nederland'.

Knowledge of tropical and equatorial climates has greatly increased in recent years. B. J. Garnier has begun a series of *Climatic abservations at University College Ibadan* including regular measurements of evapotranspiration. The new colleges of Ghana and Makerere, the University of Malaya, the School of Tropical Agriculture in Trinidad and Research Institutes such as that at Namulonghe are steadily investigating the influence of climate on crops, forestry, soil conservation, etc. In meteorology the network of airways is accumulating valuable information, some of which is summarized, as in I. E. M. Watts, *Equatorial weather, with particular reference to S. E. Asia*, 1955. The founding in 1945 of the Institute of Tropical Meteorology, as a co-operative project of the Universities of Chicago and Porto Rico has given rise to innumerable papers in scientific journals and the main results have been summarized in H. Riehl's *Tropical Meteorology*, 1954.

For the consequences of temperature and the seasonal distribution of rain, see F. Fournier, *Climat et Erosion*, Paris, 1960.

TEMPERATURE

Station	Lat.	Long.	Alt. (ft.)	J	F	M	A	M	J	J	A	S	O	N
HONOLULU .	21°N.	158°W.	38	71	71	71	73	75	77	78	78	78	77	75
PORT AU PRINCE	19°N.	72°W.	120	76	77	78	79	80	81	82	81	81	80	78
KINGSTON . .	18°N.	76°W.	24	77	77	77	78	80	81	82	82	82	81	79
KEY WEST . .	25°N.	82°W.	22	69	71	73	76	79	82	84	84	82	79	74
COLON . . .	9°N.	79°W.	36	80	80	80	81	81	80	80	80	80	80	80
RIO DE JANEIRO	23°S.	43°W.	201	78	78	77	74	71	69	68	69	70	71	73
ST. HELENA . .	16°S.	6°W.	1,900	64	65	66	65	63	60	58	57	57	58	59
DURBAN . . .	30°S.	31°E.	250	76	77	75	71	68	64	64	65	67	69	72
KAYES . . .	14°N.	12°W.	197	77	81	89	94	96	91	84	82	82	85	83
BATHURST . .	13°N.	17°W.	6	74	75	76	76	77	80	80	79	80	81	79
MONGALLA . .	5°N.	32°E.	1,440	80	82	83	81	79	77	76	76	77	78	79
GONDOKORO. .	5°N.	31°E.	1,500	83	85	86	84	81	79	78	77	78	78	79
MOGADISCIO . .	2°N.	45°E.	59	77	79	81	82	80	77	75	76	76	77	76
LINDI . . .	10°S.	40°E.	268	82	81	81	81	80	77	78	77	78	79	81
FORT JOHNSON .	15°S.	35°E.	1,558	79	77	78	77	73	69	68	71	75	80	82
SALISBURY . .	18°S.	31°E.	4,856	70	69	68	66	61	57	56	60	66	71	71
BULAWAYO . .	21°S.	29°E.	4,435	71	70	69	66	61	58	57	61	67	71	72
CARACAS . .	10°N.	67°W.	3,419	65	65	66	68	70	69	68	68	69	68	67
CUYABA . . .	16°S.	56°W.	541	81	81	81	80	78	75	76	78	82	82	82
ASUNCION . .	25°S.	58°W.	383	81	80	78	72	67	63	64	66	70	73	76
MEXICO CITY .	19°N.	99°W.	7,411	54	57	61	64	65	64	62	62	61	59	56
PUEBLA . .	19°N.	98°W.	6,987	54	57	61	64	65	64	63	63	62	61	58
OAXACA . . .	16°N.	97°W.	5,080	63	66	70	73	74	72	70	70	69	67	66
SUCRE . . .	19°S.	64°W.	9,190	55	55	56	54	51	49	49	52	56	56	57

RAINFALL

D	Yr.	Ra.	J	F	M	A	M	J	J	A	S	O	N	D	Total
72	75	7·7	3·7	4·3	3·8	2·3	1·9	1·1	1·3	1·5	1·5	1·9	4·2	4·1	31·6
77	79	5·9	1·2	2·5	3·7	6·5	9·4	4·1	2·7	5·4	7·3	6·6	3·4	1·3	54·1
78	79	4·9	1·0	0·6	1·0	1·2	4·3	4·1	1·7	3·7	4·1	7·5	3·1	1·0	33·9
70	77	14·4	1·9	1·5	1·3	1·3	3·5	4·3	3·4	4·5	6·8	5·6	2·3	1·7	38·1
79	80	1·6	3·7	1·6	1·6	4·3	12·4	13·3	16·0	14·8	12·5	15·1	20·7	11·4	127·4
76	73	10·1	5·0	4·5	5·3	4·2	3·2	2·3	1·7	1·7	2·6	3·3	4·1	5·5	43·4
61	61	8·8	3·0	3·8	4·9	3·9	4·2	4·1	4·0	3·7	3·0	1·9	1·7	2·0	40·2
74	70	12·9	4·6	4·9	5·4	3·4	1·9	1·2	1·2	1·7	3·2	5·1	5·0	5·1	42·7
77	85	19·2	—	—	—	—	0·6	3·9	8·3	8·3	5·6	1·9	0·3	0·2	29·1
75	78	7·2	—	—	—	—	0·2	2·9	10·9	19·6	10·0	3·7	0·2	0·1	47·6
79	79	6·9	0·1	0·8	1·5	4·2	5·4	4·6	5·2	5·8	4·9	4·3	1·8	0·3	38·9
80	81	8·7	0·1	0·8	2·0	3·5	6·5	3·9	5·0	4·9	4·4	4·7	1·9	0·4	38·1
77	78	6·9	—	—	—	0·7	2·2	3·5	2·0	0·6	0·5	0·7	0·4	—	16·9
82	80	5·7	6·1	4·1	7·5	4·7	0·1	—	0·3	0·5	0·6	0·7	1·8	4·8	32·0
79	76	13·0	8·5	7·0	4·0	2·9	6·3	0·1	—	0·1	0·2	2·1	1·9	6·4	33·5
70	65	14·6	7·5	7·4	4·5	1·0	0·5	—	—	0·1	0·3	1·1	3·7	5·8	31·9
72	66	15·0	5·5	4·8	4·5	0·8	0·6	—	—	—	0·3	0·6	3·1	4·9	25·1
65	67	4·0	0·9	0·3	0·6	1·6	2·8	4·0	4·3	4·2	4·1	3·8	3·3	1·8	31·7
81	80	6·6	9·8	8·3	8·3	4·0	2·1	0·3	0·2	1·1	2·0	4·5	5·9	8·1	54·6
80	72	18·0	5·5	5·1	4·3	5·2	4·6	2·7	2·2	1·6	3·1	5·5	5·9	6·2	50·9
54	60	10·9	0·2	0·2	0·5	0·8	1·9	3·9	4·5	4·6	3·9	1·6	0·5	0·2	22·8
54	61	11·0	0·4	0·3	0·4	1·1	3·5	6·8	5·8	6·0	5·5	2·5	0·9	0·4	33·6
63	69	11·0	0·1	0·2	0·6	1·4	3·4	6·3	3·7	4·5	4·3	2·1	0·3	0·1	27·0
57	54	8·0	6·5	4·8	3·6	2·0	0·2	0·2	0·2	0·2	0·9	1·3	2·7	4·6	27·2

VIII

TROPICAL MONSOON CLIMATES

The geographical peculiarity of the monsoon region is the spacing of the great continents opposed on either side of the Equator whose natural low pressure is overweighted by the still lower pressure developed over land in summer. The air flows from a high pressure in one hemisphere to a low pressure in the other, alternating with the seasons. It is true that the equatorial low is the destination for a brief time in April and May, but by June air is drawn towards the Asiatic low and the monsoon 'bursts', a continuous air stream having been diverted to a S.W. monsoon as it crosses the Equator.

In January the flow of air is direct from the great Asiatic high to the Australian low near Pilbarra and is again deflected on crossing the Equator. The air flow is a shallow one, only 12 to 18,000 ft. in depth, and carries no rain beyond the Himalaya. The normal planetary régime thus disappears and its place is taken by a seasonal and regional régime in which the contrast between north and south, which normally controls the pressure belts, yields place to a contrast between land and sea. Resulting from this seasonal reversal of wind direction there occurs a seasonal contrast of marine and continental influences, of excessive rain and excessive drought, of abnormal heat and, usually, abnormal cold, the change from one to the other being always distinct and frequently abrupt.

With certain exceptions, namely, those stations such as the west coast of Japan and Annam, whose situations cause them to receive the winter monsoon over the sea, the maximum rainfall is received during summer; in fact the winter is characterized by partial or complete drought. In this respect the monsoon climates resemble normal tropical climates, but they differ markedly in the amount of rain received and in the incidence of such rain within the rainy season. They are also peculiar in that the heaviest rainfall generally occurs on west coasts, especially when mountainous, e.g. the Western Ghats, Arakan and Tenasserim, thus providing an exception to the general rule that west coasts in the trade-wind belt are dry. A few regions with peculiarly favourable situations receive rainfall during the prevalence of both monsoons; thus Madras has nine months of adequate rainfall.

Rainfall and Vegetation. The length of the dry season thus varies according to local factors, especially the relation of relief to wind direction, and this variation is reflected in a variety of vegetation types, ranging from desert to dense forest. The monsoon rain forest rivals the selvas of the equatorial zone in density and wealth of foliage, and it yields many trees of commercial value; where the dry season is longer, a deciduous forest is found, e.g. teak, adapted to more or less prolonged drought, a period of enforced inactivity.

The coincidence of heat with adequate, or even abundant rain in summer makes these climates agriculturally very productive, rice, oil-seeds, tea and jute being characteristic monsoon products. A luxuriant vegetation of tropical aspect flourishes far to the north, while rice is grown successfully in Niigata (38°N.) and tea is an important crop in Honshu.

The success of agriculture is dependent on adequate rainfall from the monsoon, adequate both in amount and distribution. In this connection it is important to realize the phenomenal rate of fall of much of the monsoon rain, downpours being usual in which run-off is so rapid that only a small proportion is effective for vegetation or irrigation. The average fall per rainy day at Cherrapunji is 2·6 inches, in Bengal 0·7 inch, and in the Deccan 0·3 inch or 0·4 inch. These should be compared with an average intensity in the British Isles of 0·1 inch per rainy day. The torrential falls implied by these figures are powerful agents of denudation, causing great soil waste on slopes and sometimes washing away roads, railways and bridges. Much of the rain which falls is thus unnecessary and incapable of utilization, but, given storage, it provides a margin of safety in the event of partial failure, while it is those areas which have just enough which are the most subject to drought and famine when the rains are below normal.

Rainfall and Irrigation. Since, for example, three-quarters of the population of India is agricultural, and since rainfall means crops and crops mean food, the rainfall distribution determines the supporting power of the land. Where rainfall is definitely deficient there is a reliance on artificial water supplies, and all cultivated land with less than about 12 inches must be irrigated. But where more than 20 inches occur there is a tendency to rely on rainfall, a proceeding which is only really justified by a rainfall of about 50 inches.

Reliability of Rainfall. It is precisely in this zone of just adequate rainfall that havoc may be wrought by deficiency, and the danger is greater or less in these areas according to the reliability of the rain. The percentage variability is naturally greatest where the rainfall is least, and stations in Sind sometimes receive three years' rainfall in 24 hours; but between the critical figures of 12 inches and 50 inches there is a considerable range of reliability. Thus:

Lahore (15 inches June–September) has a percentage variation of 38.

Delhi (24 inches June–September) has a percentage variation of 29.

Calcutta (51 inches June–September) has a percentage variation of 16.

A deficiency of 25 per cent will cause injury to crops, a deficiency of 40 per cent will spell ruin—and this is true whether the normal rainfall is 15 inches or 50 inches, for the crop grown will have been selected with a view to a normal rainfall.

THE ASIATIC MONSOON

The interference with the planetary circulation exercised by land masses is everywhere evident, but the extent of the interference is proportional to the size and compactness of the continental block. It is natural, therefore, that Asia should be the continent of monsoons *par excellence*. Its enormous bulk, comprising about half the land mass of the world, its east-and-west grain (see p. 40), its barren interior plateaux, its temperate and sub-tropical latitudes, the huge expanse of warm seas which flank it on the south and east, all combine to produce that contrast between continental and oceanic influences which lies at the root of monsoonal phenomena.

Throughout the south-eastern angle of Asia, from Karachi to Vladivostok, and extending far into the interior, this seasonal reversal of controls is the dominant feature of the climate, but nowhere is the contrast so marked as in India. Lying between the greatest land mass and the warmest sea, shut in by the loftiest mountains, backed by the highest plateaux of the world, it provides the best conditions for a monsoon. Although not a complete barrier between India and Central Asia, the Himalaya, the Sulaimans and the Burmese chains exclude external influences to such a degree that the great summer low pressure of Sind must be filled by winds from the south, and only from the south. India, and especially North-West India, thus becomes virtually the terminus for ocean winds in summer and the centre of dispersal of land winds in winter.

Within an area of varied relief embracing $1\frac{3}{4}$ million square miles and extending through 30° of latitude there must naturally occur a great variety of climatic types, ranging from continental to marine and from arid to humid; but its almost complete detachment, by virtue of its natural frontiers, from the rest of Monsoon Asia makes India, as a climatic unit, capable of independent treatment.

INDIA

The Seasons of India. Throughout most of India the year can be divided into three seasons:

(1) The Cold Season, from mid-December to the end of February.
(2) The Hot Season, from March to the end of May.
(3) The Wet Season, from June to mid-December.

The first of these is the season of the land monsoon and the last that of the sea monsoons, while the second is in many respects transitional. Rain falls, of course, in all these seasons, and some districts actually have their maxima outside the 'Wet Season', but the term 'Wet Season' is well deserved, for 85 per cent of the rainfall of India is recorded during the summer monsoon.[1]

The Wet Season is usually further divided into two: (1) the season of the advancing monsoon, from June to mid-September; and (2) the season of the retreating monsoon from mid-September to mid-December. Since the rains leave the north-west plains in mid-September, but linger in the southern peninsula until mid-December, it is clear that no fixed date can be named for the change from the Wet Season to the Cold Season, for this event occurs later and later towards the south.

Thus we recognize a fourfold division of the Indian climatic year, and it will be convenient to consider each of these in turn and trace in each the changes of temperature, pressure and rainfall and their inter-relations.

THE COLD SEASON

January Pressure and Winds. The main centre of the Asiatic high pressure in January, exceeding 30·5 inches, is centred over Mongolia, and from it cold, heavy air is dispersed; but very little of this reaches India, and such air as does descend from the plateau is well warmed adiabatically in its descent of the Himalaya. India, therefore, never experiences those waves of intense cold which visit China, even as far south as Canton. A secondary high-pressure system is over Kashmir and the Punjab (see Fig. 45), from which centre pressure decreases fairly uniformly southward to the low-pressure belt now south of the Equator. It is this Punjab high pressure which controls the prevailing wind direction and causes north-west winds down the Ganges valley, recurving to north-east in the Carnatic and becoming almost easterly in Cochin. In Sind and Gujarat winds are north-easterly, still controlled by the Punjab high pressure, and in Central India they are light and variable. But nowhere during the Cold Season are the winds strong or constant in direction, light breezes being charac-teristic. Under this anticyclonic régime the air is dry, fresh and in-vigorating, the sky is clear and of a perfect blue and this is the most pleasant time of the year.

Rainfall. The stability of the air during the Cold Season, and the off-shore direction of the wind are factors unfavourable to rainfall, and this season is, with few exceptions, a dry one. These exceptions are:

[1] The terms 'summer' and 'winter' monsoon are used here in preference to South-west' and 'North-east' because these descriptions are only true over limited areas, the so-called S.W. Monsoon, for example, blowing up the Ganges valley is a south-east wind.

(1) the tip of the peninsula and Ceylon; and (2) along the foothills of the Himalaya and the margin of the plains in the Punjab and extending into the United Provinces. Rainfall in these two areas is fundamentally different in origin, that of Ceylon being convectional, that of the Punjab cyclonic.

Owing to its low latitude, Ceylon is not far removed from the equatorial rainfall belt, and slight northward aberrations of this bring rains to the island and to the tip of the peninsula. This rainfall is particularly heavy on the east side of the island, on which the N.E.

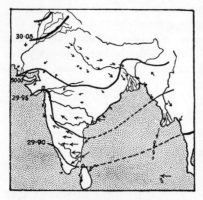

Fig. 45.—Mean Pressure and Winds (January)
The length of the arrow is proportional to the steadiness of the wind

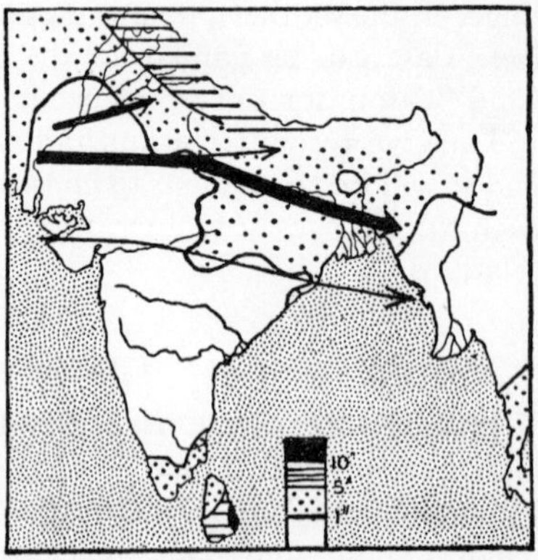

Fig. 46.—Cold Weather Rainfall (Jan. and Feb.) and Cold Weather Storm-tracks

monsoon impinges, moisture-laden from a long journey over the Bay of Bengal.

The rainfall of the Punjab and the United Provinces seems to be associated with depressions where the cold winds from the Asiatic high meet the westerlies which have come from the Mediterranean. They can occasionally be traced via Palestine and Afghanistan. The Asiatic high pressure persuades them to take a course well to the south, and the Himalaya later restrict their movements and steer them into the plains. They travel eastwards bringing one or two inches of rain to the northern plains and heavier falls to the higher slopes of the mountains; Peshawar receives more rain at this time than from the summer monsoon. Murree receives 6 inches in January and February and Simla 7 inches, while heavy falls of snow occur higher up, this being the chief time of renewal in the zone of permanent snow in the Western Himalaya. Southwards this rain extends into Rajputana, Central India and the Central Provinces, but the amount decreases markedly in this direction and the storms become more violent; frequently they are accompanied by thunder and hail which does much harm to crops. They die out eastwards and bring very little rain to the Ganges valley.

Small in amount though the rain is, it is of the greatest importance to the wheat and barley crops of the Punjab, and great distress is caused by its occasional failure. The comparative lightness of the fall increases its value inch for inch, as run-off is slight and a high proportion is available for crops.

The barometric gradients of these storms are very slight, but remarkable temperature changes are experienced during their passage. The

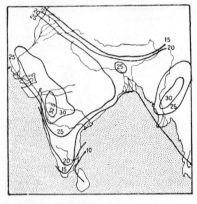
Fig. 47.—Diurnal Range of Temperature (January)

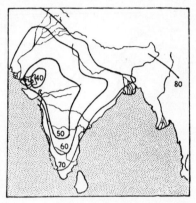
Fig. 48.—Mean Daily Relative Humidity (January)

approach of the storm is heralded by a shift of the wind to the south-east, bringing a current of warm, moist air, whose temperature is further raised by the liberated heat of condensation and by the blanket of cloud. With the passage of the storm centre the wind veers to north-west and the temperature may drop as much as 20° in 46 hours. The trajectory of this north-west wind brings it from the snow-clad slopes on to the plains, and the temperature is further lowered by evaporation. It is during these cold waves in rear of depressions that temperatures below freezing are recorded in the Punjab, and it is under these conditions that severe damage is frequently done to fruit and crops.

Cold Season Temperature. Latitude is the principal control of temperature at this time, the January isotherms being directed almost due east and west; the mean temperature of day in Ceylon is about 80°, in the Punjab about 50°. But more significant is the diurnal range of temperature, strikingly related to marine influence through the medium of relative humidity.

The thermometer often reaches 80° in the Punjab during the day, while frosts at night are common occurrences, and fires are lit every evening well into February. This rapid nocturnal radiation of heat is the direct result of the anticyclonic conditions with concomitant out-flowing winds, dry air and clear skies. Round the coasts the diurnal range is less and frosts are unknown.

Temperature. From January onwards, as the sun continues its northward march, the temperature begins to rise, and a tendency begins to be manifest for land and sea to assume control and for isotherms

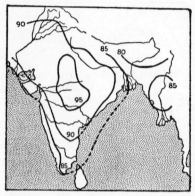

Fig. 49.—Mean Temperature (May)   Fig. 50.—Mean Maximum Temperature (May)

to curve round over the land. Thus the February chart shows a circular isotherm of 80° in the centre of the peninsula, and by March the peninsula is almost defined by the closed isotherm of 82·5°, in the centre of which is a circular area exceeding 87·5°. In April the 90° isotherm is found parallel to the coast and about 100 miles inland and in May a large area in Central India and the Central Provinces exceeds 95°. By now the heat is terrific, shade temperatures of 120° are frequently recorded in Sind, and Jacobabad, which has the evil reputation of being the hottest station in India, records a mean daily maximum above 110°.

The heat and glare of the sun, the wafts of suffocating heat reflected from the hard-baked ground or from walls compel all outdoor life to cease while the sun is above the horizon, and interiors are only kept bearable by shutting all windows and doors to keep out the stifling heat, and by the use of tatties—matting sprinkled with water through which air is drawn and cooled in passage by evaporation. Outdoor exercise is taken in what is by courtesy called 'the cool of the evening', but sunstroke is the probable penalty for venturing out during the day without sunshades or ample covering for the head and spine. All vegetation is withered and scorched by the desiccating heat, all water is dried up and the rivers have shrunk to thin threads lost in a wilderness of dazzling, shimmering sand and stones.

On the coast such excessive maximum temperatures are not recorded, but the heat is no more bearable, because it is moister. Relative humidities as low as 1 per cent may occur in Sind, but in the south the inflow of sea breezes keeps the air fairly humid. For the same reason the loss of heat at night by radiation is less, and nights are intolerably stuffy and airless. The diurnal range in Sind in May exceeds 30°, while in Madras it is less than 20° and in Bombay only 15°.

Pressure and Winds. The gradual and steady rise of temperature in the interior of India sets up convection currents, which, as would be expected, attain a maximum during the afternoon. The Punjab high pressure of January gradually breaks down, and by April has entirely disappeared, its place being taken by a low pressure extending throughout the plains. By May there is a deep low (29·6 inches) centring round Multan, and during the daytime this intensifies to 29·5 inches. A further intense low is established round Jubbulpore, and these induce a system of on-shore winds round the whole of the Indian coast.

It is during the climax of the Hot Season, more than at any other time of the year, that India is an independent climatic unit. This is a transition period between the winter and summer monsoons, a time when local factors are able to make themselves felt above the planetary controls. Air circulation is cyclonic and feeble, and conditions are ideal for the generation of local thunderstorms. Temperature and winds are locally controlled, and such storms as occur are of local origin, being due to the great changes of humidity and temperature rather than to outside influences. In this they contrast with the cold-weather storms which enter from the western plateaux, and with the storms of the rainy season, which originate in the Bay of Bengal or in the Arabian Sea.

Storms and Rainfall. Storms are fairly frequent and widely distributed in the plains during the Hot Season, but it is not everywhere that these storms bring rain. The duststorms of the north-west, the 'Nor'westers' of Bengal and the rainstorms of Assam all appear to be similar in origin, since all originate at the junction of two currents, a dry, cool upper current from the north-west riding over the shallow surface flow of air drawn in by the low pressure of the plain.

The arrangement is extremely unstable, and sudden inversions are apt to occur, there is an uprush of warm air and thunderstorms, often with hail, are the result. In the Punjab, Sind, Rajputana and the western plains these storms bring no rain; they are merely duststorms which, though bringing a welcome cooling wind, darken the sky almost like night, filling the air with blinding dust. In Bengal the sea breezes bring more moisture because nearer to the sea; the ascensional movement here frequently results in rapid condensation in the form of heavy rain or hail with lightning and thunder. The hailstones are sometimes of prodigious size, they often ruin the standing wheat and have been known to kill men and animals. But it is in Eastern Bengal, and especially in Assam, that these storms are valuable as rain-bearers; Dacca gets 18 inches in this season, and parts of Assam as much as 20 inches, chiefly in the form of afternoon thunderstorms. The economic value of this rain is incalculable, it nourishes the spring

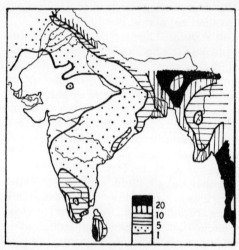

Fig. 51.—Hot Weather Rainfall (March–May)

rice crop in Bengal and it brings the first flushes of green leaves to the tea plantations of Assam. The storms become more frequent as the heat increases, as the Indian low pressure becomes more intense, and as the inflow of sea breezes becomes stronger and their humidity greater. Thus Calcutta receives 1·4 inches in March, 2 inches in April, and 6 inches in May; Cherrapunji has 9 inches in March, 30 inches in April, and 50 inches in May.

The only other region which experiences considerable rain at this period is Ceylon and the tip of the peninsula, especially on the western side. Colombo has 5 inches in March, 8 inches in April, and 13 inches in May; in fact May is the wettest month of the year, considerably exceeding the 'monsoon' months of June and July. A second maximum is reached in October and November, during the retreat of the monsoon. These rainy seasons may be considered as an invasion by the equatorial rainfall belt. In contrast, Cochin, 3° farther north, shows a more typical monsoonal régime with a pronounced maximum in June, but there is also a considerable fall (19·2 inches) in the three months of March, April and May. These rains are known as 'mango showers' in Mysore and as 'blossom showers' in coffee districts, and, as their names imply, their agricultural importance is great.

THE RAINY SEASON

Pressure and Winds. The prolonged heat and drought from which the greater part of India has been suffering for some months has generated increasingly low pressures over the plains until June and July. Throughout all this time the form of the isobars has changed

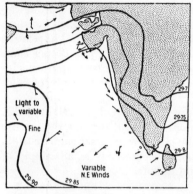

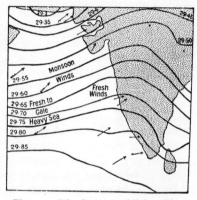

Fig. 52.—Weather on 14th May, 1897   Fig. 53.—Weather on 3rd July, 1889

scarcely at all, yet the great climatic event of the year has arrived, the 'bursting' of the monsoon in early June. In July, as in May, the isobars are arranged concentrically around the north-western plains, and the wind direction, controlled by this local low-pressure system, is to all appearances the same in each month. Yet Bombay, which has had five months without any measurable rain, and which records only ½ inch in May, is suddenly drenched with 20 inches in June and 24 inches in July. Examination of the isobaric map reveals the information that although the direction of the pressure gradient is the same in July as in May, there is a marked difference in intensity. The lowest mean pressure of May is 29·55 inches, but in July it has intensified to 29·4 inches; the gradient from Sind to Ceylon in May is 0·2 inch as compared with 0·4 inch in July. Thus winds are stronger and more constant and bring in more moisture. But more significant than the pressure distribution over India itself is that over the Indian Ocean. Figs. 52 and 53 show typical charts for May and July, and the essential differences are at once revealed. In the May chart it is clear that India's weather is the product of local conditions. There is still a tendency to an equatorial (doldrum) low-pressure zone, which attracts degenerate trade winds from both hemispheres. Between this and the

Indian low pressure is an ill-defined area of higher pressure which is the source of India's air supply. This air current is bringing no appreciable rain, such falls as occur being the products of local storms. Turning to the July chart it is noticeable that the equatorial low pressure has vanished, that the gradient over the ocean is steady and steep and that there is therefore a continuous stream of air from the southern Indian Ocean to the Indian focus. The explanation, then, of the sudden burst of the monsoon lies in the eventual overcoming of that light, unsteady circulation in the Indian Ocean and the union of the S.E. trades, deflected right-handedly on crossing the Equator, with the local winds of India. This reinforcing current from the southern hemisphere brings with it enormous reserves of moisture collected in its 4,000-mile journey over the warmest ocean. The relative humidity at Bombay in May is 74 per cent, and in July 86 per cent. Further, the increased pressure gradient gives increased wind velocity and therefore a continuous and constantly renewed supply of rain; the mean wind velocity at Bombay is 7·4 in May and 14·2 in July. It is the sudden arrival in India of this moisture-laden current which constitutes the 'burst' of the monsoon. Up to this point the intense heat and dryness of the air under the cloudless summer skies in India have been capable of absorbing the comparatively small supply of moisture which the sea breezes could import.

Breaks in the Monsoon. If, as sometimes happens, there is a temporary reversion, after the arrival of the monsoon, to the pressure conditions of May, there will also be a reversion to the May rainfall, that is, a break in the rains. The rain-bearing winds will be lost in the vague pressures over the ocean and drought and famine will result in India if the conditions are prolonged. It was a pressure distribution of this type which led to the disastrous famine of 1899.

The 'Burst' of the Monsoon. Although there may be a few days of light rain preceding the burst of the monsoon, this is usually in the form of an intense cyclonic storm with thunder and lightning and torrential rain, arriving with dramatic suddenness to relieve the parched and thirsty earth. At sea the winds attain great violence, native shipping does not venture out to sea, while larger ships experience some very rough passages. All the west-coast harbours, poor and few as they are, are closed, with the exception of Bombay; sea-level rises all along the coast, the water is driven up the flat shores of Cutch and Cambay and the Ranns of Cutch are flooded. The opening storm passes away and may be followed by longer or shorter periods of fine weather, to be followed in turn by further cyclonic storms. This pulsatory nature of the rain is extremely characteristic. It is to be attributed partly to the essentially cyclonic nature of much of the rain, and partly to a kind of self-feeding action, whereby the heat liberated by the heavy precipitation lends further energy to the system, increasing the ascensional movement and leading to a crescendo of violence. This violence soon exhausts the moisture reserve and further rainfall has to await the arrival of fresh moisture-laden currents from the sea. Slowly fresh energy accumulates as the humid currents come pressing inland and the process is repeated.

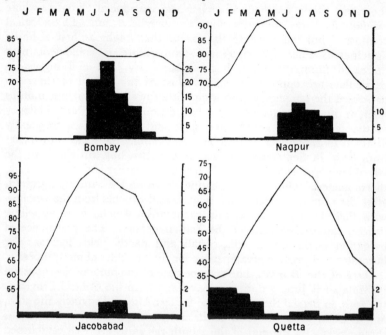

Fig. 54.—Annual March of Temperature and Rainfall in India
(Note that the scale of rainfall used in the graphs for Jacobabad and Quetta is five times that used for Bombay and Nagpur)

Effect of the Rains on Temperature. The immediate effect of the rains is to lower the temperature, both by conduction and evaporation, and to make the climate of the plains endurable again. The mean temperature at Bombay in May is 85·8°, and in July 81·4°, a fall of more than 4°; at Nagpur there is a fall from 95° in May to 81·7° in July, more than 13°, and falls of 6° or 7° are usual. Fig. 54 gives some idea of the relief which this drop of temperature affords. The line of the temperature curve in April and May for Bombay and Nagpur may be continued in imagination to a hypothetical maximum in July, such as would be reached were it not for the cooling effect of the rain. An approximation to this hypothetical curve is seen in the graph for Jacobabad which receives only just over 3 inches of monsoon rain. The curve continues to rise into June, when the mean temperature reaches nearly to 100°, and even here there is a slight depression of the crest of the curve when the rains come. At Quetta (whose climate is, of course, not monsoonal) there is practically no June–September rain and therefore no cooling effect. The curve continues to rise until a normal maximum is reached in July, nearly 80° in spite of the altitude (5,502 feet). This flattening of the top of the temperature curve in the rainy season has been observed also in tropical climates, where rainfall, as here, coincides with the overhead sun. Physiological comfort is not increased to the extent that might at first sight be anticipated, since the decreased temperature is coincident with increased humidity; moist heat replaces dry heat and nights, in particular, are oppressively sultry. Just as the arrival of the rains depressed the temperature, so their withdrawal, three or four months later, is often accompanied by a reassertion of the upward tendency of the temperature graph, and a second maximum is frequently attained in September or October. At Bombay the temperature rises from 80·9° in September to 82·4° in October, but this check in the fall of temperature is not so universal as the earlier check to the rising thermometer. The hot sun beating down on the saturated ground, after the rains have left, draws up mists and vapours from the swamps, with their attendant fevers and rheumatism which, to an appalling degree, sap the energies of the Indian peasant. The prevalence of irrigation works, tanks, wells, canals and paddy fields increase the humidity and provide greater space for the breeding of mosquitoes.

Date of the 'Burst'. The average date of the burst of the monsoon in Malabar is June 3rd, at Bombay the 5th, in the Central Provinces the 10th, in Bengal the 15th, in the Eastern United Provinces the 20th, and at Delhi about the 30th; but its arrival may be delayed as much as three weeks. By the end of the month the rains are well established everywhere, maximum falls are generally recorded in July and August and by the first week in September the rains begin to diminish. The monsoon current appears to be a shallow one, since the course it takes is profoundly influenced by the shape of peninsular India, whose relief is not, in fact, extremely mountainous. Coming from the south-west, the air-flow divides into two branches, an Arabian Sea branch and a Bay of Bengal branch, and it is the relation between the courses taken by these currents on the one hand and the relief of the land on the other that determines the distribution of rainfall at this season. The rainfall is due almost entirely to ascent of moisture-laden air, and this ascent is due to one of three causes:

1. Relief—whereby there appear striking contrasts between wind-ward stations and stations in the rain shadow in lees of mountains.

2. Convection—whereby there is a tendency to afternoon storms.

3. Cyclonic storms—which give the most evenly distributed rainfall of the three.

Advance of the Monsoon. The two currents into which the monsoon divides (see Fig. 55) may be treated separately and the rainfall due to each described in turn. The goal in each case is the low pressure of the Indo-Gangetic Plain, culminating in the intensely low pressure over Sind. But the heart of the system varies in position from day to day and so influences the trajectory of the winds and the courses of cyclonic storms.

The Arabian Sea Branch expends its full force on the Western Ghats which stand directly athwart its path. Here it must rise through

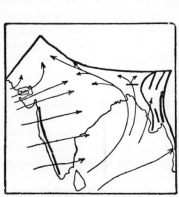 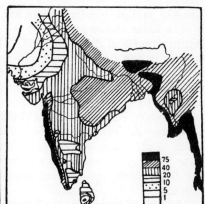

Fig. 55.—Streamlines during
Monsoon
(*After* Simpson)

Fig. 56.—Monsoon Rainfall
(June–Oct.)

a height of from 3,000 to 7,000 feet and the resultant precipitation is enormous. More than 30 inches fall in June along the coastal plain and more than 40 inches in July; higher up the slopes the rainfall is still heavier. Mahabaleshwar (4,540 feet) records an average of over 100 inches in July and over 250 inches during the five months of the monsoon. Here rain falls, on an average, on 116 out of the 122 days in June–September and nearly 2½ inches fall on each rainy day—more than London gets in the whole of July. But these heavy falls are experienced only along a narrow strip, in fact along the western slopes; for after crossing the summit the wind descends the eastern slopes as a foehn wind. Thus there exists a remarkable rain shadow immediately in lee; Gokak, only 65 miles from Mahabaleshwar, has only 22 inches a year; Dhulia has 22 inches; Bellary has only 18 inches, of which only 9 inches fall in June–September and only 1·3 inch in July. The rain shadow effect is still more pronounced in the south where the mountains are higher—Tinnevelly, in lee of the Cardamom Hills, which rise to 10,000 feet, has less than 1 inch in July and only 2·7 inches from

June to September. At their northern end the Western Ghats decrease
rapidly in height and the rainfall decreases coincidently (Surat 40 inches,
June–September). At the same time the contrast between windward
and leeward ceases to exist, as there is no obstacle to the rain-bearing
winds; 40 inches of rain from the monsoon is recorded far inland,
while 30 inches is recorded in a continuous strip from the Gulf of
Cambay to the Gangetic Plain at Cawnpore.

Northern Limit of the Rains. Northwards from the Gulf of
Cambay the monsoon rainfall decreases still further, 29 inches at
Ahmedabad, 15 inches at Bhuj in the island of Cutch, 7 inches at
Karachi, and practically none in Baluchistan. The explanation of this
is to be found in two factors: (1) the trajectory of the air currents and
(2) the dryness of the air of Sind. The eastward deflection of the air
current in the Arabian Sea, seeking the focus of Sind, has become
greater and greater farther north, until it has practically lost its northward component and the air is moving east or even south-east. The
rain-bearing winds thus reach only to the south-west tip of Arabia and
fail to reach Persia and Baluchistan whose aridity at this season is
complete. An east to west line drawn near Karachi divides those
winds to the south which have travelled over the Arabian Sea from
those to the north which have travelled over the arid lands of Arabia
and Baluchistan, and this line practically marks the limit of monsoon
rainfall.

The Bay of Bengal Branch flows up between Ceylon and Sumatra,
swinging northward and westward under the attraction of the Sind
low pressure, and travelling up the Gangetic Plain towards this destination. A branch of this current impinges on the mountainous shores of
Arakan and Tenasserim in a manner analogous to the impact of the
Arabian Sea against the Western Ghats, and in a like manner there
results torrential rain along a narrow coastal zone. Akyab has a mean
annual precipitation of over 300 inches, of which about 170 inches fall
in the four monsoon months from June to September. As would be
expected, there is a region of pronounced rain shadow in the middle
Irrawaddy valley near Mandalay, which receives only 19 inches in the
four months from June to September; but a branch of the Bengal
current penetrates some distance up the Irrawaddy from the delta,
and brings 40 inches of monsoon rainfall as far north as Thayetmyo.
Under circumstances in general similar to those of Arakan some of
the heaviest rainfall in the world is recorded in the re-entrant, between
the Khasi and Lushai Hills in Assam, where the sudden forced ascent
of air currents, confined between the convergent mountain walls, is
responsible for the prodigious fall at Cherrapunji (4,455 feet) of 450
inches a year, of which nearly 350 inches occur in the four months
June–September. Over 900 inches (25 yards) of rain have been recorded
here in one year and 40·8 inches in one day (June 14th, 1876). Beyond

the summit of the hills this figure falls off rapidly, and Shillong, 25 miles away on the plateau-like top of the Khasi Hills, has only 55 inches in June–September.

Rainfall of Gangetic Plain. But the most valuable rain in India results from that branch of the Bengal current which advances up the Gangetic Plain towards the Punjab. The very gradual rise in altitude along this track allows an even distribution of rain, infinitely more favourable to man than the torrential falls on the Western Ghats with their inevitable zone of deficiency in lee. This rainfall is partly due to relief, but more especially to cyclonic storms which follow the track of low relief and low pressure, particularly along the southern margin of the plains. It should be noticed that this current is actually a south-east wind, although a direct continuation of the S.W. monsoon. The rainfall from this current decreases westwards and southwards; westwards because farther away from the source of moisture, southwards because farther away from the precipitating agent, the Himalaya. There is a tendency for this current to bear against the Himalayan foothills and so to bring heavier rains there at this season; compare Benares (35 inches June–September) with Gorakhpur (42 inches), Agra (23 inches) with Bareilly (36 inches) and with Naini Tal (81 inches). The hill stations not only get heavy rain (Darjeeling 102 inches June–September), but are also enveloped in cloud throughout much of the rainy season. The zone of maximum precipitation lies at about 5,000 feet, beyond which rainfall decreases, while beyond the first ranges the precipitation is very slight. Shrinagar has only 8 inches and Leh only 2 inches. The decrease from east to west is still more pronounced: Calcutta has 46 inches from June to September, Patna 41 inches, Allahabad 33 inches, Delhi 22 inches, Lahore 16 inches, and Multan 5 inches. The rainfall in the extreme north-west is extremely variable, and its arrival or failure depends on the strength of the Ganges valley branch of monsoon. This in turn depends on the pressure gradient in the Gangetic Plain, the focus, as in the case of the Arabian Sea branch, being Sind. When the low pressure of Sind is well developed, rains extend far to the north-west, but it occasionally happens that relatively high pressure develops in the Punjab, and reverses temporarily the wind direction in the United Provinces and the surrounding country, which now receives north-west winds of an anticyclonic type. There is, in effect, a reversion to the Cold Season conditions, and drought results. These dry north-west winds, which might be considered as a prolongation of the Arabian Sea branch, which has travelled over Baluchistan and Sind, sometimes possess the plains as far as Bengal and extend southwards into the Deccan as far as Hyderabad. It was a pressure distribution such as this which was responsible for the disastrous droughts of 1876, 1877, 1880 and 1883.

The tracks of the two principal currents of the monsoon have now been traced, but there remains an interesting area at the junction of the two.

Storm Tracks. Throughout the peninsula the south-west direction of the Arabian Sea holds sway, while in the Gangetic Plain the air movement is from south-east to north-west; the debatable ground between the two comprises the southern margin of the Gangetic Plain from Orissa, through Allahabad and Agra to Patiala. It is a favourite track of cyclonic storms which originate in the head of the Bay of Bengal and pass westward along the foot of the peninsular hills, accompanied, especially on their southern side, by heavy rainfall, which often causes sudden floods on the Narbada and Tapti. This rain is of value since the southern side of the plains would be deficient without it, and it is especially valuable to the rice-growing districts of the Central Provinces.

The Sind Focus. Bearing in mind that the goal of both monsoon currents is the low pressure of Sind, and that this is the temporary equivalent of the equatorial low-pressure belt, it is not a little surprising that the ascensional movement here does not give rise to heavy rainfall. But actually the site is one of the world's most complete deserts, and to this condition a variety of factors contribute. In the first place, of the air currents which arrive here, those from the south-east have already shed nearly all their moisture in their passage up the Gangetic Plain; those from the south-west have passed from a warm sea to an intensely hot land, and so decreased their relative humidity from 90 per cent to about 55 per cent. But even this air would become saturated and clouds would form at about 3,000 feet were it not for another important factor, namely, the existence of an upper current of hot dry air from the arid plateau of Baluchistan, which greedily absorbs all available moisture and keeps the skies free from the slightest cloud. Finally, the clear, cloudless air and the scorching rays of the vertical sun on the sandy desert conspire to maintain that dryness of the air which makes rainfall almost an impossibility.

Retreat of the Monsoon. Within three weeks of their appearance off the west coast in early June the rains have spread over the whole of India, but the retreat is a much slower process and occupies about three months. The rains slacken and cease in the northern plains towards the end of September, but they linger in the tip of the peninsula until the middle of December. Slowly the northerly winds extend southwards, driving back the southerly rain-bearing monsoon, and gradually from north to south the Cold Season conditions are re-established. There is none of that suddenness about the change which marked the passage from the Hot to the Wet Season; for the burst of the monsoon was due to the irruption of the rainy winds from the southern hemisphere impatiently overcoming the resistance offered by the feeble high pressure north of the Equator; but the retreat is a long struggle between two well-matched but feeble currents, the one becoming steadily stronger and the other steadily weaker as the sun retreats towards the southern tropic. Thus dry winds succeed to damp winds, anticyclonic conditions succeed to cyclonic and diurnal range of temperature increases. The advance of the monsoon is northwards and its retreat, from the middle of September onwards, is to the south; the length of the rainy season therefore decreases, as in normal tropical climates, away from the Equator.

Rainfall of the Retreat. The same causes produce rain in the same way as during the advance, but the area affected grows less and less from north to south. The monsoon leaves the Punjab about 15th September, the United Provinces about 1st October, and Bengal about 15th October. A premature departure of the monsoon is a catastrophe greater than its delayed appearance; for while the latter only prolongs the discomfort and expectant inactivity of the Hot Season, the former dries up crops which have relied on a continuance of the rains. In 1883, for example, the monsoon left Bengal a month too early; the rice crop was ruined and serious famine resulted.

A change of wind direction from south to north-east accompanies the southward retreat of the low-pressure trough, and it is these north-east winds which bring the October and November rains to the Carnatic. These rain-bearing winds are not to be considered as the N.E. monsoon, but as the retreating edge of the S.W. monsoon; not as winds originating from the new continental high pressure but as the northern periphery of cyclonic circulation associated with the retiring low-pressure trough. These winds, coming off a warm sea, bring Madras 25 inches of rain in October and November, rainfall which is essential for the rice crop and which fills the irrigation tanks against the dry months of January–April.

Hurricanes. The isobaric chart for October shows a strikingly uniform pressure distribution, varying little on either side of 29·9 inches. There is a tendency to a low pressure in the Bay of Bengal (29·85 inches), and tendencies to high pressures in Peshawar, Rajputana and Khandesh (29·9 inches), but small gradients resulting in feeble winds or calms are characteristic. This is the time when violent storms brew in the Bay and travel against the coast, doing great damage. The place of origin of these storms is located farther and farther south as the year advances and as the calms move southwards. They follow the recurved track, characteristic of tropical revolving storms, travelling first westwards and then swinging round to the north and north-east. Storms at this time are of common occurrence, but fortunately severe ones are not frequent. The greatest catastrophes occur when an intense cyclone arrives at the coast while the tide is high. Then the sea water is piled against the coast and floods alone may account for 100,000 lives, while the epidemics which follow are even more destructive.

The Rainfall of India. Reviewing broadly the rainfall of India, it is noticeable that while relief rains give the heaviest falls, it is the cyclonic type which, on account of its more even distribution, is of greater economic value, as, for example, in the Gangetic Plain and the Deccan during the rainy season, and in the Punjab and the United Provinces during the Cold Season. At no time of the year is India entirely without rain. Taking India as a whole January and February are the driest months, but rain is enjoyed in the north-west, and to a less extent in the Ganges valley and Assam, from cyclonic storms, and in the southern peninsula and Ceylon from the equatorial belt. In March, April and May there is a dry triangle with its base from Bombay to Karachi and its apex at Benares, but the peninsula has 10 inches on its western side and Assam gets about 30 per cent of its yearly rain from thunderstorms. After the harvesting of the 'rabi' (spring) crop, the heat and drought make this a period of compulsory fallow in most parts, comparable to 'Charaqui' in Egypt. But Assam and Bengal grow rice and jute, relying on the rains. From June to October there is rain everywhere, though less in the north-west than elsewhere. This is the chief growing season of India, when the land, cleansed by the Hot Season fallow, and watered by the monsoon, yields fruitfully rice, millets, maize or cotton according to rainfall or soil conditions. These are the 'kharif' (autumn) crops, watered by the monsoon rains and ripened by the hot, sunny weather which follows their departure. In November and December there is a little rain all round the shores of the Bay of Bengal, while the Carnatic and eastern Ceylon gets more than 15 inches. This is a highly favoured region, where abundant and well-distributed rain gives great fertility, with resultant prosperity and leisure.

ABYSSINIA, SOMALILAND, YEMEN

In the preceding pages India has been treated as a climatic province virtually self-contained, but the western fringe of the summer monsoon current is recognizable in the climate of East Africa (see p. 111), Abyssinia, Somaliland and Yemen. In examining the earlier course of this current we obtain some useful and instructive insight into certain aspects of the Indian climate which may often be used for long-range forecasting. The July pressure map shows that the Indian low-pressure system extends westward about as far as the Nile, where winds are chiefly northerly or north-easterly, while on the south side of the low-pressure trough there blows the S.W. monsoon, the continuation of the S.E. trades. In Somaliland (see p. 130), mainly owing to the dryness of the air, it brings little or no rain, but the hot winds give rise to violent sandstorms, which harass man and beast and even kill off the mosquitoes. While it blows off-shore it makes the Arabian Sea so stormy that all native shipping is laid up; this is the season of 'Bat

Hiddan' or sea closed, the N.E. monsoon (November to March) being the season of 'Bat Furan' or sea open. In the coffee country of Yemen, the mountains, opposed to the wind, cause the precipitation of some 20 inches of valuable rain, although the rest of Arabia, Persia and Baluchistan are beyond the influence of this current (see p. 150). A further branch of the current supplies Abyssinia with 30 inches or 40 inches on the highlands.

Forecasting the Monsoon. The S.E. trades, the Abyssinian monsoon and the Indian monsoon are clearly all part of the same great air flow and would be expected to wax and wane together. It should be possible, therefore, by tracing this current backwards along its course to obtain valuable foreknowledge of the all-important monsoon rains of India and Abyssinia (the latter so vital to agriculture in Egypt). A weakening of the system, from causes largely unknown, will bring about drought and famine in East Africa and in India, and a low Nile flood, with possibly serious consequences. The following correlation table[1] shows how close is the interdependence:

Year	Nile Flood	Rainfall Variation		
		India (excluding Burma)	Burma	N.W. India
1891	Below Normal	− 3·54	+ 2·48	− 2·32
1892	Above Normal	+ 5·09	− 7·28	+ 6·88
1893	High	+ 9·07	+ 7·04	+ 7·53
1894	High	+ 6·47	+11·47	+ 8·84
1895	Normal	− 2·19	−11·63	− 5·20
1896	Below Normal	− 4·83	+ 3·79	− 2·87
1897	Normal	− 0·15	− 0·13	− 2·03
1898	Normal	+ 0·43	+ 0·40	− 2·88
1899	Far below Normal . . .	−11·14	+ 6·33	−15·56
1900	Above Normal . . .	− 0·57	− 0·91	− 1·08
1901	Above Normal . . .	− 4·13	+ 0·07	− 8·18
1902	Below Normal	− 2·05	− 7·21	− 5·17

It will be observed that Burma tends to vary in the opposite direction from the rest of India, especially from N.W. India, a fact which, though not fully understood, is of some value in forecasting. Observations over many years show also that a high pressure in May over the Indian Ocean, as represented by Mauritius, is followed by deficient monsoon rains in India. This appears, at first sight, unreasonable, as a high pressure here should strengthen the gradient and reinforce the

[1] *Imperial Gazetteer of India*, vol. I, p. 128.

monsoon, but experience has shown that high pressure here coincides with high pressures in India, which therefore offers less attraction for the monsoon.

Again it has been found that heavy rains at Zanzibar, and in British East Africa generally, in April and May are a prelude to a poor monsoon. These rains are the convectional rains of the doldrum belt at the change-over of the monsoons and they bear witness to strong ascensional movement here, accompanied by a low pressure. Now the arrival of the monsoon in India follows the elimination of the low-pressure belt (see p. 145) and will be delayed if this will not give way. In other words, the doldrum belt is waylaying the S.E. trades and preventing their arrival at the Indian focus.

THE INDO-CHINESE PENINSULA AND THE PHILIPPINES

The position of this peninsula is somewhat analogous to that of peninsular India and the climate presents many similarities. The west coast and the interior receive their heaviest rain from the S.W. monsoon, but the air currents have parted with their rain in passing over the Malay Peninsula, and the Gulf of Siam is not wide enough to replenish them; the rainfall is therefore less in amount than in peninsular India. On the other hand, the relief of the coast is much more favourable and the rain is much more evenly distributed. The west coast of Siam from Cape Cambodia to Bangkok has 10 or 12 inches in July, compared with more than 16 inches along the Malabar coast, but 8 inches is recorded far into the interior, whereas in India the 4-inch July isohyet is not far from the divide of the Western Ghats.

Except on the Annam coast the dry season is the season of the N.E. monsoon and the duration of the dry season is greatest in the north, where the rains arrive later and depart earlier. In Cambodia there are only two or three months with less than 1 inch of rain and it is a region of dense forest; the interior of Thailand, where the drought is five or six months long, is a region of tropical grassland.

The Annam coast lies at right angles to the N.E. monsoon and has its chief rains while this blows, but the rainfall régime is not wholly accounted for by the on-shore wind, since the maximum of rain (October and November) does not coincide with the maximum force of the monsoon (January). By comparison with the similarly placed Carnatic coast (p. 153) this rain would appear to be due to cyclones associated with the retreating trough of low pressure, but two other influences are at work: (1) The higher temperature of the sea with consequently greater moisture content of the winds, and (2) the occurrence of typhoons. These typhoons reach great violence in the Philippines which lie across their favourite track; they are most frequent in September, October and November, that is, at the change of the monsoon, and add appreciably to the autumn rainfall here, e.g. Baguio (see p. 16). Chiefly owing to its more mountainous hinterland Hué (102 inches) has more than double the rainfall of Madras (50 inches).

The great bend of the coast in the Gulf of Tongking brings its northern shore at right angles to the summer monsoon, and the more normal régime returns with a rainy season from May to October and a maximum in July. But from this point northwards the climate shows a strong continental influence, especially in winter, and the effects of continental cyclones are felt. The mean temperature range at

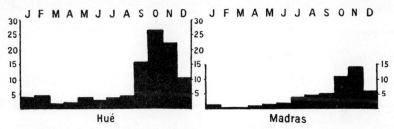

Fig. 57.—Rainfall at Hué and Madras

Hanoi is 22° and extreme temperatures exceed 100° and fall below 40°. These extremes ally the region with sub-tropical China rather than with tropical Annam, and it is more properly considered under that head (see p. 192).

AUSTRALIA NORTH OF THE TROPIC

Matching the Asiatic continent across the Equator is the island continent of Australia, whose monsoon is the complement of Asia's, but the lesser bulk of the southern continent has its sequel in a less clearly defined monsoon. Yet the climate of Australia north of this tropic is essentially monsoonal and bears a marked resemblance to India in the following respects:

1. A seasonal reversal of wind direction.
2. Clearly marked wet and dry seasons, with local exceptions where the direction of the coast ensures rain from both monsoons.
3. Rainfall decreasing steadily towards the focus of the monsoon, which, in each case, is a desert area beneath the tropic, arid in spite of being the region of greatest convection.

The difference arises from the contrasts of size, position and relief, as a result of which:

1. The air flow is weaker and the rainfall therefore less.

TEMPERATURE

Station	Lat.	Long.	Alt. (ft.)	J	F	M	A	M	J	J	A	S	O	N
LAHORE . . .	32°N.	74°E.	702	53	57	69	81	89	93	89	87	85	76	63
JAIPUR . . .	27°N.	76°E.	1,431	61	65	75	85	92	93	86	84	84	80	70
BOMBAY . . .	19°N.	73°E.	37	76	76	80	83	86	84	81	81	81	82	81
DELHI . . .	29°N.	77°E.	718	58	62	74	86	92	92	86	85	84	78	67
BENARES. . .	25°N.	83°E.	267	60	65	77	87	91	89	84	83	83	78	68
CALCUTTA . .	23°N.	88°E.	21	67	71	80	85	86	85	84	83	83	81	73
CHERRAPUNJI .	25°N.	92°E.	4,309	53	55	61	64	66	68	69	69	69	66	61
MADRAS . . .	13°N.	80°E.	22	76	78	81	85	90	90	88	86	85	82	79
AKYAB . . .	20°N.	93°E.	20	70	73	79	83	84	82	81	81	82	82	78
RANGOON . .	17°N.	93°E.	18	77	79	84	87	84	81	80	80	81	82	80
MANDALAY . .	22°N.	96°E.	250	70	75	83	90	89	87	87	86	85	83	76
NHATRANG . .	12°N.	109°E.	11	75	77	79	82	83	84	84	84	82	80	78
MONCAY . .	22°N.	108°E.	29	61	61	66	73	80	82	83	83	82	77	70
HONG KONG .	22°N.	115°E.	108	60	59	63	70	77	81	82	82	81	76	69
MANILA . . .	15°N.	121°E.	47	77	78	80	83	83	82	81	81	80	80	78
KUPANG . . .	10°S.	124°E.	48	79	79	79	79	79	78	77	78	79	80	81
DARWIN . . .	12°S.	131°E.	97	84	83	84	84	82	79	77	79	83	85	86
CLONCURRY .	20°S.	141°E.	696	87	85	83	78	71	64	61	67	72	83	85
WYNDHAM . .	16°S.	128°E.	23	88	88	88	87	82	77	76	79	85	89	90
BROOME . . .	18°S.	122°E.	63	86	85	85	83	76	71	70	73	77	81	85
CAIRNS . . .	17°S.	146°E.	Coast	82	81	80	77	74	71	70	70	73	76	79
SIMLA . .	31°N.	77°E.	7,232	42	42	50	59	64	68	65	64	62	58	51
DARJEELING .	27°N.	88°E.	7,376	40	42	50	56	58	60	62	61	59	55	48
KODIAKANAL .	10°N.	77°E.	7,688	55	56	59	61	62	59	58	58	58	57	55
ADDIS ABABA .	9°N.	39°E.	8,000	60	62	65	64	66	64	62	61	61	62	59

RAINFALL

D	Yr.	Ra.	J	F	M	A	M	J	J	A	S	O	N	D	Total
55	75	40	0·9	1·0	0·8	0·5	0·7	1·4	5·1	4·7	2·3	0·3	0·1	0·4	18·1
63	78	32	0·4	0·3	0·4	0·2	0·6	2·6	8·3	7·3	3·2	0·3	0·1	0·3	24·0
77	81	10	0·1	0·1	—	—	0·7	19·9	24·0	14·5	10·6	1·9	0·4	—	72·4
60	77	34	1·0	0·6	0·5	0·4	0·7	2·9	7·6	7·0	4·7	0·5	0·1	0·4	26·2
60	77	31	0·7	0·6	0·4	0·2	0·6	4·8	12·1	11·6	7·1	2·1	0·2	0·2	40·6
67	79	19	0·4	1·0	1·4	2·2	5·6	11·9	12·7	13·4	10·0	4·9	0·6	0·2	64·3
55	63	16	0·7	2·3	10·6	31·3	50·9	103·6	107·5	81·5	49·4	16·8	2·3	0·3	457·3
77	83	14	1·1	0·3	0·3	0·6	1·8	2·0	3·8	4·5	4·9	11·2	13·6	5·4	49·5
72	79	14	0·1	0·2	0·5	2·0	13·7	49·4	53·7	42·5	24·6	11·6	5·0	0·6	203·8
77	81	10	0·2	0·2	0·3	1·4	12·1	18·4	21·5	19·7	15·4	7·3	2·8	0·3	99·6
71	82	20	0·1	0·1	0·2	1·1	5·8	5·5	3·3	4·6	5·7	4·7	1·6	0·4	35·1
76	80	9	2·4	1·1	0·9	0·9	2·4	2·2	2·0	1·5	6·9	10·6	13·9	9·6	54·4
63	73	22	1·4	2·0	3·1	4·4	10·8	17·9	20·7	23·6	12·0	5·4	3·1	1·4	105·8
63	72	23	1·3	1·8	2·7	5·3	12·0	15·8	14·0	14·6	9·7	5·1	1·7	1·1	85·1
77	80	6	0·8	0·4	0·8	1·3	4·5	9·2	17·3	16·0	14·3	6·7	5·2	3·1	79·6
81	79	4	15·7	14·8	8·7	2·5	1·2	0·4	0·2	0·1	—	0·8	3·4	10·0	57·8
85	83	9	15·9	12·9	10·1	4·1	0·7	0·1	0·1	0·1	0·5	2·2	4·8	10·3	61·8
88	77	27	5·1	4·9	2·7	0·9	0·4	0·3	0·5	0·1	0·5	0·5	1·1	3·0	20·0
90	85	14	9·7	5·9	4·3	1·0	0·4	0·1	—	—	0·1	0·5	2·2	4·2	28·4
86	80	16	6·2	6·1	3·8	1·4	0·6	1·0	0·2	0·2	0·1	—	0·9	3·7	24·2
81	76	12	15·8	16·4	17·7	12·1	4·3	2·8	1·6	1·7	1·7	1·8	4·0	8·6	88·5
46	56	26	3·6	3·7	3·3	2·7	3·9	8·8	21·1	20·7	7·5	1·4	0·5	1·3	79·3
42	53	22	0·6	1·1	1·8	3·8	8·7	24·9	32·3	26·1	18·4	4·5	0·8	0·2	122·7
55	58	7	2·9	1·4	2·0	4·3	6·0	4·1	5·0	7·0	7·3	9·7	8·2	4·4	62·3
59	62	7	0·6	1·9	2·8	3·4	3·0	5·7	11·0	12·1	7·6	0·8	0·5	0·2	49·6

2. The trajectory of the air current is such that for much of its journey it is travelling over an archipelago (contrast the Indian monsoon which comes entirely over a warm ocean).

3. Northern Australia is a table-land sloping gently to the north-west coast. No barrier such as the Western Ghats is offered to the monsoon current, and rainfall, in consequence, is cyclonic rather than orographic, with a more even distribution accordingly.

It will be convenient to approach the study of the Australian monsoon climates in the height of the dry season, and to trace the successive changes of temperature, pressure and winds throughout the year.

The Dry Season. In July, when the Indian monsoon is at its height, there is a continuous pressure gradient from the southern hemisphere high-pressure ridge to the Asiatic low. The axis of highest pressure, the source of the monsoon, crosses the continent of Australia from west to east some distance south of the tropic, the highest average pressure (30·15 inches) being shown in the upper Darling basin. The outflowing air from this continental high pressure is characteristically dry, and no rain falls except on the east coast of Queensland where the S.E. trade is an off-sea wind. Even here the rainfall is small in winter— from Rockhampton to Townsville the coast has less than 2 inches in July, but at Halifax Bay the coast changes direction slightly, the wind meets the coast at a higher angle and the rainfall is heavier; Harvey Creek has over 4 inches. Apart from this coastal strip the rest of tropical Australia is at the height of the dry season with cloudless skies and a steady trade wind blowing, often without interruption for many weeks on end.

Temperature is virtually governed by latitude, the isotherms running east and west, 75° in Cape York and Arnhem Land, 60° along the tropic (cf. India in the cold season, p. 141). The daily range of temperature is considerable on account of the dryness of the air, and night frosts are apt to occur along the southern border of the zone.

The Wet Season. As the sun returns to the southern hemisphere temperatures steadily rise; Cloncurry, for example, records 72° in September, 85° in November, and 88° in December, while at Wyndham both November and December have mean temperatures above 90°. Temperature is now governed by continentality, the isotherms being arranged in loops round the heart of Northern Territory (cf. India in the hot season, Fig. 49).

The continental anticyclone weakens coincidently and by November has been replaced by a well-marked low centred over the northern half of Western Australia where temperatures are highest. Rain-bearing winds blow into this trough and gradually the rains spread southwards, as shown in Fig. 58.

There is nothing here resembling the 'burst' of the monsoon in

India, much more closely does it resemble the steady advance of the equatorial rains into the savanna of Africa or South America.

The monsoon reaches its height in January when the rains extend almost to the tropic (compare the limit of the Indian rain-bearing current at Karachi, p. 150). The length of the wet season as well as the actual amount of rain is thus greatest in the north and decreases steadily towards the margin of the Australian desert. The rain is mainly cyclonic in origin, associated with shallow slow-moving depressions, thus resembling the rainfall of the Gangetic Plain. But along the Queensland coast is a province characterized by rainfall which is mainly orographic, thus resembling the rainfall of the Western Ghats. The trade winds, drawn in by the Pilbarra-Cloncurry low, here become

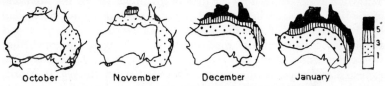

October   November   December   January

Fig. 58.—Advance of the Monsoon Rains into Australia

more easterly, impinging on the coast ranges and bestowing 10 inches of rain in January throughout a considerable length of coast. Harvey Creek, backed by the Atherton Plateau and Mount Bartle Frere, has more than 30 inches in January and well over 100 inches in the four months January to April.

The temperature at the height of the wet season everywhere exceeds 80°, and reaches 90° round Pilbarra, where the air is clear and the sky almost cloudless. But it is in the somewhat cooler but more humid localities farther north that the heat is particularly oppressive; the wet-bulb temperature exceeds 80° in January over a considerable area round Wyndham (see Fig. 3, p. 13).

Tropical Cyclones. The weather of northern Australia during the rainy season is essentially cyclonic, the monsoonal lows moving slowly south-eastward from the low-pressure centres of Pilbarra and Cloncurry in obedience to the drift of the upper air currents. More rarely rain-bearing cyclones travel south-westward along the north-west coast of Western Australia, again following the drift of the upper air. As a rule, though not invariably, the depressions are of feeble intensity, but occasionally there occur tornadoes, similar to those of the United States, highly localized, but very violent. In addition to the regular procession of monsoonal lows there are two provinces subject to invasion by tropical hurricanes; these are:

1. The Queensland coast.
2. The northern coast of Western Australia.

The former group originate in the neighbourhood of the Solomon Islands and travel south-westwards towards the Queensland coast. Usually they swing round to the south and then to the south-east before reaching the coast, though their outer fringes may bring rain to coastal stations, but when they do reach the coast they often do considerable damage and bring torrential rain; 36 inches have fallen in 24 hours at Crohamhurst and 20 inches has frequently been recorded from stations in the coastal strip.

The west-coast hurricanes, known as 'Willy Willies', appear off the north-west coast and follow a track similar to that of the monsoonal lows, recurving round the Pilbarra low and coming inland in the neighbourhood of Onslow and the mouth of the Fortescue River. This track is determined by the drift of the upper air, by the repulsion of the high pressure which lies above the West-Australian current and by the attraction of the Pilbarra low. In Northern Territory and Kimberley they give east winds gradually veering to the north, but where the centre of the storm strikes the coast great damage is to be expected and copious rain results. Once it has left the sea, its source of moisture, the cyclone rapidly loses its violence, but brings valuable and welcome rain along its track to the Great Australian Bight. In both localities the storms are restricted to the hottest months of the year and reach a maximum in late summer. Between 1877 and 1912 they were distributed as follows:

	Nov.	Dec.	Jan.	Feb.	Mar.	April
Queensland	—	—	6	2	8	1
West Coast	1	3	9	6	6	6

SUGGESTIONS FOR FURTHER READING

The Climatological Atlas of India, issued by the Indian Meteorological Department under the direction of Sir John Eliot in 1906, is a mine of information on the climates of India. There are good summaries, too, in vol. 1 of the *Imperial Gazetteer of India*, the *Oxford Survey of the British Empire*, and the *Géographie Universelle* (Tome IX, Asie des Moussons, J. Sion). See also G. C. Simpson, 'The South-west Monsoon', *Q. J. Roy. Met. Soc.*, 1921. A full and readable treatment with many illustrations is H. F. Blandford's *A Practical Guide to the Climates and Weather of India, Ceylon and Burma*, 1889.

For Somaliland and Abyssinia there are good summaries in the *Oxford Survey of the British Empire*, and for Indo-China in the *Géographie Universelle*.

For Australia, in addition to the *Oxford Survey of the British Empire*, see Griffith Taylor's *Australian Meteorology*, 1920, and Hunt, Quayle and Taylor's *Climate and Weather of Australia*, Melbourne, 1913. The rainfall is treated by Wallis in the *Scot. Geog. Mag.*, 1914.

IX

WESTERN MARGIN WARM-TEMPERATE CLIMATES

The Warm-Temperate Climates. Situated in the latitude of the oscillating front of divergence which divides the spheres of influence of the trade winds and the westerlies, these climates are characteristically transitional in nature, enjoying for part of the year a climate which is typically 'tropical' in its constancy and for part of the year weather which in its changeability is more closely allied with the temperate zone. In the simplest form the warm-temperate (or sub-tropical) climates derive their summer influence from the east and their winter influences from the west; summer is therefore continental on western margins and marine on eastern; winter conversely is marine on western margins and continental on eastern. But since the westerly circulation is less constant than the easterly the continentality of the eastern margin winter is less pronounced than that of the western margin summer. Clearly eastern or western marginal situation must be a fundamental criterion of subdivision of these climates.

This ideal simplicity is, however, not always realized, owing to the disturbing influence of the continental masses on the planetary wind circulation. Especially is this the case on the eastern margins of large land masses where the trade wind of summer is distorted into a monsoon and the westerlies of winter are reinforced to become outflowing winds of a markedly continental nature. This interference is least marked in the southern hemisphere owing to the small size of the land masses, and it is here that the simplest form of eastern margin sub-tropical climate is found. Both the great land masses of the northern hemisphere generate monsoons and in the Eurasian block this brings about so profound a modification of these climates as to justify the creation of a monsoonal sub-type.

THE WESTERN MARGIN TYPE (MEDITERRANEAN)

The wide extension of this type of climate round the Mediterranean Sea and the early familiarity with it in this region have led to the general adoption of the name to denote the climatic type. The term is a convenient abbreviation of the somewhat lengthy full title, but the Mediterranean area actually furnishes a number of highly complex varieties of the simple type which appears in the New World in a form much more closely approximating to the ideal. It is the complexity of relief and the confused intermixture of land and sea, peninsula, island and gulf in the Old World which makes for the complexity of climate, while the regularity and simplicity of the coast-line in the New World allows the planetary régime to be established with the minimum of interference. Yet the numerous variants in the Mediterranean basin all agree in certain essential respects, especially:

1. A winter incidence of rainfall and a more or less complete summer drought.

2. Hot summers (warmest month usually above 70°) and mild winters (cold month usually above 43°).

3. A high sunshine amount, especially in summer.

Distribution and Transitional Types. Since it is an essential characteristic of these climates that the winter is maritime, receiving its influence from the oceans to the west, the type has only a limited extension on the western margins of continents, except in Europe where the Mediterranean Sea extends its range for 2,000 miles into the heart of the land. Traced eastward there is a rapid degeneration into steppe and desert by the progressive diminution of the winter rainfall. The spring rainfall survives farthest (steppe type), but eventually both spring and winter rain dies out completely and such scanty rain as does occur is the convectional summer rain of continental interiors. In South Africa and in South Australia and Victoria the land is too narrow to support an interior steppe climate, the western margin warm-temperate passing directly into the eastern margin variety by a gradual increase of summer rain, as the accompanying table and graphs show:

	Percentage of Rain in			
	D. J. F.	M. A. M.	J. J. A.	S. O. N.
Cape Town	8	27	45	20
Knysna	22	21	25	32
Port Elizabeth . . .	17	25	24	34
Durban	34	24	9	33

Equatorwards the winter rains begin later and later and cease earlier and earlier until there can scarcely be said to be a wet season at all; this is the trade-wind desert.

Polewards the rainy season lengthens at each end until the dry season can no longer be said to exist; this is the western margin cool-temperate climate. Tunis has five months with less than one inch of rain, Palermo three, Naples one, and Genoa none.

Temperature. The mean temperature of the coldest month is usually between 43° and 50° and of the hottest between 70° and 80°, so that the mean annual range is about 30°—considerably greater than that of tropical climates but somewhat less than that of most cool-temperate climates. The range, however, increases with distance from the sea: at Mogador it is only 12° but at Morocco it is 34°, at Naples 29° but at Rome 33°, at San Francisco 10° but at Sacramento 27°. The highest temperatures are recorded at the eastern end of the Mediterranean farthest away from the Atlantic influence, the July mean at Athens exceeds 80°, and Beirut, though on the coast, reaches 83°. Stations on ocean margins usually have abnormally low summer temperatures on account of the cold currents which set equatorwards

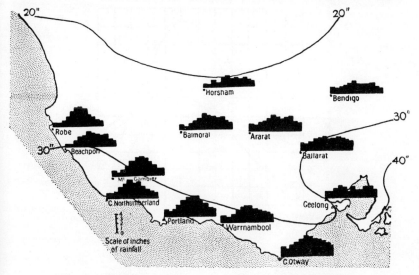

Fig. 59.—Transition of Climates in South Australia and Victoria

along the western coasts in these latitudes; the hottest month at Mogador is only 68·5°, reached in September, and San Francisco only reaches 59·3°, also in September. At Cape Town and in Swanland, situated at the extremities of continents, the cold currents are not so pronounced and summer temperatures rise higher (Cape Leeuwin 69°, Cape Town 70°). Insular stations, since they are unaffected by cold currents, do not show quite such low summer temperatures, but the range is very small, as the winters are remarkably warm; Las Palmas (Grand Canary) has a minimum temperature of 63° and a range of 11°, Funchal (Madeira) of 59° and 13°.

The extreme range, too, is small at marine stations; frost is unknown on the oceanic islands and is rare on the small islands of the Mediterranean, but is fairly frequent, though not often severe on the shores of the Mediterranean; 15° of frost have been recorded at Rome. The rarity of frost encourages the cultivation of delicate fruits, especially citrus fruits, but its occasional occurrence is a constant menace, met locally by extensive and expensive heating devices. At the other extreme the highest maximum temperatures occur in the continental variety and on the margins of deserts which lie equatorwards and to the east. Algiers has recorded 112°, while inland in Tripoli temperatures exceeding 130° occur.

The daily range of temperature is considerable, especially in the dry summer months when the conditions approximate to those of the hot deserts; 10° or 15° in winter and 15° or 20° in summer are usual. In

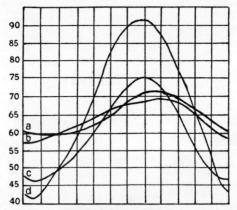

Fig. 60.—Yearly March of Temperature in Mediterranean Climates

a. Funchal (Insular)
b. Mogador (Oceanic)
c. Nice (Marine)
d. Mosul (Continental)

the brilliant sunshine the heat of the summer day is excessive and the midday siesta has become confirmed as a habit. Houses are provided with shutters to keep out the heat and glare, and the design in general aims at coolness in summer rather than warmth in winter. There is, however, often a breeze to temper the heat, especially near the sea or mountains where sea-breezes begin to blow about midday or where valley winds occur. The heat, too, is dry and not so enervating as the moist heat which sometimes afflicts the later months of summer and early autumn. The marked fall in temperature after sundown causes a feeling of chilliness even though the temperatures are not actually low. It also brings about a great increase in the relative humidity, dew is copious and heavy mists occur, which, however, quickly melt away before the rays of the morning sun. Dew and mist are most frequent in late autumn when damp airs are entering from the Atlantic and when temperatures are beginning to fall rapidly. Temperature inversions make the Lombardy plain (which is not, however, properly Mediterranean) a particular victim to dense fogs. Rheumatism here is a common complaint and the death-rate from pneumonia is exceptionally high.

The high humidity of autumn while the air temperature still remains high sometimes makes this season unpleasantly muggy and oppressive; conversely, spring with its delightful freshness and warmth of the air is the most pleasant season; to the charm of the climate at this time is added the beauty of green fields and crops and a wealth of flowers and blossom.

Rainfall. The occurrence of rain in Mediterranean climates is usually associated with the passage of cyclonic storms and its seasonal incidence corresponds to the equatorward migration of the westerly circulation in which these occur, i.e. to the winter months. Where mountains intercept rain-bearing winds orographic rain and even snow may make an important contribution, as on the Sierra Nevada (=snowy mountains) which supply California with valuable perennial water supply for irrigation and power. But the orographic rain, like the cyclonic, has mainly a winter incidence, since west winds (ocean winds) are mainly winter winds and because the land, being cooler, acts more efficiently as a precipitating agent.

The simplest rainfall régime occurs where the planetary circulation is least affected by the configuration of the land, e.g. California as well as other areas with regular coasts, e.g. Chile, South Africa and Australia. Here the rainfall graph is a simple curve with a January maximum and (at San Francisco) a five-month summer drought centring on July. In the Mediterranean basin this simple type is best seen on the southern shores in Morocco and Tripoli. But in the northern Mediterranean, where the peninsulas and seas introduce complications, there is a tendency to a double maximum in spring and autumn. Except in the Meseta, the Lombardy plain and the Balkan peninsula, whose climates are not really Mediterranean but continental, the autumn maximum tends to be the larger, for the seas are still warm and nourish local storms, as well as yielding a good supply of moisture for the air: e.g. Rome, Seville, Athens.

The yearly amount of rainfall is not large, 15 inches to 35 inches, but westward-facing shores, backed by mountains, may receive 50 inches or more (e.g. Ragusa, 59 inches); the Dalmatian highlands, behind Catarro, with over 180 inches, comprise one of the wettest spots in Europe. In general, the rain decreases eastwards, away from the oceans, and equatorwards, away from the cyclones. The drought becomes longer in these directions; Alexandria has only 8 inches, nearly all of which falls in the three months, November to January. A little barley is grown just west of here but is insecure without irrigation; southwards of this is sheer desert.

Although winter is the rainy season it is by no means cloudy or damp. Storms are not so numerous as in western Europe, but such as occur usually bring heavy rain. The Scilly Isles have rather less rain than Rome, but have more than twice as many days with rain; Cape Town has more rain than Kew, but the winter cloud amount is 5·1 compared with 7·4 at Kew and the summer cloud is only 3, compared with 6·7 at Kew. Mediterranean climates are famed for their blue and cloudless skies, they are, in fact, some of the sunniest parts of the world, and are naturally suited to be the health and pleasure resorts of every continent in which they occur. But the concentration of the rains into a comparatively few heavy showers separated by long intervals of fine weather is not an unmixed blessing. The heavy falls cause considerable soil waste on the steep hill-sides with their thin plant cover and this has been aggravated by the serious deforestation of the past; rivers are subject to considerable fluctuations of level, prone to flood, and then to dwindle away. Thanks to the rapidity with which water drains away, the soil is generally dry and healthy, but wherever soil or relief conditions are such that water stands and stagnates, e.g. on the impervious volcanic floor of the Campagna, mosquitoes breed and malaria is rife.

The Mediterranean climates lie near the limit of the cyclonic rains and the precipitation in consequence is somewhat unreliable. The rainfall at Santiago is below normal in seven years out of ten, the deficiency being made good by a few excessively wet years; the range of variability here is from 18 per cent to 226 per cent. In consequence irrigation is necessary everywhere north of 37°S. even with a rainfall of 40 inches. In the Mediterranean basins, where the course of the cyclones is guided by the natural features, the rainfall has a much higher reliability.

Snow, though by no means unknown, is sufficiently rare to call for comment when it falls, though on the mountains, of course, even in North Africa, there is a considerable yearly fall which provides a valuable source of water for irrigation and power.

Vegetation. The dominant characteristic of the Mediterranean climates is the marked rhythmic recurrence of rain and drought, and this naturally finds its echo in a marked rhythm of plant growth. But the contrasts of growth and rest are not nearly so startling as in that other essentially periodic type, the savanna, since in the latter heat and moisture coincide, while in the Mediterranean the rainy season is the coolest. Winter temperatures are scarcely low enough to forbid growth altogether, but they are too low for growth to be very vigorous. Summer drought, while in general limiting activity, is not always complete and is sometimes circumvented by local supplies of ground-water, either natural or artificial. Autumn and spring, with their moderate temperatures and yet adequate rain, are the seasons of greatest vigour. Growth is therefore steady, though sometimes slow, throughout autumn, winter and spring, but except under locally advantageous circumstances is checked in summer by the excessive drought. Even where the rainfall figures show a little summer rain this avails nothing, for in the fierce heat and dry air evaporation is far in excess of any rain which may fall. The summer landscape, dancing in the shimmering heat haze, is baked bare and brown, crudely coloured with the tints of naked soil and rock, with the dazzling white walls of houses and the grey of the olive trees—a scene of universal drought, dust and glare.

The survival of plants is dependent on their ability to withstand the intense desiccation and the whole vegetation is characteristically xerophytic. More hygrophilous plants can grow only where local conditions, generally edaphic but rarely climatic, give a supply of summer moisture; e.g. the riparian groves of oleander, the irrigated orange groves and the Californian redwoods nourished on fog during the summer drought.

Structures to escape the drought are numerous and varied. Almost every device for reducing transpiration is present, thickened cuticle, spiny leaves, wax and hair coverings, etc. Annuals run through their cycle quickly in spring and seeding is over when the drought appears; bulbous and tuberous rooted plants, many of them with beautiful flowers, e.g. tulips, gladioli, lilies, narcissi and irises bloom in the early spring and die down, only the bulb surviving.

Main Types of Vegetation. Where conditions are most favourable, where rainfall is heaviest and there has been least interference, the vegetation consists of evergreen woodland with pine, cedar and evergreen oak; more rarely, where the drought is less severe, deciduous oak may occur. The occurrence of the cork oak in the midst of a wine industry is a fortunate circumstance; while the acorns provide valuable food for pigs. In Western Australia the eucalyptus forests provide the durable jarrah and karri woods. But the distribution of such forest is very limited and has been further seriously reduced by deforestation both by man and goats. Where conditions are less favourable the forest degenerates into a low scrub, the *macquis* or *macchia* of the Mediterranean, the *chaparral* of California, the *Mallee scrub* of Australia, a dense tangle of thicket made up of low-growing evergreens, arbutus, laurel, myrtle, rosemary, etc., with occasional taller trees. The degeneration of forest into scrub is frequently to be attributed to deforestation, especially in the Mediterranean lands, but it is more usually the result of lower rainfall or poor soil. In extreme cases of drought or soil poverty the vegetation consists of scattered low-growing scrubs with bare soil showing between, the *garigue*; it is especially characteristic of limestone soils. The plants which make up the garigue are highly xerophilous with bright but short-lived flowers, e.g. broom and gorse, and many of them are highly aromatic, e.g. lavender, sage and thyme.

Difficult to clear for cultivation, the macquis is of little use to man,

though the Mallee scrub, when cleared, makes good wheat land in
South and Western Australia. The garigue is practically worthless and
serves only to provide sustenance for the ubiquitous and omnivorous
goat.

The increase of rainfall with altitude brings about a mountain
forest zone, but here the trees are broad-leaved and deciduous, beech
and sweet chestnut, for the orographic rains are not so restricted to the
winter months and the winters are colder. Above the forest is a zone
of alpine pasture, available during the summer only, and therefore,
to be useful, requiring to be supplemented by winter pasture on the
plains, a seasonal transhumance occurring across the cultivated slopes
between. But grass is unusual in Mediterranean climates since heat
does not coincide with moisture; conditions, in fact, are generally
more favourable for woody growth. In the Rhône delta, for example
however, the strong winds (mistral) discourage trees and grass is able
to establish itself. In the absence, or rarity, of good pasture, cattle
are rare, their place being generally taken by the less fastidious goat.
Thus there is throughout the Mediterranean lands a natural deficiency
of meat, butter and milk, their place in the dietary being taken by
beans, olive-oil and fruit-juices—not altogether satisfactory substitutes,
especially for children, as witnesses the high infant mortality.

Cultivation. The climatic rhythm does not impose a rhythm on
agriculture, since fruits, either naturally drought-resistant such as the
olive, or irrigated such as the citrus fruits, give employment during the
summer, while cereals and vegetables employ the labourer during
the rainy months. The reliable summer drought offers ideal conditions
for the cereal harvest and is further valuable for the drying of fruits
such as currants, raisins and figs. The long summer is a valuable asset
for fruit growing and especially for the vine; maximum temperatures
at many stations are delayed until August and even in September the
temperature is still high in the sixties.

The native fruits are all well adapted to stand the drought, especially
the olive, the fig and the vine with their long root systems; though the
latter may require careful pruning and wide spacing to economize
moisture. The imported fruits, however, are less suited and generally
require artificial watering, e.g. peaches, oranges, lemons, limes. These
are so thoroughly established in Mediterranean climates all over the
world today that they have come to be considered as characteristic,
yet most of them belong more properly to the eastern margin warm-
temperate or even tropical climates and are accustomed to summer
rain. The long sunshine hours of summer and the high temperatures,
however, ensure the success of the importation, and the general
uniformity of the climatic type all over the world has made their
transference from one Mediterranean region and their establishment
in another a comparatively simple matter.

REGIONAL TYPES: THE MEDITERRANEAN BASIN

The limits of the Mediterranean climates on the north and west are practically coincident with the mountains which enclose the basin, but the plateaux and basins of the Iberian and Balkan peninsulas, together with the Lombardy plain, while retaining certain characteristics of the Mediterranean climates, have adequate summer rain which allies them more closely with Central Europe. Furthermore, they fail to satisfy the requirements of the typical Mediterranean vegetation, of which the olive is the most sensitive test. Southwards and eastwards the boundary is vague, the Mediterranean climates grading in this direction imperceptibly into desert and steppe.

Winter Conditions. During the winter months the warm waters of the Mediterranean intrude a 'lake' of relatively low pressure between the North Atlantic high, which now extends over the Sahara, and the Eurasian winter high which stretches down the central axis of Europe to the Meseta. It becomes a much used track for cyclones of the westerly circulation, but is virtually separated from the more frequented path up the west coast of Europe by the westward extension of the continental high. Winds on the northern shore of the Mediterranean are therefore northerly and in the southern Mediterranean westerly in obedience to this pressure distribution. The actual direction of the wind varies during winter from day to day in accordance with the position of cyclonic centres in their passage, but it is in general from land to sea, that is, from high pressure to low.

There is a tendency for the three peninsulas to intrude tongues of high pressure southwards while the intervening seas tend to generate local lows. As a result of these pressure distributions there is a tendency for warm, wet south-west winds to occur on the west coasts of the peninsulas and cold, dry north-east winds on their eastern shores. Thus Lisbon has 30 inches while Murcia has only 15 inches, Ragusa has 59 inches and a January temperature of 48°, while Athens has only 15 inches and a January temperature of 46°.

Depressions. Although depressions of local origin affect the weather of the Mediterranean they are usually feeble, the dominant type belonging to the westerly circulation and entering the area either by the Straits of Gibraltar or from the Bay of Biscay via the Gate of Carcassonne. Their passage from west to east is connected with many important local phenomena. On the advancing front the winds are southerly, coming from the deserts of North Africa and often excessively hot and dry, sometimes laden with red, penetrating dust. This is the *Sirocco* of Algeria, the *Leveche* of Spain, the *Khamsin* of Egypt, and it receives other names in other parts. Its hot breath dries and cracks the skin, bringing considerable bodily discomfort and sorely trying the nerves and will; it withers the vegetation, often doing permanent damage, especially if it comes when the vines and olives are in blossom. Where it is a descending wind, e.g. on the Algerian coast or on the north coast of Sicily, its heat and aridity are further increased and maximum temperatures of over 110° often occur. In its passage across the warm seas it sucks up water, and is consequently less desiccating on the northern shores, but the high humidity, combined with the heat, make it now enervating and depressing in the extreme.

The rear of the depression is associated with northerly winds from the cold interior of Europe, reinforced to give rise to the *mistral* when a high pressure over the continent combines with a low pressure in the Ligurian Sea to bring a torrent of bitterly cold, dry air from the plateau of the Cevennes and concentrates it in the narrow passage of the Rhône Valley. In spite of the warming by its descent it is a biting wind with a temperature often below freezing. The mean minimum at Marseilles is 22° and the extreme minimum 11°, figures which are not matched elsewhere in Mediterranean climates. Hedges of cypress are planted to protect orchards and gardens from its icy breath and the sites of houses are chosen for their protection from it. The *Bora* (see p. 40) is a similar wind in the Adriatic.

Summer Conditions. In summer the Azores high extends along the now relatively cool waters of the Mediterranean which is thus largely responsible for the maintenance of rainless conditions. Winds are northerly over the whole area and are equivalent to the trade winds, being directed towards the great trough of low pressure which extends westward from India and Arabia. In the eastern Mediterranean where the gradient is steepest these winds reach great force and constancy; these are the *Etesian winds*, well known to the ancient Greeks. From mid-May to mid-October they blow with great regularity with a velocity of 10 to 30 and sometimes 45 miles an hour; they increase in force during the day as the daily convection in the hot lands to the south and east temporarily increase the pressure gradient, but at night they weaken and often die away. On land they bring clouds of dust which make the summer at Athens an unpleasant season, at sea they raise foam-crested waves strangely contrasting in their storminess with the deep blue dome of a cloudless sky. For the Etesian winds, travelling towards warmer lands, are dry (relative humidity 20 to 30 per cent) and in spite of their heat are physically cooling and refreshing. Sailing is dangerous, especially to windward of exposed rocky coasts, but the wind is reliable, the conditions are familiar and accidents are less frequent than during the squally cyclonic storms of winter. The force and dryness of the wind is sufficient to prohibit tree-growth in exposed places, and orchards have to be protected by a row of cypresses as wind-breaks on the north side.

In their passage over the Mediterranean the north winds pick up moisture and although they seldom cause rain they bring mist and fog which may last for days in Algeria and Tunis. They assist, too, in moderating the temperature along the African coast; should the wind blow from the south, the heat increases at once. At Benghazi (Tripoli) September is as hot as July (78°) and October as June (75°) for the summer wind blows from the north into the Arabian low (71 per cent from the north), but in autumn the Arabian low is weakening and the Mediterranean low begins to draw south winds off the still hot Sahara (34 per cent from the south).

SPAIN

The climate of the Iberian Peninsula is a curious variety in which the Mediterranean type is struggling against the effects of continentality. The size of the peninsula is sufficient to produce appreciable monsoons which supersede the planetary winds. In winter the temperatures on the plateau are abnormally low, in places below 40°, and frost is common; the resultant land monsoon partially excludes rain-bearing winds and the winter fall of rain on the plateau is well below normal.

In summer temperatures are abnormally high. Madrid, though more than 2,000 feet above sea-level, has a July mean of over 75°. The days are excessively hot and dusty, the fierce rays of the sun scorching through the thin air. Strong convection is thus set up, and a steady flow of air, strengthened by day, moves in towards the centre of the plateau. Yet this inflow of air brings no rain, July and August being almost completely rainless; for the heat of the plateau is so great that the moisture capacity of the air is still further increased and the rising air is carried away by an upper current before saturation is reached (cf. Sind, p. 152). It is during spring and autumn, when the air currents are humid, and while the land monsoon is not strong enough to exclude them, that most of the rain falls. This is augmented by local thunderstorms, especially in the spring. These features, a large temperature range, strong winds, spring rains and hot summers are steppe characters and much of the vegetation is of the steppe type.

THE NEW WORLD

The uninterrupted line of the west coast of the Americas in these altitudes enormously simplifies the relationship of land and sea influences, the resultant climates being correspondingly simple. The rainfall régime shows a single winter maximum and a single summer minimum, a steady decrease in amount equatorwards and a steady increase in that direction of the duration of the drought. The presence of a coastal chain of mountains causes the rainfall to increase inland up to a point, after which it diminishes.

California, Winter Conditions. The wind system of California is closely bound up with the circulation over the North Pacific, being virtually independent of conditions east of the Sierras, which act as an efficient climatic barrier, especially in winter. In January the chief influences are the Aleutian low, the North Pacific high and the continental high. The westerly winds on the north flank of the Pacific high impinge on the coast in about latitude 40°N. and are turned north and south by the double obstruction of the mountain barrier and the continental high. The northern air stream is drawn past British Columbia into the Aleutian low, the southern turns to the south-east and then to the south and finally to the south-west to become the N.E. trade wind. The prevailing wind of California is thus north-west or north, but the passage of storms causes considerable variations from day to day. The majority of the cyclones pass some way to the north, over Oregon and Washington (see p. 213) and only their southern edge affects California. They are felt less and less farther south and rainfall diminishes in this direction, as the following table shows:

Station	Lat.	Jan. Rain Inches	Winter Rain (Oct.-March) Inches	Dry Months (less than 1 inch)
Eureka . . .	41°N.	8	36	3
San Francisco . .	38°N.	5	19	5
Los Angeles . .	34°N.	3	15	6
San Diego . .	33°N.	2	9	8

The typical temperate depressions, in fact, scarcely reach southern California and most of the storms are thunderstorms (*Sonoras*).

As in the European Mediterranean, the passage of the depressions often draws hot winds off the deserts. The *Santa Annas* of southern California and the *Northers* of the Sacramento Valley are hot, dry winds, charged with dust, and, like the sirocco at its worst, owe their high temperatures largely to adiabatic heating during their descent of the mountain slopes. The effects are the same as those of the sirocco, acute discomfort to man and disastrous drying up of the vegetation. Being due to cyclones they are chiefly winter phenomena, but the worst damage is done if they come in spring when the fruit-trees are in blossom or when the young fruit is formed.

Mean temperatures in winter increase steadily from north to south (San Diego is 5° warmer than San Francisco). More extreme temperatures occur inland, especially in the valley bottoms where inversions occur.

Summer Conditions. Although coastal temperatures are kept low by the cold current, inland temperatures rise rapidly in spring, and by April there is a conspicuous northward bend of the isotherms. In July the isotherms run north and south; continent and ocean are the controls. The temperature of the North Pacific, thanks to the cold current, is only 57° and, the wind being on-shore, these low temperatures are carried on to the land. Temperatures along the coast are below 60° (e.g. Eureka 56°, San Francisco 57·3°), but these phenomenally low temperatures are confined to a narrow coastal strip and rise rapidly inland: Mount Tamalpais, although 2,375 feet above sea-level, is 13° warmer than San Francisco, while in the Great Valley mean temperatures exceed 80°. San Francisco thus lies between a cold sea and a hot land; while the sea breeze blows, the temperature may fall to 50°, when the land breeze blows it may rise above 90°. The very low mean temperature shows that the sea breeze blows nearly all the while, in point of fact 98 per cent of the July winds are westerly (S.W., W., or N.W.). The striking persistence of this wind is due to the suction effect of the heated interior, by which air is drawn into the Great Valley and concentrated on San Francisco by the relief of the Coast Ranges and the funnel of the Golden Gate. The westerly wind is only a shallow current, easterly return winds are frequent on the summit of Mount Tamalpais.

Temperature in the Great Valley itself is governed by proximity to the funnel entrance; thus Sacramento, opposite the gap, is 9° cooler than Red Bluff which actually lies 1° farther north. Not until the heating of the Great Valley diminishes does this ventilation of San Francisco cease; but by September the conditions are weakening, westerly winds now make up only 88 per cent of the total and their velocity has decreased. San Francisco, freed from the cooling draught, begins to warm up and reaches its maximum temperature (59·9°) in September.

The south-west wind at San Francisco has a high humidity (85 per cent), yet during the six summer months rain is very rare, which, at first sight, is surprising; but a combination of conditions prevents precipitation: (1) The sea off which the wind comes is cold; (2) the land on to which it blows is hot; (3) the air is extremely stable since the temperature lapse is inverted. But though no rain falls, fog is most persistent from May to October, rolling in from the sea every afternoon as the strength of the sea breeze increases; Port Reyes has 1,860 hours of fog a year. It affects only a limited area where the air current flows and the fog-free areas are the popular residential districts (Berkeley and Oakland). The fog supplies moisture during the summer months for a hygrophilous vegetation (the Californian redwood and the Californian laurel), contrasting strongly with the bracken and berry shrubs of the hill-sides and inland districts which lie beyond its reach.

CHILE

From Coquimbo (30° S.) to Concepcion (37° S.) the climate of Chile qualifies as Mediterranean, the type thus occurring at rather lower latitudes than in the northern hemisphere. The same controls are at work here as in California, namely:

1. The sub-tropical ridge of high pressure, swinging south in summer and becoming an independent anticyclone over the sea as the continental low develops to the east.

2. A cold current off-shore.

3. A straight coast-line backed by a mountain range. The same simplicity is therefore noticeable especially as regards the rainfall régime, thus:

	Lat.	Yearly Rainfall Inches	No. of Months with less than 1 inch
La Serena . . .	30°S.	4·3	9
Valparaiso . . .	33°S.	20	8
Puerto Carranza . .	35°S.	28	5
Concepcion . . .	37°S.	53	1
Valdivia	40°S.	105	0

But unlike California, where the only gap in the coast range is the Golden Gate, the coast ranges of Chile are low and broken, thus permitting marine influence to penetrate much more freely inland. Thus although on account of the cold current coastal temperatures are low in summer (Valparaiso 69°) and maxima are delayed, there is nothing comparable to the conditions at San Francisco. The inland temperature gradient is also much less steep and the longitudinal valley has nothing approaching the furnace heat of the Sacramento-Joaquin valley. Santiago (1,703 feet) has a January temperature below 70°. The greatest heat, as in California, occurs under foehn conditions; with winds blowing down from the Sierras over 100° has been recorded at Punta Tumbez, near Concepcion. Although in 37°S. latitude the extreme temperatures here are considerably higher than in tropical Chile, where conditions are not so favourable to foehn winds.

Rainfall, low on the coast on account of the cold current (Valparaiso 20 inches), increases up the slopes of the coast ranges (Quilpue 27 inches), decreases again in the rain-shadow of the longitudinal valley (Santiago 14 inches), increases again up the slopes of the Andes (Portillo 60 inches) and finally decreases and practically ceases at high altitudes and across the watershed in Argentina. (All those stations are in the same latitude, namely, 33°S.) The scanty rainfall in the valley makes irrigation necessary for most crops and fruits, but the Andean slopes, with their heavier rainfall and their snow, are a valuable source of water.

With its dry bracing air and with the heat of summer tempered by altitude and the cold current, the climate of this heart-land of Chile is almost ideal for man; and to the advantages of health and comfort it adds considerable agricultural and horticultural productivity. Good crops of wheat, barley, lucerne, etc., are grown and fruit flourishes, those varieties being especially grown which lend themselves to long-distance transport, e.g. vines (as wine) and nuts (especially walnuts).

CAPE TOWN

The area with a Mediterranean climate in South Africa is very small —from the Olifants River in the north to the Breede River in the east —but small as it is its importance is great owing to its suitability for wheat and fruit growing. As in California, Chile, Morocco and Western Australia, the summer temperatures of the coastal zone are kept down by a cold current (the Benguela current), off which blows the prevailing wind. Cape Town has a January mean below 70°, but temperatures are higher inland and a shift of the wind to this quarter means a marked rise in temperature, especially when winds blow off the high plateau. These winds, known as *Berg Winds* (see p. 254), are especially characteristic of the winter months when there is a strong anticyclone on the plateau and low pressure out to sea from a depression passing to the south. As a result of adiabatic heating their temperature may exceed 100°, the winter temperatures thus temporarily exceeding those of midsummer. They are analogous to the Santa Annas of California and are equally damaging. They occur on all sides of the plateau edge, but the season of their occurrence varies from place to place.

Less than 6 inches out of a total of 25 inches falls in the six summer months at Cape Town and four months each have less than 1 inch. The depressions of winter generally pass south of the continent, and it is the south-west wind in rear which brings most of the rain. The rainfall is extraordinarily variable within small distances, different stations within Cape Town itself recording means of 18 inches and 40 inches. There is a marked increase on hill slopes; parts of Table Mountain have 80 inches, while over 200 inches is recorded in small areas near by.

AUSTRALIA

Swanland and the south Australian littoral may be included among the Mediterranean climates, but Adelaide is much less characteristic than Perth. Swanland is subject to controls almost identical with those of Cape Town and the climates and resultant cultivation are very similar in the two cases. The west Australian plateau is, however, much lower and less steep-sided than the South African and there is no 'Berg Wind'. In Australia, too, the climatic type has a much wider extension, the 15-inch isohyet cutting off the corner of the continent from Geraldton in 28°S. on the west coast (cf. 33°S. in South Africa) to Esperance on the Bight. The rainfall, too, is extremely reliable.

TEMPERATURE

Station	Lat.	Long.	Alt. (ft.)	J	F	M	A	M	J	J	A	S	O	N
PUNTA DEL GADA	38°N.	26°W.	73	58	58	57	59	62	66	70	72	70	66	62
FUNCHAL . .	33°N.	17°W.	82	59	59	60	61	64	67	70	72	71	68	64
LA LAGUNA .	28°N.	16°W.	5	54	55	57	58	62	65	69	72	70	66	60
LISBON . . .	39°N.	9°W.	209	51	52	54	58	60	67	70	71	68	62	57
GIBRALTAR . .	36°N.	5°W.	53	55	56	57	61	65	70	73	75	72	67	60
*MOROCCO . .	32°N.	7°W.	1,542	52	55	59	67	69	77	82	85	76	70	62
MOGADOR . .	32°N.	9°W.	33	57	59	60	63	65	68	68	68	69	67	63
ALGIERS . . .	37°N.	3°E.	72	53	55	58	61	66	71	77	78	75	68	62
MARSEILLES . .	43°N.	5°E.	246	44	46	50	55	61	68	72	71	66	59	51
ROME . . .	42°N.	12°E.	207	45	47	51	57	64	71	76	76	70	62	53
PALERMO . .	38°N.	13°E.	230	51	52	55	58	64	71	76	77	73	67	59
ATHENS . . .	38°N.	24°E.	351	48	49	52	59	66	74	80	80	73	66	57
*ALEXANDRIA .	31°N.	30°E.	105	58	60	63	67	72	76	79	81	79	75	68
SMYRNA . . .	38°N.	27°E.	65	46	48	53	59	68	75	80	79	72	66	56
JERUSALEM . .	32°N.	35°E.	2,454	44	48	51	59	66	70	73	73	71	67	56
SAN LUIS OBISPO .	35°N.	121°W.	201	52	53	54	55	57	61	63	64	64	61	56
SAN FRANCISCO .	38°N.	122°W.	155	49	51	53	54	56	57	57	58	60	59	56
SACRAMENTO .	39°N.	121°W.	71	46	50	54	58	63	69	73	72	69	62	53
RED BLUFF . .	40°N.	122°W.	332	45	49	53	59	66	74	81	79	72	63	53
SANTIAGO . .	33°S.	71°W.	1,703	67	66	62	56	51	46	46	48	52	56	61
VALPARAISO .	33°S.	72°W.	135	67	66	65	61	59	56	55	56	58	59	62
CAPE TOWN . .	34°S.	18°E.	40	70	70	68	63	59	56	55	56	58	61	64
GERALDTON .	29°S.	115°E.	13	74	75	73	69	64	60	59	59	61	64	68
PERTH . . .	32°S.	116°E.	25	74	74	71	67	61	57	55	56	58	61	66
EUCLA . . .	32°S.	129°E.	30	71	71	69	66	61	56	54	56	59	63	66
ADELAIDE . .	35°S.	139°E.	140	74	74	70	64	58	54	52	54	57	62	67
ROBE . . .	37°S.	140°E.	Coast	65	65	62	59	56	52	51	52	54	57	60

* Morocco $\left(\frac{T}{R}=7\right)$ and Alexandria $\left(\frac{T}{R}=9\right)$ are really desert climates, but are degener
Mediterranean type.

RAINFALL

D	Yr.	Ra.	J	F	M	A	M	J	J	A	S	O	N	D	Total
60	63	14	2·8	3·1	2·3	2·0	2·2	1·3	0·8	1·5	2·4	3·3	3·6	3·1	28·4
61	65	13	3·4	3·6	3·4	2·0	1·1	0·4	0·1	0·1	1·2	4·0	4·7	3·2	27·2
56	62	18	4·1	3·5	2·9	1·8	0·6	0·2	0·2	0·1	0·5	2·4	3·4	3·8	23·5
52	60	20	3·6	3·5	3·4	2·6	2·0	0·8	0·2	0·2	1·4	3·3	4·3	4·1	29·4
56	64	20	5·1	4·2	4·8	2·7	1·7	0·5	—	0·1	1·4	3·3	6·4	5·5	35·7
54	67	33	1·3	1·2	1·4	1·1	0·7	0·3	0·2	—	0·3	0·5	1·5	0·9	9·4
59	64	12	2·2	1·5	2·2	0·7	0·6	0·1	—	—	0·2	1·3	2·4	2·0	13·2
57	65	25	4·2	3·5	3·5	2·3	1·3	0·6	0·1	0·3	1·1	3·1	4·6	5·4	30·0
46	57	28	1·7	1·4	1·9	2·2	1·7	1·1	0·7	0·8	2·4	3·8	2·8	2·1	22·6
46	60	31	3·2	2·7	2·9	2·6	2·2	1·6	0·7	1·0	2·5	5·0	4·4	3·9	32·7
53	63	26	3·2	2·7	2·8	1·9	1·1	0·7	0·2	0·4	1·8	3·2	3·3	3·6	25·0
52	63	32	2·0	1·7	1·2	0·9	0·8	0·7	0·3	0·5	0·6	1·6	2·6	2·6	15·5
61	70	23	2·2	0·9	0·5	0·2	—	—	—	—	—	0·3	1·4	2·6	8·1
49	63	34	4·3	3·3	3·2	1·7	1·3	0·6	0·1	—	0·7	1·7	3·6	5·2	25·7
46	61	29	6·2	4·6	3·5	1·5	0·3	—	—	—	—	0·4	2·5	5·7	24·7
53	58	12	5·0	3·9	3·5	1·4	0·6	0·1	—	0·3	0·3	0·9	1·6	3·8	21·4
51	55	11	4·8	3·6	3·1	1·0	0·7	0·1	—	—	0·3	1·0	2·4	4·6	22·2
46	57	27	3·8	2·9	3·0	1·6	0·8	0·1	—	—	0·2	0·9	2·1	4·0	19·4
46	62	36	4·8	3·9	3·4	1·6	1·1	0·5	—	—	0·8	1·4	2·8	4·4	24·7
66	56	21	—	0·1	0·2	0·6	2·6	3·2	3·2	2·1	1·2	0·5	0·3	0·2	14·2
64	61	12	—	—	0·6	0·2	3·5	5·8	4·8	3·2	0·8	0·4	0·1	0·3	19·7
68	62	15	0·7	0·6	0·9	1·9	3·8	4·5	3·7	3·4	2·3	1·6	1·1	0·8	25·3
72	67	16	0·2	0·2	0·4	1·1	2·6	4·6	3·6	2·9	1·1	0·7	0·3	0·1	17·8
71	64	19	0·3	0·5	0·7	1·6	4·9	6·9	6·5	5·7	3·3	2·1	0·8	0·6	33·9
69	64	17	0·7	0·5	0·9	1·2	1·2	1·1	0·9	1·0	0·8	0·7	0·7	0·4	10·1
71	63	22	0·7	0·7	1·0	1·8	2·8	3·1	2·7	2·5	2·0	1·7	1·2	1·0	21·2
62	58	14	0·8	0·7	1·2	1·9	3·0	4·0	4·0	3·6	2·2	1·7	1·1	1·0	24·7

The winter temperature of Perth is 55° (equivalent to May in London) and frost is unknown. The heat of summer is tempered during the day by a regular sea-breeze known as 'the Doctor' and cloud amount in summer is less than 3. This is one of the most favoured regions of Australia with a healthy and enjoyable climate admirably suited to dairy farming, wheat cultivation and fruit growing.

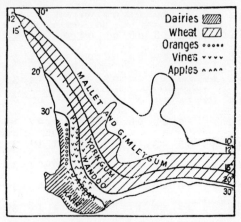

Fig. 61.—The Rainfall Control of Vegetation and Crops in Swanland
(Griffith Taylor)

The shores of the Great Australian Bight, lying parallel to the rain-bearing winds, have only about 10 inches of rain and the desert reaches down to the coast, but the Eyre's Peninsula has 15 inches and the isohyets are carried far to the north by the Flinders Range. Adelaide has 21 inches, of which 70 per cent falls in the six winter months, but there is a small fall in spring and autumn and even a little in summer; December, January and February each have less than 1 inch, but are not quite dry. The winter rain is due to the depressions of the temperate zone, that of the summer half of the year chiefly to tongues of low pressure intruded from the tropics through the sub-tropical high-pressure belt (see p. 186).

Eastwards the summer rains begin to increase in importance; Robe has 25 inches, of which 78 per cent falls in the winter half of the year and may still be referred to the Mediterranean régime, but by Melbourne the incidence of rain is uniformly distributed over the year; the dividing line coincides roughly with the north and south range of the Grampians (see Fig. 59).

SUGGESTIONS FOR FURTHER READING

For the Mediterranean basin see A. Philippson, *Das Mittelmeergebiet*, 1904; *Atlas of Normal Monthly Values of the Meteorological Elements of the Mediterranean Sea and the Adjacent Lands*, Met. Off., 224, London, 1919; H. A. Matthews, 'Mediterranean Climates of Eurasia and the Americas', *Scot. Geog. Mag.*, 1924; W. W. Jervis, 'The Mediterranean Climate and its Variants', *Geog. Teacher*, 1925; *Notes on the Climates of the Eastern Mediterranean and Adjacent Countries*, I.D. 1117, and M.O. 391, *Weather in the Mediterranean*, 1937, both published by H.M.S.O.; E. G. Mariolopoulos, *Etude sur le Climat de la Grèce*, Paris, 1925; R. de C. Ward, 'Climatic Notes on Palestine, Mesopotamia and Sinai', *Nature*, 1918; 'Bewolkung und Sonnenschien des Mittelmeergebietes', Hamburg, *Arch. D. Seewarte*, 35, 1912, No. 2.

For California, see Matthews, *op. sup. cit.*; C. E. P. Brooks, 'Variations of Temperature at San Francisco', *Geog. Teacher*, 1927; R. de. C. Ward, *Climates of the United States*; R. J. Russell, 'Climates of California', *Univ. of Cal. pubs. in Geog.*, vol. 2, No. 4.

For Chile, see Matthews, *op. sup. cit.*; R. C. Mossman, 'Climate of Chile', *J. Scot. Met. Soc.*, 1911; M. Jefferson, 'Rainfall of Chile', *Am. Geog. Soc. Research Ser.*, No. 7.

For South Africa, see *Oxford Survey of the British Empire*; and for Australia, in addition to the above, see G. Taylor, *Australian Meteorology*, and Hunt, Quayle and Taylor, *Climate and Weather of Australia*, Melbourne, 1913.

X

EASTERN MARGIN WARM-TEMPERATE CLIMATES

The transition zone between the trade wind and westerly circulations is represented on the eastern margins of continents by a type of climate which, while generally sharing the mild winters and hot summers of the Mediterranean climates, differs fundamentally in the amount and distribution of rainfall. The trade winds, or winds closely related to them, which bring summer drought to western margins are here rain-bearing, because on-shore the westerlies which bring cyclonic rain to western margins are here continental and depressions are less vigorous though by no means extinct. Winter rain is, for this reason, relatively less than in Mediterranean climates, but is amply compensated by generous summer rains, as the following comparable figures show:

	Lat.	Temperature		Rainfall		
		Jan.	July	6 Summer Months	6 Winter Months	Total
Valparaiso . .	33°S.	69	53	2	18	20
Montevideo . .	35°S.	72	50	18	21	39
Cape Town . .	34°S.	70	55	6	19	25
Port Elizabeth . .	34°S.	69	58	10	13	23
Perth . . .	32°S.	74	55	4	29	33
Sydney . . .	34°S.	72	52	22	26	48

Weather Influences. As might be expected in a region of transition, wind direction is somewhat variable, especially during winter when passing depressions cause frequent and rapid veering and backing, accompanied by changeable weather. Both temperate and tropical storms are elements of some importance in humid sub-tropical climates which lie within a zone exposed to both types. Visits by the latter are comparatively rare in the southern hemisphere, but the West Indian hurricane and the Chinese typhoon are fairly frequent visitors to the extra-tropical coasts of the United States and China and Japan. The depressions of the temperate zone bring some remarkable weather types, with temperature changes of remarkable amplitude and suddenness and with correspondingly important influences on life conditions. Sudden irruptions of polar air may lower the temperature by 30° or 40°

in 24 hours, damaging fruit and crops and making fires and warm clothing a real necessity for the shivering inhabitants, the more so as they are, as a general rule, unprepared for a phenomenon so at variance with the average climatic conditions and therefore inadequately provided against it. Such are the *Southerly Burster* of New South Wales, the *Pampero* of the Argentine and the *Norther* of the Gulf-Atlantic States. Their arrival is usually sudden and squally and is often accompanied by violent hailstorms and thunder, destructive on land and a menace to shipping at sea. At the other extreme invasions of equatorial air bring spells of unpleasantly hot weather, especially during the summer months. Dust-laden winds with daily temperatures above 100°, often prolonged for days at a time, are an unpleasant feature of the climate of Melbourne; the early settlers called them *Brickfielders*. The *Zonda* of the Argentine is a hot wind which brings a feeling of complete prostration, leaving the body an easy victim of disease and the mind in a state of dejection and depression. In South Africa and South China their temperature is further raised by a foehn effect (see pp. 188 and 197).

Temperature. Winters are mild, the mean temperatures being about 50°, but there are considerable departures from the mean figures, as described above, and frosts, though rare, occasionally occur, especially inland. The temperature rises steadily as the sun rises higher, the spring temperatures being about those of a London summer. The maximum, usually between 70° and 75° in the southern hemisphere but from 75° to 85° in the more continental northern hemisphere, is reached rather late, especially near the coast, owing to the strong marine influence in summer; autumn, for the same reason, is always considerably warmer than spring. The summer heat is oppressive since the humidity is high and there is little wind to bring relief. The thermometer rises above 90° almost daily at midsummer and sometimes exceeds 100°. This is the most unhealthy season of the year, the mortality from dysentery and malaria rises rapidly as heat and humidity increase and it continues to rise into late summer and autumn, for the humidity remains high after the highest temperatures are past. Even when health does not actually suffer there is a loss of energy which makes such climates unsuited to manual labour by white men, and it has been found more satisfactory, if not absolutely necessary, to employ coloured labour in, for example, the cotton-fields of the Gulf-Atlantic States and in the tea plantations of Natal.

Rainfall. The rainfall is characteristically adequate, but not excessive and its incidence is well distributed over the year. The ideally uniform rainfall is realized, for example, at Dubbo in New South Wales, where a difference of only half an inch separates the wettest from the driest month. But though it is characteristic that the monthly totals are more or less equal, the nature of the rain and its efficacy vary considerably between the seasons. The winter rain is chiefly of the cyclonic type, occurring as light showers or prolonged drizzles; the summer rain, on the other hand, is either orographic or instability rain, occurring as heavy downpours, an hour of which may produce as much in the gauge as a week of the winter type. Much of the summer rain is therefore incapable of utilization, being lost by run-off and evaporation, while the winter rain possesses a high degree of efficiency. The rainfall of the summer half of the year usually slightly exceeds that of the winter half, but the maximum occurs at different times in different places.

Away from the marine influence which prevails along the eastern coasts there is a tendency for the maximum to occur in spring (see Fig. 62, *a* and *b*), thereby showing a transition into the steppe type of

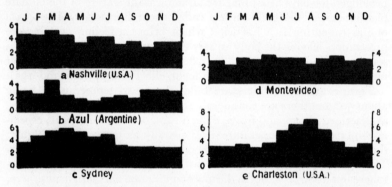

Fig. 62.—Rainfall Régimes in Eastern Margin Warm-Temperate Climates
(For explanation see text)

the interiors of continents in these latitudes. Often a distinct minimum occurs in autumn (Fig. 62, *a*), an arrangement very favourable for the harvest. Along the eastern littoral, however, the heaviest fall tends to occur in autumn, during which season the off-sea winds are most humid (see Fig. 62, *c* and *d*). Yet a third rainfall régime is found at such stations as New Orleans and Charleston (see Fig. 62, *e*) where there is a pronounced summer maximum, the result of a summer monsoon. This type will be met with in a more pronounced development in the monsoonal variety of South China, to be discussed presently.

Vegetation and Cultivation. As in the Mediterranean climates, the temperature, even of the coldest month, is not so low as to check entirely the growth of the plant, and perennial activity is possible. Owing to the similarity of temperature conditions, many species of conifers (e.g. cypress), shrubs (e.g. laurel) and other evergreens (e.g. evergreen oak) are common to both climates, but the absence of a dry season makes the conditions favourable for many handsome and important species, such as the tree ferns, bamboos, lianas, tulip-tree, etc., which are excluded from western margins by the summer drought.

The regular rainfall supports a forest vegetation which is usually broad-leaved evergreen, but sometimes deciduous, while coniferous forests also occur. Many of the trees are of considerable economic value as timbers and especially cabinet woods, e.g. oak, maple, walnut, hickory, tulip-wood, etc., while the mulberry and tea grow abundant crops of valuable leaves during the long moist summer.

There are, however, considerable areas, e.g. the Argentine Pampas, which, though well enough watered for forest growth, have a grassland vegetation. The reason for this is a matter for speculation; it may be due to deforestation, or perhaps to strong winds or perhaps to the geological history—an earlier episode of greater aridity and a lack of later colonization by forest trees from areas outside. If given protection in early growth trees can be readily established here and are frequently planted round estancias, acting as wind-breaks.

But typically these climates are agricultural and horticultural in their economy, growing valuable crops of tobacco, cotton, maize, rice, tea, sugar-cane, oranges, etc. Their great virtue in this connection lies in the length of the summer, combined with its constant humidity; a growing season, free from frost, of 200 days or more is ensured, while the promise of a cool season to come induces heavier fruiting in the autumn than in the perennially hot climates of tropical latitudes. The typical cereal is maize which finds here the summer rain it requires, and, in the wetter parts, rice. Other cereals do not as a rule thrive, because of the absence of a hot dry season for ripening and harvest, though there are many local exceptions to this general statement. The warm winters make autumn sowing possible and two or more crops may be raised should the density of the population demand it; but the more usual practice is to grow a single crop which may safely have a long growing season. Numerous varieties of fruit can be raised, irrigation being generally unnecessary; but there is a certain risk in damage to orchards from the cold snaps which are to be expected in the winter; this applies especially to the delicate citrus fruits.

Their high productive capacity makes these climates capable of supporting a population of considerable density, but with the exception of the monsoonal variety they are as yet not fully developed. Where the stress of a dense population demands that two or more crops shall be taken off the land the routine of the agriculturist is arduous in the extreme; but where one crop only is aimed at the yearly round allows more leisure, and anxiety is reduced to a minimum by the reliability of the climate.

REGIONAL TYPES:
NEW SOUTH WALES AND VICTORIA

Rainfall. The south-eastern angle of Australia from Port Macquarie to Cape Otway enjoys uniformly distributed rains, transitional between the tropical summer rains of Queensland and the cyclonic winter rains of Tasmania. Bass Strait and the seas to the south of the continent offer the easiest route for the cyclonic storms of the westerlies and the storm centres seldom cross the land, the rain being borne chiefly on the southerly and westerly winds in rear of the centres. Such rain does not often penetrate far inland unless the cyclonic centre is connected by a trough through the high-pressure ridge with the tropical low pressures lying to the north. The heavy autumn rains of Sydney (31 per cent in the three autumn months) come chiefly on south-easterly winds connected with anticyclones over Bass Strait or the Tasman sea. Similar

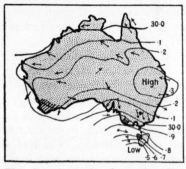 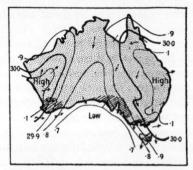

Fig. 63.—Antarctic Depression Unfavourable to Inland Rains. No Tropical Connection

Fig. 64.—Antarctic Depression Favourable to Good Inland Rains. Trough Connection with Tropical Low

anticyclones follow similar tracks in spring but bring less rain, the decisive factor being the temperature difference between sea and land. The high humidities which often obtain in autumn make that season unpleasantly muggy at Sydney.

The rainfall decreases rapidly inland from 40 inches along the coastal strip to 20 inches beyond the crest of the Australian Alps, but there are numerous examples of local increase or decrease due to relief. The Gourock Range has 30 inches, which decreases to less than 20 inches in the trough of the upper Murrumbidgee, but the Australian Alps beyond have more than 50 inches.

Storms. It is along the coast of New South Wales from Cape Howe to Port Macquarie that the *Southerly Burster* is felt at its worst. The weather chart of one of these storms shows a V-shaped depression extending northwards from a cyclone over the seas to the south of Australia; winds in front are northerly and warm, then, as the trough approaches, from the south or south-west there appears a typical long dark roll of heavy cloud, often with a front 30 miles long. The wind drops to an ominous calm and then, without warning, returns as a violent cold blast from the south, whirling before it a cloud of blinding dust, and accompanied by thunder and lightning and, more rarely, rain or hail. The drop in temperature is sudden and considerable, a fall of 15° in five minutes is not unknown and the total fall often exceeds 30°, though 20° is a more usual figure. Two factors encourage the best development of *Southerly Bursters* along the New South Wales coast: (1) The depressions acquire renewed vigour on reaching the warm eastern seas; and (2) the meridional trend of the mountains exercises a selective influence over wind direction favouring particularly south or north winds. The storms are most numerous in spring and summer, November and December being the favourite months.

ARGENTINE, URUGUAY AND SOUTH BRAZIL

During the winter months the high-pressure ridge extends almost continuously across South America just south of the tropic; south of this the weather is affected by the procession of lows which enter from the Pacific, cross the pampean plains and pass out into the South Atlantic in the region of the Plate estuary, accompanied on their south-western margin by south-easterly rain-bearing winds. South-easterly winds also occur on the advancing edge of moving anticyclones and these also are rain-bearing (cf. New South Wales, p. 186). The passage of storms during the summer months and especially from October to January, is often characterized by the sudden arrival of a cold, dry, stormy wind of gale force (the *Pampero*), exactly comparable to the *Southerly Burster*, though rather less violent on account of the less favourable relief.

During the summer the continental low splits the high-pressure belt into separate Atlantic and Pacific anticyclones, and winds, though by no means regular, are prevailingly easterly, winds from the N.E., E. and S.E. constituting about 60 per cent of the total from December to February. These are the equivalent of the trade winds and are rain-bearing; thus there is well-distributed rain at all seasons.

Much of this region is excellent wheat and maize country, the wheat being autumn sown and utilizing the winter and spring rain (12 inches at Santa Fé), the maize being spring or early summer sown and utilizing the summer and autumn rainfall (20 inches at Santa Fé). The percentage of the rain falling during the summer months decreases from north to south, and this, taken in conjunction with the progressive southward decrease of temperature, gradually sways the balance of conditions from maize to wheat. Rosario, with 24 inches in the six summer months and a January temperature of 76°, has a maize climate; Bahia Blanca, with only 12 inches in the six summer months and a January temperature of 71°, is well suited to wheat. There is, moreover, an increasing tendency towards the south for a second minimum of rainfall to occur at midsummer, thus providing a dry season for harvest.

At Bahia Blanca, November and March are the wettest months, while December and January each have less than 2 inches.

SOUTH AFRICA

The type has a very small distribution in South Africa since the continent does not extend south of 35°, but the coastal zone from Cape Agulhas to Port St. John's may be placed there. The driest month at Knysna has nearly 2 inches and the wettest less than 3·5 inches. Westward along the south coast there is a gradual passage to the Mediterranean type with summer drought (see p. 164), northwards along the east coast to the tropical marine type with a distinct summer maximum (Durban) and inland to a steppe type on the plateau (Aliwal).

Owing partly to the Agulhas current winter temperatures are rather high; Port Elizabeth is 4° warmer in July than Cape Town in the same latitude, while Berg winds, similar to those of the Cape Peninsula, bring unpleasantly high temperatures at times.

THE GULF-ATLANTIC STATES OF U.S.A.

North America introduces a slight monsoonal complication into the normal Eastern Margin Warm Temperate type, but not sufficient to justify its inclusion in the monsoonal sub-type. Charts of mean wind direction do actually show a complete seasonal wind reversal, but this is rather misleading as the direction is by no means constant, but manifests from day to day a high degree of variability and the weather types vary accordingly. The pressure gradient is never comparable with that of Asia (o·2 inch between Nebraska and Cuba compared with o·5 inch between Ordos and Luzon), so that the influence of, for example, passing cyclonic storms is sufficient to bring about a complete obliteration of the 'prevailing' wind. The prevailing wind direction brings to bear, however, a strong influence which makes itself felt especially in the strong summer maximum of rain.

Winter Conditions. Winter winds are northerly along the Gulf Coast and north-westerly along the Atlantic littoral, but it is not until December that these winds are firmly established. Mean temperatures are rendered somewhat low by this prevailing continental influence, as the following pairs of stations show. In each case the latitude, relation to surrounding relief and proximity to sea make the stations comparable:

Coldest month	Charleston:	49·3°.	Pt. Macquarie	54·5°.
„ „	Vicksburg:	47°.	Paraná	54°.
„ „	New Orleans	54°.	Durban	64·6°.

These mean temperatures are the resultants of considerable variations of daily temperature, usually connected with passing cyclonic storms which frequently travel along the 'southern circuit' (see p. 213) during the winter months. The southerly winds of the advancing edge bring spells of warm muggy weather, followed by cold dry spells as the wind swings to the north in rear. The northerly winds are often strong since they reinforce the prevailing wind and sometimes, under special conditions, they are gales of bitterly cold air, the *Northers* of Texas in particular and of the Gulf States in general. The special conditions consist of a steep gradient between a deep cyclone and a following anticyclone advancing across the plains; it is a 'line-squall' phenomenon exactly comparable to the *Pampero* and the *Southerly Burster* and is accompanied by the same roll of heavy cloud, the same suddenness of onslaught and the same rapid drop of temperature, often 50° or 60° in a few hours. The sufferings of man, beast and plant are all the more acute on account of the muggy warmth of the weather immediately preceding.

The rainfall of the winter season is essentially cyclonic and it is the southerly wind off the Gulf (the exact opposite of the prevailing wind) which brings most of the rain. Some stations show a clearly defined tendency to a secondary maximum in late winter when cyclonic activity is greatest.

As winter gives way to spring and summer the cyclonic control weakens, and though depressions passing along more northerly routes may still draw occasional wet southerly winds off the Gulf, the prevailing wind, still northerly, is less frequently interrupted; hence the rainfall generally decreases until by April and May a minimum has been reached.

Summer Conditions. The steady rise of temperature in the continental interior weakens the high-pressure system and ultimately substitutes a low pressure into which a southerly monsoonal wind blows off the Gulf and up the Mississippi Valley, gradually increasing in strength and reliability. It is a current of high temperature and humidity, bringing abundant rain to the cotton-fields of the southern states and to the maize belt farther north. Temperature mounts steadily to July maxima of 80° which extend as far north as the Ohio confluence, and this, combined with a relative humidity of 80 per cent or more, while encouraging vigorous plant growth, makes the summer weather most unpleasantly enervating.

The prevalence of a sea wind, combined with thunderstorms, is the main cause of heavy rainfall throughout the summer months with a maximum coinciding with the greatest strength and regularity of the monsoonal wind. Along the Atlantic coast and especially in Florida, the maximum is delayed into September by the occurrence of violent hurricanes with torrential rain (see Fig. 65, Miami). They originate in the west Atlantic between 10° and 20°N. and follow the characteristically curved path passing up the east coast of Florida, Georgia and

Carolina, but seldom penetrating far inland. In the Miami hurricane of September, 1926, the barometer fell to 27·6 inches, a record for U.S.A., wind velocities of over 100 m.p.h. were registered, causing great havoc among fruit groves, crops and buildings, and 15 inches of rain fell, followed, naturally, by serious floods.

Autumn conditions are broadly similar to those of spring and, like spring, the late autumn is generally rather dry; in fact, irrigation is sometimes necessary at this season.

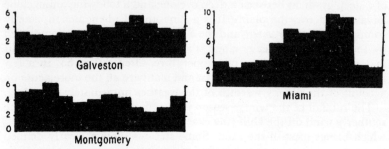

Fig. 65.—Rainfall Régimes in the Gulf-Atlantic States

Rainfall Régimes. Reviewing broadly the rainfall of this southeast quadrant of the United States two main types may be recognized:

1. Cyclonic rain with a winter maximum.
2. Monsoonal rain with a summer maximum.

To these may be added two subsidiary types:

3. Thunderstorm rain with a summer maximum.
4. Hurricane rain with a late summer maximum.

Types 2, 3 and 4 combine to give a predominant summer maximum at nearly all stations, while Type 1 is generally recognizable as a secondary winter maximum. Spring and autumn are usually rather dry, occupying positions intermediate between the cyclonic rains of winter and the monsoon rains of summer. This is the normal Gulf Coast type (see Fig. 65, Galveston).

Stations round the southern end of the Appalachians show a maximum in late winter or early spring which corresponds to a maximum of cyclonic activity (see Fig. 65, Montgomery). This last shows a close relationship to the rainfall régime over the greater part of the continental interior (Greely's Missouri Type).[1]

Crops. If the January isotherm of 43° is taken as the northern limit, the boundary of the climatic province practically coincides with that of

[1] A. W. Greely, 'Rainfall Types of the United States', *Nat. Geog. Mag.*, 1893.

the Cotton Belt as defined by O. E. Baker,[1] including the Carolinas, Georgia, Florida, Alabama, Mississippi, Louisiana, Arkansas and much of Oklahoma and Texas. This is, however, mainly coincidence, since it is summer temperature (the July isotherm of 77°) which delimits the Cotton Belt. Cotton of the Sea Island type along the coast and of the American Upland type inland, is by far the most important crop, the long, moist, hot summer giving suitable growing conditions, while the somewhat dry and cool autumn ensures good quality and reliable conditions for picking. Farther west, in Texas, the rainfall decreases and Egyptian type is grown, often with irrigation. Sugar, rice and other semi-tropical crops are cultivated, while the especially mild area round the Gulf is noted for fruit and early vegetable farming on a huge scale. Florida, projecting 5° farther south and with winter temperatures 10° higher than the Gulf coast, is the chief centre of the citrus fruit industry.

THE MONSOONAL SUB-TYPE

It has been laid down as one of the characteristics of the normal type that the wind is variable in force and direction owing to the essentially transitional latitude, and it is in this fundamental respect that the monsoon type varies from the normal; for its winds, under the strong continental influence, exhibit a markedly greater degree of reliability and regularity. In consequence of this the seasons are much more clearly differentiated, the summer being typically marine and the winter steadily continental. This finds expression particularly in the following ways:

1. A considerably greater range of temperature, due especially to the lower winter temperatures. Shanghai is 17° colder in winter and 7° warmer in summer than Port Macquarie.

2. The rainfall is much more seasonal in nature: at Amoy 74 per cent of the rain falls in the six summer months, at Port Macquarie only 54 per cent.

3. The arrival and departure of summer are much more sudden and in general the passage from one season to another is much more clearly defined.

Limits. The element which perhaps most clearly dissociates the monsoonal from the normal climates is the winter temperature; for this falls below freezing everywhere north of Shantung; Peking, in almost the same latitude as Melbourne, has four months with mean temperatures below freezing. Such a climate is clearly not 'Warm-Temperate' even though its high summer temperature allows the cultivation of rice to the north of 35°. But rice is a monsoonal, not a warm-temperate cereal, its growing season being comparatively short and its prime

[1] U.S. Dept. of Agriculture, Year-book for 1921.

requirements not winter mildness (the criterion of warm-temperate)
but adequate moisture and heat during the short growing season.

The Northern Limit of the Type.[1] The January isotherm of 43°
has been taken as the poleward limit of the warm-temperate climates
(see p. 89) and this runs just south of the Yangtse; China north of this
line would belong, by virtue of its cold winters, to the cool-temperate
climates. But in practice this line is not very satisfactory since it bisects
the natural unit of the Yangtse valley which is uniform in climate,
vegetation and agriculture. The winters here, though in the northern
part of the basin they are as cold as those of England, are much shorter;
too short, in fact, to impose an effective check on the growth of trees, the
majority of which are broad-leaved evergreens, though mixed with
deciduous species belonging to the more northerly climatic province
(chestnut and maple). It is the mountain rim enclosing the Yangtse
Valley on the north, the Tapa-shan and the Hawaiyang-shan, which
function as the real climatic and vegetational divide. The water-
parting, in fact, being more palpable and better defined, forms a more
significant limit than the isothermal line and may, with advantage, be
adopted as the boundary between the warm- and cool-temperate
climates. The climate usually called 'Chinese' is thus considered here
under two heads:

1. A warm-temperate type with a long growing season in which two
or three crops a year may be obtained; mulberry, tea, sugar and other
sub-tropical crops are grown.

2. A cool-temperate type with cold winters and a relatively short
growing season, only one or two crops a year being possible as a rule,
and these are cool-temperate crops, e.g. wheat, barley, maize, beans, etc.

The warm-temperate type, as delimited above, has a comparatively
small extension in southern China only and does not extend north of
about 32°. Such a latitude elsewhere corresponds more closely with the
equatorward margin of warm-temperate climates since it roughly coincides with the lower margin of the transition zone between the trades and
the westerlies. Here, however, these elements of the planetary circulation have no significance, being entirely obliterated by the alternating
monsoons. The pronounced continentality of the winter, in fact, forces
the winter isotherms farther south here than anywhere else in the
world and repels all the climatic zones towards lower latitudes (see the
northern limit of the tropical monsoon, p. 157).

Winter Conditions. According to the latitude, South China should
lie in the high-pressure belt and its winds should be directed equatorwards; this tendency gives an added impulse to the winter monsoon

---

[1] For a discussion on the climatic provinces of China, see Coching Chu, *The Climatic
Provinces of China*, *Mem. Inst. Met.*, *No. 1*, National Research Institute, Nanking,
April 1929.

which consequently exceeds, both in strength and duration, the summer monsoon which is established in the face of the planetary circulation. Winds over southern China are strong (10–15 m.p.h.) and northerly in direction, swinging to north-easterly in the extreme south. The steepest pressure gradient occurs between Japan and the Philippines (0·5 inch in the 1,500 miles between Shanghai and Mindanao) and here are found the strongest winds. Their arrival is sudden and violent; rough seas confine native shipping to the harbours, and even the fastest mail steamers may take five days to cover the 850 miles from Hong Kong to Shanghai in the teeth of the gales concentrated in the Formosa Strait. To the northern shores of the mountainous island of Formosa they bring 8 inches of relief rain in January, but elsewhere the January rainfall is less than 3 inches and mainly cyclonic in origin.

The Continental Cyclones introduce a second main factor into the winter climate, temporarily interrupting the dominance of the monsoon and upsetting the ideal regularity of the winds. On the relative vigour of these on the one hand and of the Siberian anticyclone on the other depends the weather experienced. With an excessive development of the latter to the exclusion of the former are associated seasons of excessive cold and drought, while an excessive development of the cyclones brings a winter of rain and mist with mild periods interspersed with cold waves.

These cold-weather storms may be compared with those of Northern India (p. 140) and, like these, their place of origin is uncertain. It is inconceivable that they should travel across the whole width of Europe and Asia from the Atlantic; it is, in fact, a matter of observation that the Atlantic storms very rarely reach as far as the Urals, while the rare survivors pass eastward by the valleys of the Ob and Yenisei, and die out near Lake Baikal. Very exceptionally, storms may survive into Mongolia, but these are much too far north to affect southern China. Most would appear to originate within the continent and to follow fairly well-defined tracks (see Fig. 70, p. 220). The storms are frequent throughout the cold season, but reach a maximum in spring, when the influence of the anticyclone is waning, January to April being the favourite months. Along the most southerly track, following the Si-kiang Valley, the frequency is as follows: October 2, November 2, December 7, January 10, February 13, March 18.

The temperatures in front and in rear of these storms are, as usual, strongly contrasted; warmth comparable to that of the summer months and often exceeding 80° is often brought by the southerly winds on the advancing edge and may endure for many days since the storms usually travel but slowly. The northerly winds in rear are reinforced by the monsoonal pressure gradient and blow with added force, bringing temperatures below freezing almost as far south as Hong Kong.

TEMPERATURE

Station	Lat.	Long.	Alt. (ft.)	J	F	M	A	M	J	J	A	S	O	N
DUBBO . . .	32°S.	149°E.	870	79	77	71	64	55	49	47	51	56	63	71
WILCANNIA . .	31°S.	143°E.	267	81	80	74	65	58	52	50	54	60	68	75
PORT MACQUARIE	31°S.	153°E.	49	73	73	71	66	61	56	55	57	60	64	68
SYDNEY . . .	34°S.	151°E.	138	72	71	69	65	59	55	53	55	59	64	67
MELBOURNE .	38°S.	145°E.	115	68	68	65	60	54	51	49	51	54	58	61
AUCKLAND . .	37°S.	175°E.	260	67	67	66	61	57	54	52	52	55	57	60
WELLINGTON .	41°S.	175°E.	10	63	63	61	57	53	49	48	49	52	54	57
PORT ELIZABETH	34°S.	26°E.	181	69	70	68	65	62	59	58	58	60	62	65
MONTEVIDEO .	35°S.	56°W.	96	72	72	69	63	57	51	51	51	55	58	65
PARANÁ . . .	32°S.	61°W.	66	77	77	73	66	59	54	54	57	61	66	72
ROSARIO . .	33°S.	61°W.	95	77	76	70	62	56	49	51	52	57	62	69
BUENOS AIRES .	35°S.	58°W.	82	74	73	69	61	55	50	49	51	55	60	66
BAHIA BLANCA .	39°S.	62°W.	82	74	72	67	60	53	47	47	49	54	59	66
NEW ORLEANS .	30°N.	90°W.	53	54	57	63	69	75	80	82	81	78	69	6
GALVESTON .	29°N.	95°W.	69	54	56	63	70	76	82	84	83	80	73	6
MOBILE . . .	31°N.	88°W.	57	51	54	60	66	73	79	80	80	77	68	5
PINE BLUFF .	34°N.	92°W.	215	43	45	55	64	72	79	83	81	76	63	5
VICKSBURG . .	32°N.	91°W.	247	47	51	58	65	73	79	80	80	75	65	5
HATTERAS . .	35°N.	76°W.	11	46	46	51	58	67	74	78	77	74	65	5
ABILENE . .	32°N.	100°W.	1,738	45	47	56	64	72	79	82	82	75	65	5
FUCHAU . . .	26°N.	119°E.	66	53	52	56	64	76	80	84	84	80	73	6
CHUNGKING . .	30°N.	107°E.	750	48	50	58	68	74	80	83	86	77	68	5
KAGOSHIMA . .	35°N.	131°E.	394	45	45	51	60	65	71	78	80	75	66	5

RAINFALL

D	Yr.	Ra.	J	F	M	A	M	J	J	A	S	O	N	D	Total
76	63	32	2·1	1·9	1·8	1·9	1·9	1·9	2·0	1·8	1·9	1·6	1·8	2·0	22·6
79	66	31	1·0	0·8	1·1	0·7	1·0	1·1	0·6	0·8	0·7	0·9	0·7	0·8	10·2
71	64	18	5·9	7·5	6·5	5·9	5·6	4·6	4·5	3·8	3·9	3·2	4·1	5·9	61·4
70	63	19	3·7	4·3	4·8	5·6	5·1	4·8	4·8	3·0	2·9	3·2	2·8	2·9	47·9
65	59	19	1·9	1·7	2·2	2·3	2·2	2·1	1·8	1·8	2·4	2·6	2·2	2·3	25·5
64	59	15	2·6	3·0	3·1	3·3	4·4	4·8	5·0	4·2	3·6	3·6	3·3	2·9	43·8
60	55	15	3·3	3·1	3·3	3·9	4·7	4·8	5·6	4·5	4·0	4·1	3·5	3·2	48·0
68	64	11	1·2	1·3	1·8	2·0	2·4	1·7	1·9	2·1	2·2	2·1	2·1	1·7	22·5
69	61	21	2·7	2·8	3·2	4·5	3·5	3·2	2·5	3·6	3·4	2·6	3·2	3·6	38·8
76	66	23	4·4	3·8	3·7	3·8	1·8	0·9	1·0	1·3	1·9	4·4	3·6	4·8	35·4
75	63	28	3·7	3·2	5·3	3·1	1·8	1·5	1·0	1·5	1·6	3·5	3·4	5·3	34·9
71	61	25	3·1	2·7	4·4	3·5	2·9	2·5	2·2	2·5	3·0	3·5	3·1	3·9	37·3
71	60	27	2·0	2·2	2·6	2·2	1·1	0·9	1·0	1·0	1·6	2·3	2·0	2·1	21·0
55	68	28	4·5	4·3	4·6	4·5	4·1	5·4	6·5	5·7	4·5	3·2	3·8	4·5	55·6
57	70	30	3·4	3·0	2·9	3·1	3·4	4·2	4·0	4·7	5·7	4·3	3·9	3·7	46·3
52	67	29	4·7	5·2	6·4	4·9	4·4	5·4	7·0	7·1	5·3	3·5	3·7	4·9	62·5
45	63	40	5·9	3·9	5·6	4·6	5·2	4·0	3·9	2·7	3·7	2·0	4·8	4·9	51·2
49	65	33	5·2	4·8	5·5	5·0	4·3	4·0	4·6	3·4	3·3	2·6	4·3	5·0	52·0
48	62	32	4·3	4·1	4·5	3·9	3·8	4·5	5·5	5·8	5·2	5·2	3·6	4·6	55·0
46	64	37	0·9	0·9	1·2	2·6	3·9	2·7	2·0	2·4	2·6	2·5	1·4	1·2	24·3
57	68	32	1·9	3·8	4·5	4·8	5·9	8·2	6·3	4·8	8·4	2·0	1·6	1·9	54·1
50	67	38	0·7	0·9	1·3	4·0	5·3	6·7	5·3	4·4	5·8	4·6	2·0	0·9	41·9
48	62	35	3·5	3·3	6·1	9·1	9·6	13·9	11·2	7·4	8·7	5·1	3·7	3·5	85·1

Temperature. Exposure to the north-west monsoon makes China abnormally cold for its latitude in winter; Hong Kong is nearly 7° colder in January than Calcutta in the same latitude, the thermometer here has actually fallen to freezing-point and the mean minimum is 43°. February on the coast is usually cooler than January, after which temperatures rise steadily as the year advances. There is nothing in South China comparable to the 'Hot Season' of India in the same latitudes, since winds from the heart of Asia continue to cool the country right down to the coast; Hong Kong is 17° cooler than Calcutta in March and 15° cooler in April. Temperature is largely dependent on exposure to or protection from the cold monsoon winds; thus Szechwan, screened by the high mountains on its northern rim, is, in spite of its altitude, much warmer than the Yangtse delta. The January temperature at Chungking is 11° higher than that of Shanghai. This favoured basin can grow many sub-tropical plants (e.g. sugar and tobacco) which the rigour of winter forbids elsewhere. In some cases, especially on southern slopes which fortunately predominate in the basin, three crops may be obtained in the same year, and this exceptional fecundity permits the basin to support a population of about 50 millions at a density of some 500 per square mile, with agriculture as the chief means of subsistence.

Rainfall. Winter rainfall in South China is almost entirely cyclonic and its distribution is closely related to the most frequented cyclone tracks. Central China has more than South China since the Yangtse valley is a much used route; the coast has more than the interior since the sea is the chief source of moisture. Depressions become more frequent and the rainfall increases as the season advances and as the anticyclone weakens (see data for Hong Kong and Fuchau).

Summer Conditions. Winds and Rainfall.[1] The continental anticyclone weakens in spring and by April has ceased to be an important control; winds along the coast are now easterly or south-easterly but cannot yet be called the S.E. monsoon since they are extremely light and variable and bring but little rain. By May the Asiatic low is developing strongly, the south-east winds are gaining in strength and regularity and the heavy rains have set in at Hong Kong (April, 5 inches; May, 12 inches). The effects are felt later and later northwards up the coast; at Shanghai May (3·6 inches) is no wetter than April (3·7 inches), but the rainfall of June (7·4 inches) is double that of May. Sudden though the increase of rainfall is, it is no way comparable with the 'Burst' of the monsoon in India, so well displayed along the Western Ghats, as a comparison of the rainfall of Bombay and Hong Kong clearly shows. This is explained by a variety of causes:

---

[1] The rainfall régime of the Yangtse valley is described in connection with North China on pp. 219 to 222.

1. Owing to the continental depressions there is no real dry season in South China.

2. The monsoon current is of less vigour and lower humidity; the pressure gradient in India is much steeper and the volume of air arriving is proportionately greater there.

3. The Indian monsoon has travelled over 4,000 miles of continuous hot sea, while the Chinese monsoon has parted with some of its moisture to the East Indies and other islands which lie in its path.

The prevailing south-east wind raises the level of the sea along the coast by an amount which varies from 18 inches to 5 feet. It imports high humidities and much cloud from the Pacific Ocean, but rain does not generally fall until a cyclonic disturbance, generally shallow but covering a wide area, upsets the equilibrium of the air. Thus the S.E. monsoon is not a season of persistent rain, but a time of unsettled weather in which heavy showers are interspersed with spells of fine weather. Rain falls, as a rule, at least once a week, and in some places on one day in three during the wettest months, but occasionally the fine spells are unduly prolonged and crops suffer; for though there may be no actual drought the rice crop is particularly sensitive at this stage and ten days or a fortnight without rain at the time of planting or transplanting will have a disastrous sequel at the harvest.

It should be noticed that the Gulf of Tongking renders to China a service similar to that which the Gulf of Mexico renders to North America, supplying moisture to the on-shore winds and increasing the rainfall; but the Gulf of Tongking is not so warm for its latitude and its effect is not so great, as the restricted area with 40 inches shows.

Temperature. The summer temperatures are monotonously high (Hong Kong has four months above 80°), and the humid heat is disagreeably enervating, but since the wind is off the sea unduly high figures are seldom reached along the coast; the mean maximum at Hong Kong is only 4° higher than the mean temperature of the hottest month. Inland it is hotter and the 85° isotherm makes a closed loop in the heart of central and southern China. Consequently the highest temperatures along the coast are recorded when the wind is off the land. In late autumn, for example, depressions sometimes cause north-west winds to descend the slopes from Yunnan into the delta of the Red River as a foehn wind with temperatures of 105°.

Typhoons. A characteristic element of the climate is the tropical cyclone, here known as a typhoon, of which eight or nine may be expected in a year. They are most frequent in late summer (85 per cent occurring in July, August and September) and they affect chiefly Southern China. The course followed is closely related to the pressure distribution of the season and especially to the position of the high-pressure areas. In winter they are repelled by the continental high and do not penetrate inland, but in summer they frequently invade South China and Tongking, gradually weakening as they pass inland but bringing considerable rain to central and northern China. Usually, however, they follow the typical parabolic course, skirting the Pacific anticyclone on its western margin and thus passing close to the Chinese coast. During the disastrous Swatow typhoon of 1922 the barometer fell to 27·5 inches and the wind blew at 100 m.p.h. for two hours. The shift of the wind to the south as the centre passed brought a 'tidal' wave which inundated the town and caused an estimated loss of 50,000 lives. Torrential rain accompanies these storms and materially affects the average figures for coastal stations in this zone. Hong Kong has recorded more than 27 inches in 24 hours, and nearly 20 inches in 8 hours. The importance of this rain is, however, often over-stressed and it has been shown that the maxima of rain do not coincide with the maxima of typhoons; it is very highly localized and it concerns only coastal stations; probably it does not add much to the mean monthly figure at any given place inland.

SUGGESTIONS FOR FURTHER READING

For Australia, see works quoted in earlier chapters.

For South America, in addition to the general works already quoted, there is *The Climate of the Argentine Republic*, by W. G. Davis, Buenos Aires, 2nd ed., 1910. An article by N. A. Hessling on 'Relation between Weather and the Yield of Wheat in the Argentine', *Monthly Weather Review*, 1922, may be consulted.

For the United States, read Ward's *Climate of the United States*, and frequent articles in the *Monthly Weather Review*.

For China, there is an excellent summary by Kendrew in Dudley Buxton's *China*, 1928. See also 'Climate of China', C. E. Koeppe and N. H. Bangs in the *Monthly Weather Review*, 1928; *Etude sur la pluie en Chine*, E. Gherzi, Shanghai, 1928 (reviewed and summarized in the *Monthly Weather Review*, 1929); 'Weather Types in East China', Coching Chu, *Geography*, 1928.

For Japan, see 'The Climate of Japan and Formosa', E. M. Sanders, *Monthly Weather Review*, 1920.

The climates of the sea and the coasts are described in M.O. 4042, *Weather in the China Seas*, H.M.S.O., 1938. Much good meteorological work has been done since the war in China and Japan, but is not readily available in English.

XI

COOL-TEMPERATE CLIMATES

The quality which divides the cool-temperate from the warm-temperate or sub-tropical climates is the possession of a real cold season which retards or inhibits active plant growth and places a check on agricultural activity. The severity of the cold season increases from west to east, away from the marine influence, and in this direction the seasonal control of occupation becomes correspondingly more pronounced; the autumn-sown wheat of western margins is replaced by the spring-sown wheat of continental interiors and there is, for example, a considerable exodus from the prairie provinces when the dead season of winter sets in. Winter warmth and rain come from the oceans to the west and it is the relative position of land and sea which, more than latitude, determines the seasonal distribution of temperature and pressure and the position of forest, grassland and desert.

The Cyclonic Unit which exercised such a strong influence on the winter conditions of the warm-temperate climates now plays an important role throughout the year, though it still assumes greater importance in winter than in summer. The regular periodicity of prevailing winds, pressure, temperature and rainfall is masked and obscured by the non-periodic cyclonic unit; the significance of climate decreases, and weather assumes an importance such as it enjoys in no other climatic type.

While they exercise a considerable degree of individuality in the matter, these depressions tend to frequent certain fairly clearly defined paths, especially following seas, gulfs, straits, lakes, river valleys, plains and humid areas of low relief generally. They are repelled by the high pressures which develop over continents in winter and tend, therefore, to take a course along continental margins at this season. The cyclonic control is consequently more marked in the marine than in the continental variety, whose winter conditions, under the influence of polar continental air, tend to be more reliable, more regularly periodic and more constantly faithful to the climatic type.

Continental and Marine Types. The distance to which marine influence penetrates depends on the configuration of the western margin, as a comparison of the new and old worlds well shows. It makes itself felt in the following ways:

1. A small annual range of temperature.
2. A higher humidity and rainfall.
3. Rainfall evenly distributed throughout the year, but with a

tendency to a winter maximum (cyclonic) as compared with a summer maximum (convectional) of the continental type.

4. An imperceptible gradation of the seasons with frequent relapses from, for example, spring into winter or autumn into summer.

Pressure and Winds. The prevailing westerly wind, though fairly constant in the upper air, is disturbed near the ground by the frequent passage of cyclones and anticyclones and by the yearly changes of the continental pressure systems to such an extent as to be scarcely recognizable except by records kept over a period of time. The seasonal pressure changes over the two great land masses of the northern hemisphere attempt to establish systems of outflowing winds in winter and inflowing winds in summer. In Eastern Asia this tendency is realized to a high degree, the wind direction is almost completely reversed and a palpably monsoonal régime is established. Elsewhere their success is less complete and they are able only to modify the planetary régime. During the winter of the northern hemisphere these continental highs and the oceanic lows govern a system of winds which tend to be south-west along the western margins of continents and north-west along their eastern margins. The continental interiors are under the dominance of the anticyclones, while south of the axis of highest pressure the winds are northerly or north-easterly, bitterly cold and excessively dry.

In summer the continental lows become the principal foci and the oceanic lows virtually disappear; winds are, therefore, mainly westerly on western margins and southerly, or even south-easterly on eastern margins.

In the southern hemisphere the disturbance of the planetary winds is much less; 'Roaring Forties' and the 'Brave West Winds' blow all the year round with considerable force and with a high degree of dependability. In spite of the frequent passage of depressions, westerly winds (S.W., W. and N.W.) make up 80 per cent of the total winds at Punta Arenas.

Temperature. The designation 'temperate' for these zones is a decided misnomer, for within the limits of the cool-temperate are found some of the most extreme temperatures in the world, exceeded only in the cold climates farther to the north. Mean annual ranges exceeding 80° occur in Manchuria and Mongolia, mean monthly temperatures 15° below zero at the one extreme and above 80° at the other are included in the zone, while extreme temperatures of 105° and below −20° find places in the climatic type. However, these enormous ranges are limited to extreme continental climates, the marine type well deserving the description 'temperate'. The annual range in the Scilly Isles, for example, is only 15°, frosts are rare and temperatures above 75° are exceptional.

The striking contrasts between oceanic and continental types are a feature of the zone in the northern hemisphere, for here there occur two of the largest land masses, separated by two oceans which, for their latitude, are the warmest in the world. The North Atlantic shows a positive anomaly in January of over 40° and the North Pacific of over 20°, while both Siberia and North America have negative anomalies exceeding 30° (see Fig. 1, p. 12). In the extremely oceanic southern hemisphere, by contrast, the anomalies are less than 10° and the mean annual range only very locally exceeds 20°.

Temperatures below freezing are to be expected everywhere away from the marine influence and river navigation is regularly blocked by ice. The canals in Holland are frequently ice-bound and the flowing water of the Rhine is frozen on an average for three weeks in the year at Cologne. Farther east the closed season is longer, the lower Danube is ice-bound for five or six weeks and the upper Sungari for five months.

The prolonged cold of the continental winter makes some efficient form of artificial heating necessary, and most houses, especially in towns, are centrally heated; in the less extreme winter of the marine climates there is not the same need of constant heating and fires are generally lit every day, not in stoves but in open grates whose bright glow helps to cheer the darkness of cloudy winter days.

Maxima and minima of temperature are reached in January and July, i.e. soon after the solstice, in the continental type, but there is frequently a retardation on western margins which may amount to two months (e.g. the Scilly Isles where February is the coldest month and August the hottest). Autumn is decidedly warmer than spring in the marine type, but the difference becomes less marked inland and finally almost disappears.

Mean figures, however, convey a very inadequate idea of the temperature conditions, since owing to the irregularity of wind direction resulting from cyclonic control rapid fluctuations of considerable amplitude occur on either side of the mean. This is a feature not found in hot climates since winds in low latitudes bring warm air from whatever direction they blow; but in high latitudes a change of wind direction imports air now from the warm lands to the south (in the northern hemisphere), now from the frozen north; in extreme cases a drop of 50° or 60° in 24 hours may be recorded and of 30° within an hour. This demands great adaptability on the part of plants, animals and man, and survival depends on the degree of response to the demand. Up to a point the body is able to meet these extremes and adapt itself thereto, in fact they are stimulating to both body and mind, but powers of resistance are weakened by the artificial conditions of civilized life, heavy clothing, fires and central heating. Many of the diseases characteristic of the zone, e.g. colds, influenza, pneumonia, are more or less directly traceable to these sudden temperature changes, for the lowering of vitality which results from the strain on the system makes the individual more susceptible to germ-borne diseases. Winter and early spring are generally the season of highest mortality since weather changes are most frequent and severe at this time.

Rainfall. Since winds and rain come mainly from the west it is natural that rain should be heaviest on western margins and decrease eastwards. Where high relief presents a barrier to the westerlies 80 inches or 100 inches may result and even 200 inches locally. Hokitika (New Zealand), exposed to the full force of the winds and backed by the New Zealand Alps, has 120 inches, Bahia Felix, similarly exposed in South Chile, has over 200 inches, while the Welsh mountains round Snowdon have a similar amount under similar conditions. Such phenomenal figures are mainly made up of orographic rain and are paid for by a rapid eastward decrease and often a serious deficiency in lee of the mountain barrier. But whether by coincidence or from more significant geological causes, the west coasts in these latitudes are almost everywhere highland coasts of the fjord type, of which British Columbia, Norway, Western Scotland, South Chile and New Zealand are typical. The west coast of Europe is the most fortunate in this respect, for there are numerous breaks in the fringing 'Caledonian' Mountains which, moreover, reach no great height south of the Skagerrak; thus it is only in Scandinavia that there is any approximation to the suddenness of transition from wet to dry, from marine to continental that occurs elsewhere. In the Americas the 20-inch isohyet is within 200 miles of the west coast, while in Europe 20 inches penetrate to Moscow, Kiev and Bucharest—more than 1,000 miles from the ocean to the west. But so great is the width of the Eurasian continent that rainfall eventually becomes too small to support even good grass, and poor steppe and desert eventually supervene.

The rainfall thus decreases, suddenly or gradually according to relief, to a minimum in the interior of continents. On approaching the east coasts rainfall again increases, this rain being due to the importation of moisture by easterly winds, generally in the front of cyclonic centres. Thirty or 40 inches may occur here, but again orographic influences may augment this considerably (e.g. the Appalachians). In the southern hemisphere where the westerlies are stronger, more constant and less disturbed, relief rain plays a more important part than cyclonic, and eastern margins (e.g. Patagonia) are arid up to the coast.

A further source of rain on eastern margins is the monsoon wherever this is well developed; since this is, by nature, summer rain only, the seasonal régime is totally different and many important consequences follow. This monsoonal sub-type, well developed only in North China and Japan, deserves special consideration and will be dealt with later.

Transitional Rainfall Types. In a meridional direction the rainfall decreases polewards, in which direction the air is colder and therefore carries less moisture, but equatorwards the changes are less simple.
On western margins there is a decrease by loss of summer rain as these
areas fall more and more under the influence of the high pressures of
mid-latitudes in summer, thus providing a transition to the Mediterranean type. In continental interiors there is a similar decrease, which,
though small in actual amount, is rendered more serious by the already
deficient precipitation and by the higher temperatures which diminish
the efficacy of such rain as does fall. On eastern margins, on the
contrary, there is an increase equatorwards since the frequency of
rain-bearing easterly winds increases in this direction, especially during
the summer months. In this direction there is a gradual passage to
the humid warm-temperate type. Where there is a monsoon the nature
and distribution of the rain does not change, but there is an increase in
amount towards lower latitudes where winds are warmer and carry
more moisture.

Seasonal and Diurnal Distribution of Rain. It has been pointed
out on p. 17 that each type of rain tends to have its own seasonal
distribution; now all three types are represented in the cool-temperate
climates but with varying degrees of importance in different places.

1. Relief rain, which tends to have an autumn maximum, is of
greatest importance on western margins; it is much reduced in significance in continental interiors and attains renewed, though secondary,
significance on eastern margins. Where there is a monsoon the maximum
of relief rain coincides with the greatest strength of the monsoon (i.e.
midsummer). It tends to be heaviest at night or in the morning when
the difference of temperature between land and sea is greatest.

2. Cyclonic rain attains its greatest importance in winter, at which
season the storms are most numerous and most vigorous; nearly 70
per cent of the storms over the North Atlantic between 45° and 60°N.
occur in the four months of November–February. The storms decline
in force and precipitation value as they pass inland and their chief
influence is therefore felt on the west. The almost featureless European
plain illustrates this diminution, for here there is virtually no complication by relief rain; convectional rain is, however, increasing in this
direction, and the net decrease represents the decrease of cyclonic
minus the increase of convectional. Furthermore, the cyclones, repelled
by the continental highs, follow a marginal course in winter, while in
spring and summer they penetrate farther into the heart of the
continent.

3. Convectional and thunderstorm rain, with its typical summer
maximum, is characteristically the rain of the interiors of continents.
The fall chiefly occurs in the afternoon: over 70 per cent of the rain at

Berlin occurs between the hours of noon and 8 p.m. The rain, while it falls, is heavy, but does not last so long as the persistent cyclonic and orographic rain of the west. Bucharest and London have about the same amount of rain, but rain falls on 167 days at Kew and on only 106 at Bucharest.

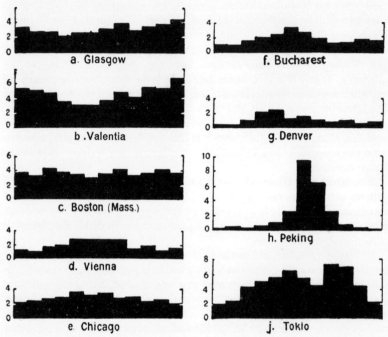

Fig. 66.—Rainfall Régimes in Cool-Temperate Climates
a, b, and c: Marine type; d and e: Continental type; f and g: Steppe type; h and j: Monsoon type

Rainfall Régimes. As a result of the different distribution in time and space of these three fundamental types we may recognize four rainfall régimes as follows:

1. A marine variety in which cyclonic and relief rain predominate. The rainfall is fairly evenly distributed over the year but there is a tendency for a winter maximum (e.g. Glasgow). At west coast stations the predominance of relief rain is testified by an autumn maximum (e.g. Valentia). This marine type, while characteristic of western margins, reappears in a modified form on eastern margins with the return of the relief and cyclonic controls (e.g. New England), but is not found in Eastern Asia on account of the monsoon.

Passing inland from west coasts convectional rain soon becomes important, and by Paris has brought a summer maximum. This becomes more and more pronounced, affording a gradual passage into the second type.

2. A continental variety in which convectional rain predominates. Winter rainfall is small since pressure is high, the air is cold and humidity is low. There is a similar régime in North America (the Missouri province of Greely).[1] Agriculturally this and the following are ideal wheat climates since the rain comes at the critical growing season and a small fall is therefore adequate for cultivation. On its equatorward side this type grades into desert by a transitional type which is:

3. A steppe type in which the slight rainfall is largely convectional but in which there is a spring or early summer maximum to which depressions contribute. These storms penetrate from the Mediterranean region as soon as the weakening of the continental high permits their ingress. The equivalent in North America is Henry's[2] 'Rocky Mountain foothill type' (see p. 215).

4. The monsoon type in which monsoon rains (relief and cyclonic) predominate, giving a summer maximum, generally somewhat suddenly reached, and a strongly marked seasonal rhythm.

Snow. Even in the extreme south and west snow falls almost every year, though the snow cover is only continuous and prolonged in the continental variety. In the marine type it does not lie many days and in cities is reduced to slush almost immediately. The number of days a year with snow decreases southwards and westwards: Warsaw has 47, Berlin 34, Paris 14, Scilly 3.

Sunshine and Cloud. These are cloudy latitudes on account of the frequency of depressions, the seasonal variation of whose paths is accompanied by a seasonal variation of cloudiness. Cloudiness increases towards higher latitudes, from five-tenths in Southern France to seven-tenths in Western Ireland and North-west Scotland; Gascony has 2,000 hours of sunshine a year, parts of Scotland have less than half this. There is also an increase of sunshine towards the east corresponding to the greater dryness of the air, but a decrease again as the east coast is approached (cf. the increase of rainfall).

Fog is frequent especially near the coast, for the air is always moist and a fall in temperature usually results in condensation. Inland, under anticyclonic conditions, radiation fogs are characteristic of early winter while the air is still moist.

Storms. The pressure gradient in the normal temperate depression is not, as a rule, sufficient to give rise to really dangerous storms, but wind velocities of over 60 m.p.h. are not uncommon in winter and do much damage to property. Some of the most serious storms are the result of V-shaped depressions or line-squalls similar to the Southerly

[1] See reference on p. 190.
[2] A. J. Henry, *Rainfall of the United States*, U.S. Weather Bureau Bull. D., 1897.

Bursters of Australia, already described. The 'wind-shift line' in the trough of these depressions is a region of great disturbance with strong squally winds and often thunder and lightning. This is, in fact, the general form of winter thunderstorm especially along western margins, but in continental interiors they are generally heat thunderstorms due to convectional overturning and are much more frequent during the summer months and especially during the afternoon. The number of days with thunder increases southward and inland: Lerwick has only 1 day a year with thunder, Dublin 8, Kew 14, Berlin 15, Vienna 18.

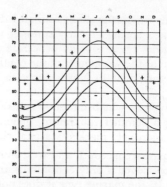

Fig. 67.—Mean Temperatures at Kew

Downpours of rain and hail accompany the storms; the hail may do considerable damage to standing crops, vineyards being especially vulnerable.

The violent tornadoes of the middle Mississippi Valley generally occur along the trough of a V-shaped depression, but have some connection with convection, as is shown by the fact that, though they may occur in any month and at any time of day or night, they are most numerous in the hottest months and just following the hottest part of the day. Though affecting a very narrow area only, they are among the most violent and destructive of meteorological phenomena.

The Seasons. Temperature is obviously the determinant of season, rainfall being of very subsidiary importance in this respect. In the extreme continental type where the annual range of temperature is huge there are virtually only two seasons, summer and winter, the change from one to the other being clear-cut and well-defined. At Warsaw, for example, April is 11° warmer than May, and November is 10° colder than October. But in the marine type, when the annual range is small, there is a gradual passage extending over two or three months, the transition seasons being spring and autumn. Owing to the considerable day-to-day variation and the small annual range, temperatures at any time during these seasons may revert to those of the season before, or anticipate those of the season to follow. In Fig. 67 the curves represent:

(a) **Mean** monthly temperature.
(b) Mean monthly maximum.
(c) Mean monthly minimum.
+ Highest daily mean on record.
− Lowest daily mean on record.

It will be seen that the daily range is greater in the relatively dry months of spring and summer than in the wetter months of autumn and winter; also it will be noticed that the mildest January days may be warmer than the chilliest July nights. This unreliability of the seasons and especially of the seasons of early growth and harvest is a serious handicap to cultivation, for killing frosts may return when blossom and shoot have been encouraged by weeks of summer-like weather, or autumn fruit may be prematurely nipped by an unseasonably early visitation of winter conditions.

VEGETATION AND CULTIVATION

In the peculiarly favoured climates of Western Europe, with their exceptionally mild winters, evergreen trees of a sub-tropical aspect flourish; arbutus thrives in south-western Ireland, myrtle, laurel, etc., in Cornwall, but over the rest of the zone the winter season, lasting from one to six months, during which temperatures are below the minimum for active growth (43°), is generally evaded by the trees by the shedding of their leaves. The deciduous forest is characteristic of the western margins and extends for a considerable distance inland, but the diminution of rain in this direction places a limit at about 50°E. in Eurasia (see p. 206). In North America the equivalent forest, here largely coniferous (the Pacific coniferous forest), does not extend beyond the Cordillera. In an easterly direction this temperate forest, usually deciduous but occasionally coniferous, gradually gives way to grassland and ultimately to semi-desert. On approaching eastern margins the forest reappears as the rainfall increases. Polewards of these vegetation types lies the northern coniferous forest (Taiga), stretching as a broad continuous band across both Eurasia and North America, belonging characteristically to the cold climates with their long winters.

The Broad-leaved Deciduous Forest. The extreme western margins do not produce the best deciduous forest, for the seasonal rhythm is scarcely well enough pronounced, and moreover salt-laden winds are inimical to tree growth; it is in the semi-oceanic climates of England, France and Germany that the forest reaches its finest development. Oak, beech, ash and maple are the chief forest trees,

but chestnut, elm, sycamore and lime are common. It is a feature of
these forests, as of the coniferous forests, that they often consist of
almost pure stands of a single species, a fact which adds considerably
to their economic value, for many of the trees make excellent timber.

Deforestation has caused the disappearance of this forest over huge
areas and its replacement by grassland and meadow, often cleared for
cultivation. Thanks to the heavy rainfall, the grass which is thus
established is extremely rich and nourishing; and thanks to the mild
winters of western margins it continues to grow practically throughout
the year, thus providing excellent fodder for stock, especially dairy
cattle. Unfortunately those same factors of high rainfall and humidity,
together with the cool summers in the extreme oceanic type, are
unfavourable to agriculture and especially to cereal cultivation. This
climatic type has a small extension in space and there is a scarcity of
economic plants adapted to life in it. Potatoes are practically the
only crop which really thrives, while oats is the most successful cereal.

Coniferous Forest. Under less favourable conditions, and especially
on sandy soils, coniferous trees often displace deciduous, in some cases
forming pure stands over very large areas. The North American
forests of this zone, for reasons which are not climatic, are mainly
coniferous. The centre of the lumbering industry in eastern Canada
lies about the northern edge of the 'Eastern Temperate Forest', along
the banks of the St. Lawrence, where almost pure stands of white pine
(on the lighter soils) and spruce and hemlock (on the heavier soils)
provide much of the timber and wood-pulp of commerce.

Deciduous trees are often mingled with the coniferous, especially
sugar maple, elm, oak, yellow birch, etc., but they become less and
less important northwards. In British Columbia, under milder and
wetter conditions, the Douglas fir and Sitka spruce grow to a prodigious
size and provide fine large timber.

Temperate Grasslands. With less than 15 or 20 inches yearly
the forest passes into grassland as is clearly brought out by a comparison
of the mean annual rainfall and vegetation maps of Eurasia or North
America. The grassland climate is the steppe variety with early summer
rain and a considerable temperature range. The heat of late summer
scorches up the grass which is a tougher, drier type than the meadow
grasslands of, say, Britain, and causes annual prairie fires whose ashes
have added fertility to the rich black soils. This grass is well suited
to stock-raising and its economy is mainly pastoral, but the better
watered parts have been largely brought under cultivation, especially
for cereals, for which the moist spring and the hot dry period of late
summer are ideal. Wheat is favoured on the poleward borders, maize
in lower latitudes where the summer is longer and hotter and where
the rains tend to be prolonged later into summer. The richer parts of
the steppe and prairie are the great granaries of the world.

REGIONAL TYPES: BRITISH ISLES

The climate of the British Isles is characteristically mild, the weather typically variable. The mildness of the climate is due to the profound marine moderation exerted by the warmth of the North Atlantic; the variability of the weather to the frequent procession of cyclones and anticyclones accompanied by veering and backing winds importing air now from the ocean, now from the continent, at one time polar, at another tropical.

The British Isles lie in the debatable ground between three great pressure and weather systems, the Azores high, the Icelandic low and the continental pressures which are high in winter but low in summer; as these systems wax and wane Britain may be enveloped entirely in one or other of them. In the hot, dry summers of 1911 and 1921 the Azores high spread northwards over us, in the cold winter of 1879 when the Thames was frozen over and when −23° was claimed at Blackadder, Berwickshire[1] it was the continental high that covered us. Normally the prevailing wind is the result of the air-flow from the Azores high towards the Icelandic low and is therefore south-west, but in summer the continental low begins to exert an attraction and winds become more westerly.[2]

Winter Conditions. The Icelandic low is now at its deepest and largest and the pressure gradient between here and the Azores is at its steepest, the south-west winds blow, therefore, with their greatest force. At the same time the continental high is strongly marked and attempts to repel the warm westerly winds from the continental margin. Between the cold outflowing continental winds and the warm, stormy westerly winds a constant struggle for mastery is fought out over the British Isles. While the ocean prevails, the weather is mild, but windy; while the continent prevails, cold snaps occur; temperatures as low as −23° have been recorded, but it is rare for the thermometer to sink below 10°. That proximity to the ocean is more important than latitude at this season is shown by the north and south run of the isotherms; the east coast has 38° as far south as the Thames estuary in January, the west coast has 42° as far north as the Outer Hebrides, while a gulf of warmth, a small-scale replica of the North Atlantic, is thrust into the Irish Sea between the relatively cold lands of Britain and Ireland.

The cyclonic storms of this season tend to follow a marginal course up the west coasts of Ireland, Scotland and Norway and to brush the western seaboard with their wet southerly fringes. Owing to the strength

---

[1] *This figure is now discredited owing to the exposure of the thermometer: −17° at Braemar on 11th February, 1895, is now the official record low temperature for Britain.*

[2] The meteorological circumstances that underlie these climatic conditions are examined in Chapter IV.

of the wind and the frequent depressions, the west coasts are receiving their heaviest rain: Ben Nevis, with 18 inches in January, has 11 per cent of its total in this month; Seathwaite, with 12 inches, has 10 per cent. On the other hand, the Midlands and East are under continental

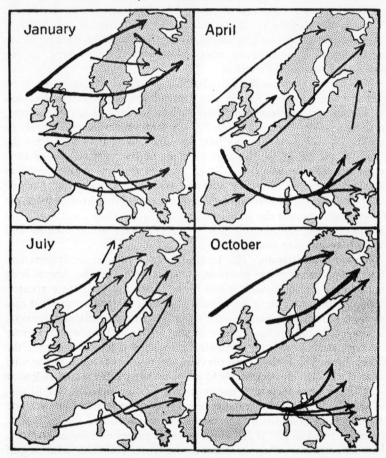

Fig. 68.—Principal Tracks of Depressions in Western Europe

influence and have their lowest rainfall at this season; East Anglia, with less than 2 inches, has only about 7 per cent of its total in January.

Spring. As the continental anticyclone weakens, especially in the south, storms begin to take a more southerly course, a favourite track at this period being the warm Mediterranean Sea. In this case the British Isles lie to the north of the cyclonic centre and suffer from frequent easterly winds off the Continent. These winds, especially in the eastern counties, are dry and cold, chapping and roughening the skin and bringing a crop of lung diseases and allied complaints. But anticyclone weather is common, with a large diurnal range of temperature and night frosts well into May. This is the driest season of the year, most places having only about 6 per cent of their total in April.

Summer. The Azores high has now spread far to the north and the Icelandic low has virtually disappeared as an individual system, owing to the growth of continental lows on either side. Winds, therefore, blow mainly from the Azores high to the continental low and are westerly with very little of the southerly component; the gradient is small and the resultant wind is generally light. The track of cyclones lies directly across the British Isles or over the seas to north and south, but they are less frequent and are not as a rule deep when they do occur. There are, however, occasional thunderstorms, often of considerable intensity, especially in the eastern counties which experience their heaviest rain at this season. The amount of rain is not large: London has 2 inches in July, but this makes up 9 per cent of the yearly total. Places in the West have five times as much (e.g. Seathwaite 10 inches), but this makes up only about 7 per cent of their yearly total. The summer rainfall of the eastern counties comes mostly in heavy showers (convectional) which may give 1 inch or 2 inches in an afternoon—a continental feature. Evaporation and run-off are rapid at this season, and it does not appear to be nearly so wet as winter although the statistics show that the rainfall is actually heavier.

By contrast with the January conditions the isotherms now run east and west, but have a tendency to form loops round the larger land areas. The South and Midlands are enclosed by the isotherm of 62°, while London is enclosed by the 64° isotherm; marine influence keeps the coast somewhat cooler. The highest temperature recorded is 100° at London in the hot summer of 1911; this, however, is exceptional and temperatures above 90° are rare, occurring only under persistent anticyclonic conditions.

Autumn. These anticyclonic conditions of summer persist into early autumn, but temperature falls and the nights are positively cold. The strong nocturnal cooling of the still moist air causes fog and mist to form in the evenings, but these are soon dispersed by the morning sun.

By November anticyclonic conditions have ceased to be common, a cyclonic control is established and the winter pressure distribution is restored. Autumn is a wet season everywhere, for the sea is still warm and the winds off it are very humid, while the land is growing cold, especially at night. Rainfall is chiefly orographic and cyclonic and is therefore heaviest in the West and at night.

EUROPE

In the absence of any climatic barrier on the European plain there is an almost perfect gradation from the humid, mild, equable marine type to the dry, extreme continental type, as the following figures for stations, all in the same latitude, show:

Station	Temperature Range	Rainfall	Wettest Month	Percentage Rain in Winter (six months)
Utrecht . .	28	29	August	47
Berlin . .	34	22	July	43
Warsaw . .	40	15	July	43
Nikolaewskoe .	61	14	June	40

But south of the plain the scattered blocks of the Hercynian Mountains and the more formidable loops of the alpine chain cause the transition to be more spasmodic, the western slopes being unduly marine, the basins to leeward being prematurely continental. Not only do the western edges of these blocks receive more abundant rain than the enclosed basins which they protect, but also they are more 'marine' as regards their temperature range and the transition of the seasons. Steppe conditions occur in the Hungarian basin some 500 miles west of the general boundary of the steppe. The heart of the Balkan peninsula is 'cool-temperate continental' far into latitudes which would normally have a Mediterranean climate, and summer rains here nourish a dense oak forest, the home of herds of swine. Even the Plain of Lombardy, shut off from the Mediterranean Sea by the Apennines, has a climate which is typically continental and has little affinity with the Mediterranean type. There are summer rains here, of great value in supporting good summer pasture for dairy cattle (rare elsewhere in the Mediterranean basin) and for the peculiar crops of the Lombardy Plain, rice and maize. Much of the rain is in the form of thunderstorms, often of great violence with large hailstones which batter the crops and damage the fruit. It is quite a general practice to insure against it.

By longitude 30°E. the west winds are parched and dry and steppe supervenes, but the Black Sea renews their moisture, and its eastern shores, backed by the lofty Caucasus, have 80 inches of rain in places and dense forest as a result.

The Caspian has no mountains on its eastern shore, which is extremely arid, but the prevailing wind here is not westerly but northerly, determined by the Mediterranean low of winter and the Sind low of summer. Thus the south shore is the lee shore, and, like the lee shore of the Black Sea, it is mountainous. Along the south Caspian littoral, backed by the Elburz Mountains, is a region with

50 to 70 inches of rain; a swamp in winter, steamy and hot in summer. Here are dense forests (from which the Romans obtained the tigers for the gladiatorial games) and sub-tropical cultures, rice, cotton, sugar and fruit.

This is the last of the cool-temperate climates in Eurasia, north-east of this point there is sandy desert, east of it there is high plateau, while to the north, where the rainfall is heavier (Saratov, Orenburg),

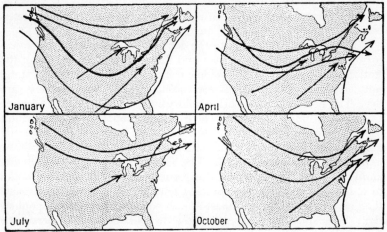

Fig. 69.—Cyclone Tracks, North America

the long winters compel their inclusion in the cold climates. When the type reappears in North China it clearly belongs to the monsoonal sub-type.

NORTH AMERICA

Storm Tracks. Unlike Europe the continent of North America presents no obvious lines of least resistance for the penetration of cyclonic storms, which are compelled to break across high mountains by whatever route they enter. Yet there are certain tracks followed with some degree of fidelity, the most important of which are shown in Fig. 69.

The northerly route, taking advantage of the Great Lakes and the St. Lawrence valley, is generally referred to as the 'Northern Circuit' and is the most frequented track during the summer when, however, storms are fewest, feeblest and least constant. During this time the southern state lies almost entirely to the south of the cyclonic influence which does not disturb the serenity of their summer.

As the winter approaches the cyclones begin to show an occasional and increasing preference for the so-called 'Southern Circuit' and by midwinter may reach to the shores of the Gulf (see p. 72). In addition to these two main routes there are intermediate routes and 'spur tracks', e.g. from Colorado and Texas, joining up with the 'Northern Circuit' in the neighbourhood of the Great Lakes. It is noticeable that most of the storms enter near Puget Sound, while nearly all leave by New England; these places, and especially the latter, are the storm centres of North America and consequently have very changeable weather. The storm tracks pay little attention to the continental winter high pressure which is, in fact, difficult to identify on the daily weather maps, and disguises rather than emphasizes the frontogenetic contrast between Arctic and Canadian air.

The weather associated with the depressions depends on the trajectory of the air masses concerned and therefore on the position of the station with relation to the cyclonic centre. Thus in the eastern states southerly and easterly winds are moist, westerly and northerly are dry. The northerly winds bring cold waves, the southerly hot waves. Cold waves, three or four of which may be expected each year, are usually sudden in their arrival, hot waves are generally more gradual and cumulative in their effect.

Distribution of Rainfall. The Cascade Mountains of Oregon and Washington and the Coast Ranges of Canada limit marine climates to a narrow coastal fringe, quickly followed to the east by a rain-shadow area whose aridity must always condemn it to scanty population. With small local exceptions due to relief this aridity persists as far as the 100th meridian, but east of this line America is highly favoured by an adequate and well-distributed rainfall increasing steadily in amount towards the east and south. This favourable treatment it owes to a large degree to the presence of a great body of warm water in the Gulf of Mexico and the flow of a warm current up its eastern coast.

As a result of the meridional trend of the relief units, the well defined climatic boundaries in North America run from north to south rather than from east to west, the changes in the former direction being in the nature of gradual transitions and changes of degree rather than of kind.

The Pacific Province. (British Columbia, Washington and Oregon.) During the winter months the three major controls, exactly analogous to those of Western Europe, are:

1. The Aleutian low (cf. the Icelandic).
2. The North Pacific high (cf. the Azores high).
3. The Continental high (cf. the Eurasian).

The resultant prevailing wind is southerly or south-westerly, warm and wet; but owing to the weaker influence of the North Pacific than the North Atlantic drift, it is neither so warm nor so wet as the corresponding European wind. Coastal temperatures are about 5° cooler, latitude for latitude, in the New than in the Old World (cf. Brest, January 44° and Victoria, B.C., 39°: Blacksod Point 43° and Masset, B.C., 35°). Inland, away from marine influence, temperatures fall somewhat rapidly, even when reduced to sea-level; the isotherms run north and south as in western Europe, but the more continuous coast-line and the sudden rise of the mountains are reflected in the straighter and more crowded isotherms.

The mountain screen behind the coast more than compensates for any inferiority of humidity and the winter rainfall is very heavy all along the coast; 8 inches in January is recorded over a long strip in Oregon and Washington.

In summer, when the Pacific high has moved north and the Aleutian low has faded and merged with the continental low, the prevailing wind is north-westerly. Partly because it is now less directly on-shore, partly because it comes from higher latitudes, partly because the land is now warmer than the sea and partly because depressions are less numerous, the summer rain is considerably less: Victoria has less than 1 inch in each of the three midsummer months. The corresponding climates in Western Europe are much wetter in summer. Farther south the summer rainfall decreases still more until in California the drought of the 'Mediterranean' summer is complete.

The Basin and the Plains. The decrease of rainfall eastward bears a very close relationship to relief: Clayoquot, on the west side of Vancouver Island, has 119 inches, Nanaimo, on the lee side, only 37 inches; just across the strait the rainfall at Vancouver has increased to 59 inches, but decreases inland to 10 inches at Kamloops and the maximum here occurs in early summer. Thus within 200 miles of the coast there is a steppe climate closely resembling that of south-eastern Europe. At Spokane the temperature range already exceeds 40°, the rainfall is only 17 inches, there is a tendency to an increased rain in spring and the autumn is strikingly dry (July, August and September have less than 1 inch each); these are steppe characteristics. A marine feature is, however, retained in a considerable winter precipitation from cyclonic storms. Thus there is good rain in winter and spring which hinders spring sowing but is useful for fall-sown wheat, the spring and early summer rains being valuable for the growing plant and the autumn drought suiting the harvest.

Across the Rockies the winter rain is no longer found, the spring maximum being dominant (e.g. Denver) and due to the penetration of cyclonic storms (cf. the steppes of the Black Sea region). This is Henry's 'Rocky Mountain foothills'[1] rainfall régime—a true steppe type. Farther east the maximum tends to occur later into summer, the rainfall being mainly convectional. This is the continental type of the plains, the 'Missouri type' of Greely.[2]

[1] See reference on p. 205.   [2] See reference on p. 190.

Winter temperatures in the plains are below freezing and there is a continuous snow cover in winter, but summers are hot (over 70°). Thunderstorms are an almost daily occurrence and the *tornado*, though a comparatively rare event, is more frequent and more destructive here than anywhere else. They are apt to occur along the trough of V-shaped depressions and are a 'squall-line' phenomenon at the meeting line of two air currents of different temperatures and humidities.

The Eastern States. Eastward again there enters a new source of winter rain, namely, winds off the Gulf and the Atlantic. At first these rains are less important than the convectional type and the rainfall graph still shows a maximum in summer (somewhat retarded) though winter is well supplied (e.g. Portsmouth, Ohio), but nearer the coast they assume greater importance and rainfall is fairly evenly distributed over the year (e.g. Boston, see p. 224). This evenly distributed rainfall confers great advantages, risks of drought and flood are minimized and the discharge of the rivers is not subject to great fluctuation, a feature of considerable value in their utilization for power. Winter precipitation is frequently in the form of snow which, in the north, lies to a considerable depth.

The climate of the eastern states is a curious mixture of marine and continental, for the prevailing westerly wind tends to carry continental conditions as far as the coast, yet the occasional south-east wind in front of depressions brings influences from the sea. The Great Lakes are a further important influence in moderating the climate: the Michigan and London peninsulas being particularly mild, and famous for their fruit, including grapes and peaches. The high humidity, a marine trait due to the proximity of ocean and lakes, makes much more unpleasant the large (though diminished) temperature range, a continental trait due to the prevailing land wind. The July mean for New York is 75°, but temperatures above 95° are common in July and August, over 100° is occasionally recorded and on 7th August, 1918, the mercury reached 104°.

Heat waves such as this are slow to arrive but are cumulative. The summer cyclones, with which they are associated, move slowly and sometimes remain stationary for days while temperatures mount higher and higher. Nor is there any relief at night, for the high humidity of the air prevents nocturnal cooling, night temperatures may not fall below 75° and sleep becomes almost impossible. In agricultural regions the heat and drought, if prolonged, may cause grave financial loss, but it is particularly in the great cities, and more especially in the overcrowded quarters that the hot waves bring the greatest suffering, and deaths from sunstroke and heat apoplexy are numerous. Light clothing is worn and quantities of ice and cold drinks are consumed but bring little relief; all who can do so leave for the seaside where

hotels and boarding-houses are full and the beaches are crowded all
night long with fugitives from the sweltering city. Not until the cyclonic
centre has passed and the wind swings round to the north is there any
relief.

THE SOUTHERN HEMISPHERE

In the ocean hemisphere the planetary winds suffer no appreciable
deflection by continental influences, and the Brave West Winds impinge
with full force throughout the year directly on to the mountainous
coasts which form the western seaboards of southern Chile, Tasmania
and New Zealand. The western coasts are deluged with rain, the
rain-shadow of the eastern slopes is profound and is reached with
striking suddenness. All three regions have considerable areas in the
west with more than 100 inches and all three have areas in the east
with less than 20 inches.

Southern Chile and Patagonia. The western slopes of the Andes
in southern Chile with their 100 inches or 200 inches of rain, are a region
of dense forests on saturated swampy ground which militates against
their economic exploitation. Rain falls on six days out of seven at
Evangelistas Island. As in North America, where the disposal of relief
is similar, there is a sudden decrease in the rain-shadow in lee; most
of Patagonia has less than 10 inches of rain fairly evenly distributed
throughout the year.

New Zealand. Chiefly owing to its oceanic situation New Zealand,
for its latitude, enjoys very mild winters; it is only in the extreme south
of South Island, in fact, that the mean temperature of the coldest
month falls below 43°. The genial winter is one of its greatest assets
as a pastoral country, for winter grazing is available and the need for
stall-feeding of cattle is largely obviated. In spite of the absence,
however, of a real cold season, a characteristic of warm-temperate
climates, it would not be legitimate to consider the island group under
that heading, for in other respects it is typically cool-temperate. With
the exception of the north of North Island, for example, the prevailing
wind throughout the year is westerly, and it is only in the Auckland
peninsula that the influence of the trade winds is felt in summer. The
weather throughout the year is thus subject to cyclonic influences and
consequently lacks that reliability of summer conditions which the
warm-temperate climates expect to enjoy. Furthermore, the oceanic
influence which kept the winters warm keeps the summers cool;
Wellington has about the same summer temperature as London,
though it is 10° nearer the Equator. The climate of New Zealand,
then, with the exception of the Auckland peninsula, is best described
as cool-temperate marine, being, in fact, comparable with England,
but being warmer in winter, windier and sunnier.

Following the line of least resistance most of the depressions follow a track to the south of New Zealand, though a few cross South Island, especially in winter, moving generally from south-west to north-east. The north winds on their advancing edge are warm and muggy, but as the centre passes the wind veers to north-west or west and, reinforcing the normal westerly winds, blows strongly against the western slopes, rising and releasing copious rain. On the eastern side it descends rapidly as a hot foehn wind which is extremely trying on the Canterbury Plains. Later the wind swings to the south and blows cool, damp and refreshing.

In South Island the rainfall is fairly evenly distributed through the year, both on the wet western and dry eastern sides, but in North Island the influence of the anticyclone begins to be noticeable in a summer decrease of rain. This becomes more and more pronounced until at Auckland nearly 66 per cent falls in the six winter months, but even here the driest month has nearly 3 inches of rain and there is no approach to a summer drought. The rainfall, too, has a high degree of reliability and New Zealand, unlike its nearest neighbour Australia (not, of course, climatically comparable), does not suffer from drought.

One of the pleasantest characters of the New Zealand climate is its sunniness. Considering that the islands lie in the cyclonic westerly circulation and that they are surrounded by wide oceans it is rather surprising that they should be able to boast a sunshine record comparable with that of the Mediterranean; yet stations on the east coast show records of 2,500 hours a year.

THE MONSOONAL SUB-TYPE

Eastward from the Atlantic we have traced a progressive degeneration of the climate by decreased rainfall into the steppe and desert of Central Asia. On approaching the Pacific the rainfall increases again in a manner similar to the increase in eastern North America (see p. 216), but whereas the eastern margin of North America is characterized by an extremely high degree of uniformity of seasonal incidence, the eastern division of Asia shows as its most striking characteristic a strongly marked seasonal rhythm. At Che-foo 20 inches, out of an annual total of 24 inches, fall in the six summer months, and half the annual total is recorded in the months of July and August alone: at Tientsin 72 per cent falls in the three summer months (cf. Washington 30 per cent) and only 2 per cent in the three winter months (cf. Washington 23 per cent).

In the face of the constant off-shore winds it is clear that the influence of the sea will not be able to make itself felt, and the monsoonal sub-type is typically continental during the winter months, the cold of the interior being carried right down to the coast. China is, in fact, some

10° or 15° colder than the United States in the same latitude during winter; Mukden (January 8°) is 15° colder than Albany (23°), Shanghai (38°) is 12° colder than Charleston (50°).

These features result from the steeper pressure gradient in eastern Asia, for which reason not only are the winds more regular, but also cyclonic disturbances are a less powerful influence. The swirl of air round these centres brings, from time to time, warmth and rain from the Atlantic to eastern North America, interrupting the reign of the continental anticyclone. The winter weather of the eastern states is therefore much more variable and much more subject to sudden changes than that of China; at Mukden the thermometer hardly ever rises above 50° in January, at Albany it frequently exceeds 60° and it is the effect of these mild spells which explains the relatively high mean temperatures in North America. Thus the difference between the two continents lies rather in the greater persistence of continentality in Asia than in its greater severity.

REGIONAL TYPES: NORTH CHINA

Although a degree of uniformity is lent by the monsoon to the climates of the whole of eastern Asia, it is necessary to recognize subdivisions on the grounds of differences of temperature. The tropical monsoon of Annam (p. 156) and the sub-tropical monsoon of South China (p. 192) have already been described. Northern China differs from these in the possession of a pronounced cold season which, farther north, increases in length and severity until, north of about 45°, the length of the winter places the climate in the category of cold climates (p. 240). The cool-temperate monsoonal climate therefore exists between the latitudes of about 32° and 45°N. and includes the whole of the great plain of China, much of the highlands which overlook it on the west and the three peninsulas of Shantung, Liao-tung and Korea.

Winter Conditions. Enclosed by mountains on three sides the air of the Asiatic high finds its chief escape to the east and, impelled by the strong barometric gradient, cold dry winds sweep over North China with great force and constancy; the mean wind velocity in January is 10 m.p.h. Though the air is very dry and the sky may be cloudless, the fine yellow dust which these winds bring from the deserts hides the sun with a heavy haze and carries bad visibility far out to sea. Closed windows and doors in Peking are impotent to keep out the penetrating grit which settles everywhere and irritates and inflames the eyes, nose and mouth. The cold winds serve, however, to cleanse and pulverize the soil of the plains and bring every year a fresh surface-dressing of fine yellow soil. In many thousands of years they and their still more powerful Pleistocene predecessors have endowed Shansi and the neighbouring provinces with a rich legacy of a mantle of loess hundreds of feet in thickness.

The temperatures which these winds import are phenomenally low and proximity to the sea brings no moderating influence since

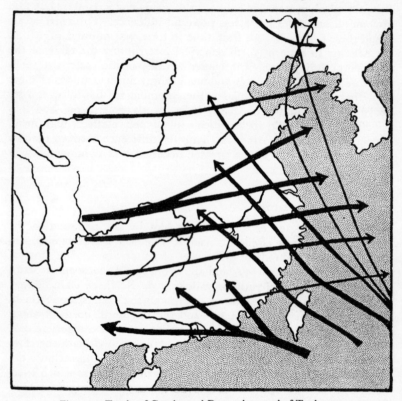

Fig. 70.—Tracks of Continental Depressions and of Typhoons
(*Mainly after* Gherzi)
Thickness of lines approximately proportional to frequency of storms

the land monsoon carries continental conditions right down to the seaboard.

Everywhere in China winter is the season of least rain, and this is particularly the case in North China where the continental influence is strongest; here the three winter months have generally less than 5 per cent of the yearly total. Such rain as does fall in winter is generally to be attributed to the passage of shallow depressions, the chief tracks of which and the relative frequency along each are shown in Fig. 70. Since this rain generally comes on south-east winds (i.e. off-sea), the heaviest incidence is in the coastal zone and there is a rapid decrease towards the interior. The Yangtse valley provides a much favoured path, and Central China, in consequence, has heavier winter rain than either north or south; Hankow has 9 per cent and Shanghai 13 per cent in the three winter months and at the latter place January is the only month with less than 2 inches. The intensity of winter rain is much lighter than that of summer and though the total amount is small the number of rainy days is greater than might be supposed.

North of the Hoang Ho and Shantung the winter precipitation is mainly in the form of snow, and severe blizzards form a most unwelcome feature of the climate of the plain.

Summer Conditions. The passage from winter to summer is remarkably sudden in North China, much more sudden than in the corresponding latitudes in America; the land is warmed with great rapidity and April at Peking is nearly 10° warmer than March. At the same time the weakening of the anticyclone allows marine influence to enter to an increasing extent, the humidity rises and rainfall begins to increase both in frequency and amount. These early rains are not the monsoon rains as yet (see p. 196), but they are of inestimable value for the growth of the spring crops (e.g. wheat) which always suffer from the disadvantage of inadequate moisture in the early stages of germination and growth.

At the height of summer the temperature exceeds 80° nearly everywhere south of Peking, and North China is nearly as hot as South. The monsoon rains have arrived and typically reach a maximum in July. The Yangtse Valley is an exception, however, to the general rule in having a double maximum, the first in June, the second in August and September, with a somewhat drier spell between. The early maximum is apparently attributable to shallow cyclones of continental origin moving slowly down the valley and later across the Yellow Sea to Japan where they cause the 'Plum Rains'. As is usually the case with the continental depressions the rain comes chiefly on south-east winds and these are followed by northerly winds which bring a marked drop in temperature, accentuated by a sudden fall in humidity. The second maximum is connected with the passage of tropical cyclones which, though they may not come ashore, nevertheless cause rain over a wide area.

Economic Effects of Rainfall in China. Summarizing the facts of rainfall, it is clear that the length of the dry season increases inland, for at one end the coast gets winter rain from the continental depressions and at the other the summer rains are prolonged by typhoons in the autumn when the monsoon begins its southward retreat. Up to a point the summer maximum of rain is a valuable asset for crop production, but in many ways the highly seasonal incidence is unfortunate, though the absence of winter rain is less serious in North China where the winter temperatures are in any case too low for cultivation. It has, on the other hand, supplied a valuable incentive towards irrigation, in its turn a powerful stimulus to the growth of a Chinese civilization.

Monsoon rainfall, largely cyclonic in origin, is characteristically variable in amount, and famine is the curse of monsoon lands. The very fertility of the lands encourages the growth of population up to the limit of supporting power, and agricultural practice is all too frequently based on the most optimistic expectation of rain. For this reason the disaster of a deficiency is all the greater, particularly in North China where the annual amount is lowest and the margin of sufficiency least. The normal rainfall at Hankow is 50 inches, but in any year the total may fall short of this by more than 50 per cent or exceed it by an even greater amount; July, 1886, was practically rainless at Shanghai; July, 1903, had 12 inches.

The heavy summer fall is often too torrential for full utilization and the loss by run-off has been aggravated by deforestation. A considerable rise of the rivers follows rapidly on every heavy fall of rain and serious floods all too often result. The flatness of the plains of Chi-li and Honan, coupled with the raising of the river banks as levees, lends itself particularly to such disasters, the distressing frequency of which is amply testified in Chinese history and legend. Central China, though its rainfall is heavier, is more fortunate in the possession of a better developed drainage and of large lakes, e.g. Poyang and Tungting, which act as reservoirs to equalize the flow.

JAPAN

Marine influence and the warm waters of the Kuro Siwo combine to give Japan an adequate and well-distributed rainfall and to moderate the extremes of temperature, but the effects of insularity are diminished by proximity to land, especially during winter when the prevailing influence is off the continent. Sakhalin, separated from the mainland only by the narrow Gulf of Tartary, shares the extreme continentality of Amuria; Yezo is a little more fortunate, but has four or five months below freezing-point; Hondo, separated by the wider Sea of Japan, has much milder winters, though even here the January isotherms are some 10° south of their mean latitude. Shikoku and Kyushu are virtually sub-tropical, there is no real winter, the January mean being everywhere above 40°, palms, camphor trees and citrus fruits flourish and three crops a year may be obtained.

Winter Conditions. The Kuro Siwo, the Asiatic counterpart of the Gulf Stream, sends a branch up each coast of Japan; the smaller westerly branch bears against the Japanese coast, but the more important easterly branch tends to swing away into the Pacific, yielding place along the coast, as far south as the angle of the Awa peninsula, to the cold Kurile current. Thus the Kuro Siwo is less effective than might be expected in raising the winter temperature of Japan; the January mean is below freezing north of Niigata which is in the same latitude as Lisbon, and frost occurs on more than 60 days in the year at Tokio. The warm current does, however, strongly influence the course of depressions, the barometric minima closely following the warm water. Fig. 71 shows clearly how the storm tracks are related to the two branches of the warm current, the more frequented track

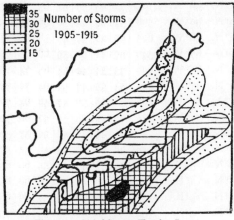

Fig. 71.—Chief Storm Tracks, Japan
(*After* Sanders)

coinciding with the greater volume of the current. The source of these storms is in the Asiatic continent (see p. 196) and their destination is the Aleutian low. On leaving the mainland they acquire renewed vigour from the warm waters and cause the Japanese climate to be much more cyclonic, much more variable and therefore much more stimulating than that of North China.

The cyclonic winds bring adequate rain to both sides of Japan, but the prevailing westerly (monsoon) wind, meeting the mountains, brings heavy relief rain to the western coast; Kanazawa has 32 inches in the three winter months and scarcely a day passes without rain or snow in the winter. Much of this precipitation is in the form of snow, especially on the mountains where it lies to a depth rarely equalled elsewhere. Though the temperature is higher than on the mainland the cold is more raw and unpleasant owing to the high humidity. The east coast, watered only by the cyclonic rains, is drier and more pleasant; Tokio has only 7 inches in the three winter months.

Summer Conditions. In summer the conditions are reversed and the east coast is the windward one, well watered by the south-east monsoon; the west coast, sheltered by the mountains, is drier though

TEMPERATURE

Station	Lat.	Long.	Alt. (ft.)	J	F	M	A	M	J	J	A	S	O	N
VANCOUVER .	49°N.	123°W.	136	36	38	42	47	54	59	63	62	56	49	43
PORTLAND . .	46°N.	123°W.	153	39	42	46	51	57	61	67	66	61	54	46
DENVER . . .	40°N.	105°W.	5,291	30	32	39	47	57	67	72	71	62	51	39
OMAHA . . .	41°N.	96°W.	1,105	22	25	37	51	63	72	77	75	66	55	39
ST. LOUIS . .	39°N.	90°W.	568	32	34	44	56	66	75	79	77	70	58	45
CHICAGO . .	42°N.	88°W.	823	26	27	37	47	58	68	74	73	66	55	42
NASHVILLE . .	36°N.	87°W.	546	39	41	50	59	68	76	79	78	72	69	49
CINCINNATI . .	39°N.	85°W.	628	33	34	44	54	65	74	78	76	69	58	45
WASHINGTON .	39°N.	77°W.	112	34	35	43	54	64	72	77	74	68	57	46
NEW HAVEN .	41°N.	73°W.	106	28	28	36	47	58	64	72	70	64	53	42
BOSTON . . .	42°N.	71°W.	125	27	28	35	45	57	68	72	69	63	52	41
VALENTIA . .	52°N.	10°W.	30	44	44	45	48	52	57	59	59	57	52	48
ABERDEEN . .	57°N.	2°W.	46	38	38	40	44	48	54	57	56	53	47	42
BREST . . .	48°N.	5°W.	213	45	45	47	50	55	60	65	64	61	56	50
PARIS . . .	48°N.	2°E.	405	37	39	43	49	56	62	65	64	58	50	43
BERLIN . . .	53°N.	13°E.	196	30	33	38	48	57	63	66	65	58	49	39
MILAN . . .	46°N.	9°E.	490	32	38	46	55	63	70	75	73	66	56	44
BRESLAU . . .	51°N.	17°E.	482	30	32	37	46	56	63	66	64	58	48	38
VIENNA . . .	48°N.	15°E.	664	29	33	40	50	59	65	68	67	60	50	39
WARSAW . .	52°N.	21°E.	436	26	29	35	46	57	63	66	64	56	46	36
BELGRADE . .	45°N.	20°E.	459	29	34	43	52	62	67	72	71	63	55	43
BUCHAREST . .	44°N.	26°E.	269	26	31	41	52	62	69	73	72	64	53	40
KIEV . . .	50°N.	31°E.	590	21	23	31	45	57	64	67	65	57	46	34
ODESSA . . .	46°N.	31°E.	210	25	28	35	48	59	68	73	71	62	52	41
HOKITIKA . .	43°S.	171°E.	9	60	61	59	55	50	47	45	46	50	53	55
DUNEDIN . .	46°S.	171°E.	40	58	58	55	52	47	44	42	44	48	51	53
HOBART . . .	43°S.	147°E.	177	62	62	59	55	51	47	46	48	51	54	57
VALDIVIA . .	39°S.	73°W.	141	60	59	57	54	51	49	46	46	49	51	53
PUNTA ARENAS .	53°S.	71°W.	92	52	51	48	44	39	36	35	37	40	44	47
ICHANG . . .	31°N.	111°E.	167	40	43	51	62	70	78	83	83	75	65	54
ZIKAWEI . .	31°N.	121°E.	23	38	39	46	56	66	73	80	80	73	63	52
PEKING . . .	40°N.	116°E.	131	24	29	41	57	68	76	79	77	68	55	39
MUKDEN . .	42°N.	123°E.	144	8	14	30	47	60	71	77	75	61	48	29
CHEMULPO . .	37°N.	127°E.	240	25	29	36	49	59	67	73	78	69	57	42
NIIGATA . . .	36°N.	139°E.	84	35	34	40	51	59	67	74	78	70	59	49
NAGASAKI . .	33°N.	130°E.	443	42	43	48	58	64	71	78	80	74	64	55
TOKIO . . .	36°N.	140°E.	70	37	39	44	55	62	69	76	78	71	60	51
MIYAKO . . .	40°N.	142°E.	100	31	32	37	47	54	61	68	72	65	55	45

RAINFALL

D	Yr.	Ra.	J	F	M	A	M	J	J	A	S	O	N	D	Total
48	48	27	8·6	6·1	5·3	3·3	3·0	2·7	1·3	1·7	4·1	5·9	10·0	7·8	59·8
41	53	28	6·7	5·5	4·8	3·1	2·3	1·6	0·6	0·6	1·9	3·3	6·5	6·9	43·8
32	50	42	0·4	0·5	1·0	2·1	2·4	1·4	1·8	1·4	1·0	1·0	0·6	0·7	14·3
27	51	55	0·7	0·9	1·3	2·8	4·1	4·7	4·0	3·2	3·0	2·3	1·1	0·9	29·0
36	56	47	2·3	2·6	3·5	3·8	4·5	4·6	3·6	3·5	3·2	2·8	2·9	2·5	39·8
30	50	48	2·1	2·1	2·6	2·9	3·6	3·3	3·4	3·0	3·1	2·6	2·4	2·1	33·2
41	59	40	4·8	4·2	5·1	4·4	3·8	4·2	4·1	3·5	3·5	2·4	3·5	3·9	47·4
36	56	45	3·5	2·9	4·0	3·1	3·6	3·7	3·4	3·3	2·6	2·5	2·9	3·1	38·6
36	55	43	3·2	3·0	3·5	3·3	3·6	3·9	4·4	4·0	3·1	3·1	2·5	3·1	40·7
32	50	44	3·8	4·0	4·0	3·5	3·9	3·2	4·3	4·5	3·7	3·7	3·6	3·7	45·9
32	49	45	3·7	3·5	4·1	3·8	3·7	3·1	3·5	4·2	3·4	4·1	4·1	3·8	44·6
46	51	15	5·5	5·2	4·5	3·7	3·2	3·2	3·8	4·8	4·1	5·6	5·5	6·6	55·7
39	46	18	2·2	2·1	2·4	1·9	2·3	1·7	2·8	2·7	2·2	3·0	3·0	3·2	29·5
46	54	20	2·6	2·4	2·2	2·1	2·4	1·5	1·3	1·9	2·5	3·4	3·1	3·7	29·1
38	50	28	1·5	1·4	1·6	1·7	1·9	2·1	2·2	2·1	1·9	2·3	1·9	2·0	22·6
33	48	36	1·7	1·4	1·6	1·5	1·9	2·3	3·0	2·3	1·7	1·7	1·7	1·9	22·7
36	55	43	2·4	2·3	2·7	3·4	4·1	3·3	2·8	3·2	3·5	4·7	4·3	3·0	39·8
32	47	36	1·3	1·0	1·6	1·5	2·4	2·4	3·4	2·8	2·0	1·5	1·5	1·5	22·9
32	49	39	1·5	1·3	1·8	2·0	2·8	2·7	3·1	2·7	2·0	1·9	1·8	1·8	25·4
30	46	40	1·2	1·1	1·3	1·5	1·9	2·6	3·0	2·9	1·9	1·6	1·5	1·5	22·1
34	52	43	1·2	1·3	1·6	2·3	2·8	3·2	2·7	1·9	1·7	2·2	1·7	1·7	24·4
31	51	47	1·3	1·1	1·6	1·7	2·5	3·5	2·7	2·0	1·6	1·7	1·9	1·6	23·1
24	44	46	1·1	0·8	1·5	1·7	1·7	2·4	3·0	2·4	1·7	1·7	1·5	1·5	21·0
31	49	48	0·9	0·7	1·1	1·1	1·3	2·3	2·1	1·2	1·4	1·1	1·6	1·3	16·0
58	53	16	9·8	7·3	9·7	9·2	9·8	9·7	9·0	9·4	9·2	11·8	10·6	10·6	116·1
56	51	16	3·4	2·7	3·0	2·7	3·2	3·2	3·0	3·1	2·8	3·0	3·3	3·5	36·9
60	54	16	1·8	1·5	1·7	1·9	1·8	2·2	2·1	1·9	2·1	2·2	2·5	2·0	23·7
57	53	14	2·9	3·2	6·4	9·3	15·3	17·5	15·4	13·5	7·3	5·0	4·4	4·8	105·0
50	44	17	1·4	1·2	1·7	1·6	1·6	1·2	1·2	1·2	1·1	0·8	1·1	1·4	15·5
44	62	43	0·8	1·1	2·0	4·2	5·0	6·2	7·8	6·8	4·0	3·6	1·5	0·6	43·6
42	60	42	2·0	2·4	3·4	3·7	3·6	7·4	5·9	5·7	4·7	3·1	2·0	1·3	45·2
27	53	55	0·1	0·2	0·2	0·6	1·4	3·0	9·4	6·3	2·6	0·6	0·3	0·1	24·9
14	44	69	0·2	0·3	0·8	1·1	2·2	3·4	6·3	6·1	3·3	1·6	1·0	0·2	26·5
30	51	53	0·8	0·5	1·0	2·6	3·2	4·3	8·6	8·2	4·2	1·6	1·7	0·7	37·4
39	55	43	7·7	4·9	4·1	4·2	3·7	5·2	6·2	5·2	7·4	5·7	7·2	9·1	70·6
46	60	38	3·1	3·5	5·2	8·1	7·4	13·2	9·3	7·3	8·6	4·6	3·3	3·3	76·9
41	57	41	2·2	2·8	4·4	4·9	5·7	6·5	5·3	5·7	8·7	7·4	4·2	2·1	59·9
36	50	41	2·7	2·6	3·4	3·9	4·7	5·0	5·3	7·0	8·5	6·7	3·2	2·5	55·5

still well watered. Kanazawa, exposed to the winter monsoon and protected from the summer monsoon, has 56 per cent winter rain and 44 per
cent summer; Tokio, exposed to the summer monsoon and protected
from the winter monsoon, has 39 per cent winter rain and 61 per cent
summer. The smallest rainfalls are recorded on the shores of the
Inland Sea, protected from both monsoons, and there are places here
with less than 30 inches.

The summer rains, like those of the Yangtse (p. 221), reach a
maximum in June or July (the Plum Rains), decrease noticeably in
August and rise to a second maximum in September. The earlier
maximum is valuable for the transplanting of the rice but is unfortunately unreliable; the June mean at Tokio is 6 inches, but there have
been years when less than half this amount was recorded and in 1917
there was less than ¾ inch. The cause of these rains is somewhat
obscure; they accompany a temporary falling off of the monsoon and
are thus clearly not orographic. They are generally ascribed to the
passage of shallow continental depressions (see p. 221) and this would
account for their variability. The autumn maximum is the greater of
the two and is due mainly to typhoons but partly to the relative
difference in temperature of land and sea. Advantage is taken of the
drier period between the maxima to harvest quick-growing varieties
of rice, in fact quick-growing varieties are in general use to ensure
harvesting before the second rains set in.

Summer temperatures are everywhere high (the 70° isotherm for
July passes only a little way south of the Tsugaru Strait), but owing
to marine influence, not excessive. Marine influence also prolongs
summer warmth far into autumn. August is nearly everywhere the
hottest month and October is almost as hot as May. The August
maximum of temperature is, however, to be ascribed in some measure
to the decrease of rainfall and humidity during this month.

Variability of the Climate. The climate of Japan shows considerable variations from year to year, for the factors which control
it are themselves subject to fluctuations: the strength of the monsoon,
the strength of the ocean currents and the frequency of storms. In
consequence agriculture is often in a precarious condition. For
example, the cultivation of rice, a cereal requiring heat and moisture,
extends in Japan, owing to other favourable conditions, somewhat
beyond its proper range (the 75° isotherm for the four months of active
growth). A cold summer therefore means a decreased yield and widespread distress. Such cold summers occur when the Kurile current
intensifies and extends its influence, which it does when, among other
causes, large quantities of ice accumulate in the Bering Sea. Excessive
ice here is thus followed by poor harvests in North Japan.

SUGGESTIONS FOR FURTHER READING

For the British Isles, see E. G. Bilham, *The Climate of the British Isles*; Gordon Manley, *Climate and the British Scene*, New Naturalist Series, 1952; M. de C. Salter, *Rainfall of the British Isles*, 1921; *British Rainfall*, published annually by the Meteorological Office; the *Rainfall Atlas of the British Isles*, published by the Royal Meteorological Society, 1926; and the *Climatological Atlas of the British Isles* (M.O. 488), 1952. See also C. E. P. Brooks, 'Weather Influences in the British Isles', *S.G.M.*, 1924; R. H. Hooker, 'Weather and Crops in England', *Q. J. Roy. Met. Soc.*, 1922; C. E. P. Brooks and J. Glasspole, 'The Drought of 1921', *Q. J. Roy. Met. Soc.*, 1922.

For Europe, see A. Angot, *Régimes de pluie de l'Europe Occidentale*, Paris, Ann. Bur. Cent. Météorol, 1895; J. Glasspole, 'The Distribution of Average Seasonal Rainfall over Europe', *Q. J. Roy. Met. Soc.*, 1929; A. Paulsen, *The Climate of Denmark*, U.S. Weather Bureau Bull. 11, 1893.

For N. America, see C. E. Kœppe, *The Canadian Climate*, 1931; Ward's *Climate of the United States*; and F. N. Denison, 'The Climate of British Columbia', *M.W.R.*, 1925. Norman Taylor, *The Climate of Long Island*, Brooklyn Botanic Gardens Contrib., No. 50. *The Climate of Minnesota*, Bull. 12, Min. Geol. Surv., 1915. *The Climate of Western U.S.A.*, Bull. Amer. Geog. Soc., 1915; *Climatic Atlas of the U.S.A.*, S. S. Visher. 'Characteristic Properties of North American Air Masses', H. C. Willett in *Air Masses and Isentropic Analysis*, J. Namias.

The distribution of ice and its influence on seasons of navigation are beautifully shown in the *Ice Atlas of the Northern Hemisphere*, Hydrographic Office of the U.S.A. Navy, 1946. The Meteorological Office publishes *Monthly Ice Charts, Arctic Seas* (M.O. 390a, 1944) and *Monthly Ice Charts, Western North Atlantic* (M.O. 478, 1944).

The Hydrographer of the Navy publishes monthly weather charts of the British Isles and adjacent waters.

XII

COLD CLIMATES

Lying within the sphere of influence of the westerly winds the cold climates are subject to controls differing little from those of the cool-temperate climates, and much that has been said of the latter is equally applicable here. There is the same eastward gradation from marine to continental type, making itself felt in the same way by an increase in annual and diurnal range of temperature, a decrease of rainfall and an increasing tendency to a summer maximum of rain. From the meteorologist's point of view there is little need for separation of the two, but from the geographer's point of view there is every need; the differences are of degree rather than of kind, but the environments they present differ widely. The essential distinction is the greater severity and longer duration of the winter season which puts a stop to agricultural duties in winter and restricts outdoor activities to trapping, fishing and lumbering.

The land masses of the southern hemisphere do not extend far enough south to experience this type of climate and only the northern continents of North America and Eurasia need be considered.

It is worthy of note that both North America and Europe in these latitudes present to the westerly winds high and rugged coasts deeply indented with fjords. The mountains confine marine influence to a narrow coastal fringe and beyond them we are plunged at once into extreme continental climates. This arrangement of relief units is unfortunate, for the life-giving rain which might spread so far across the wide and open plains which lie to the east is extravagantly squandered on the unresponsive western slopes. The marine type accordingly has a very limited distribution and is fairly clearly marked off from the continental along the line of the topographic divide and it will be convenient to describe the two types separately.

On approaching the east coast in North America a modified continental type is encountered in which the annual range of temperature is reduced and the precipitation is evenly spread over the year. But the eastern margin of Asia in these latitudes is under the influence of the monsoons and has their typical régime of temperature and rainfall. Since the two continental masses have little in common on their eastern margins they will be described separately under 'regional types'.

Pressure and Winds. The principal controls of the prevailing winds, as in the previous type, are the Icelandic and Aleutian lows and the continental winter highs and summer lows (see p. 200), but as in temperate climates, these are subject to complete temporary

228

obliteration by the passage of cyclones and anticyclones. The weather element is, however, less insistent and the climatic element more persistent than farther south; continental interiors especially lie under the influence of the winter anticyclones for long periods at a time, meanwhile enjoying cold, clear, calm and settled weather with light outward-blowing winds. From time to time, however, depressions with their anti-clockwise system of often violent winds and their sudden and severe temperature changes succeed in penetrating into the hearts of continents, but they are the exception rather than the rule, pursuing, more usually, a course marginal to the high pressures. In summer the continental interiors are regions of low pressure with light inflowing winds; subject to sudden disturbances and convectional overturnings.

MARINE (NORWEGIAN) TYPE

Temperature. Both in the North Atlantic and the North Pacific drifts of warm water bathe the eastern shores (=western coasts) and weaken the grip of winter far to the north. The North Pacific drift carries the January isotherm of 32° as far as Sitka (Alaska) in latitude 57°N., while the still warmer North Atlantic drift carries it beyond the Lofoten Islands 10° farther north still. The Norwegian coast is kept ice-free all the year round, although extending far beyond the Arctic Circle, and the January mean at Bergen is above freezing, although there are six months below 43°. This phenomenal mildness, it is true, applies only to a narrow coastal zone and the fall in temperature away from the marine influence is very rapid. Even at the heads of the fjords temperatures are 5° or 10° colder and ice forms here every year, though the mouths are ice-free.

Characteristically the minimum temperature is delayed, February being little warmer and, in N.W. Europe, generally actually colder than January. The sun, low in the sky, is slow to melt the snow; spring is late in arriving and summer is also delayed, the maximum at many coastal stations not being reached until August. In the north spring and autumn virtually do not exist; the long winter gives way suddenly about June to the cool summer and in September winter returns as suddenly. In the south winter departs earlier and returns later and there are fairly recognizable seasons of spring and autumn. The July mean is below 60° in the south and below 50° in the north, temperatures sufficient to ripen only the least exacting of crops, but the long days of summer daylight partially offset this disadvantage and permit the ripening of hardy cereals even within the Arctic Circle. At North Cape the sun does not sink below the horizon between 12th May and 29th July; at Bergen June days are nearly 19 hours long and twilight lasts throughout the night. The angle of the sun is, of course, low, and there is great advantage to be derived from a south aspect.

16

August is usually as warm as July, and at typically marine stations is actually warmer (e.g. Sitka, Alaska; Port Simpson, B.C.; Christian-sund, etc.). September remains warm, but in October the long winter sets in again.

Rainfall. The west winds, blowing over the warm waters of the North Atlantic and North Pacific drifts, are heavily charged with moisture which they precipitate on the mountainous western seaboards but there is a further important source of rainfall in the depressions whose chief tracks are shown in Figs. 68 and 69. It should be remem-bered that apart from the cooling by ascent there is a difference of some 15° or 20° in winter between the temperature of the sea and the temperatures only a short distance inland, even when reduction is made to sea-level equivalents.

Rain may be expected on two days out of three at Bergen and an umbrella is part of the equipment of every man, woman and child; in fact it is alleged that ponies in Bergen shy if they see a man without one. The high humidity of winter detracts from the benefit of the mild tem-peratures; the air is often damp and chilly, the sky leaden and cheerless and fog is a frequent occurrence; these conditions are typical only of the coastal zone—a short distance inland rain, cloud and fog decrease rapidly.

Partly because depressions are most frequent in autumn and winter and partly on account of the steep temperature gradient at that season the winter half of the year is considerably wetter than the summer half and there is a clearly marked peak of rainfall in the autumn when the differences of temperature and humidity are greatest and when the marginal cyclone track is particularly favoured (see Fig. 68 and data at end of chapter). This rainfall régime is exactly the same as that of the most typically maritime type of cool-temperate climate (p. 203) and for exactly the same reasons.

The annual amount decreases to the north since the air is colder here and its moisture capacity therefore less; it is mainly the winter precipitation that suffers the decline, as the following figures show:

	Lat.	Total Precipita-tion	Percentage June-Aug.	Percentage Dec.-Feb.
Bergen . . .	60° N.	81	21	29
Christiansund . .	63	47	20	28
Bodó . . .	67	36	22	26
Gjesvar . . .	71	29	23	25

The winter precipitation is, of course, mainly in the form of snow, which lies to a considerable depth and which is perennial above about 7,500 feet in the south and 2,500 feet in the north. As a result of the heavy winter snowfall and the low summer temperatures Norway has the largest ice-fields in Europe (Jostedalsbrae, 300 sq. miles) and glaciers which descend to within 150 feet of sea-level and actually reach the sea in the extreme north.

CONTINENTAL (SIBERIAN) TYPE

Temperature. The marine influence along western margins does not penetrate very far, especially in winter, when the continental high pressures, though acting in opposition to the prevailing westerlies, tend to keep out oceanic winds. Along eastern margins the marine type virtually does not exist, since here the winter continental winds reinforce the planetary winds and are competent to exclude completely the influence of the sea.

The January mean at Bergen is 37°, at Oslo it is 24°, and approximately along the same parallel of latitude are Helsinki (20°), Leningrad (15°), Tobolsk (—3°) and Olekminsk (—31°). To state the same thing in another way, the January isotherm of 20° runs near Vardö (lat. 70°N.), Helsinki (60°N.), Kharkov (50°N.), and Kazalinsk (45°N.).

The cold of the continental interiors is phenomenal, yet even where it is most extreme (Verkhoyansk —59°) life is not impossible for either animal or plant. The plants, and some animals, seek refuge in hibernation, while man and other animals find conditions far from unendurable, since the air, under the influence of the anticyclone, is generally still and excessively dry, so that the body loses little heat by conduction, while the bright sun, shining out of a clear blue sky, warms the skin, though the temperature of the air may be far below zero. Heavy clothing, chiefly furs which the forests of the zone furnish in abundance, must, of course, be worn, and is often not removed throughout the winter; the cold discourages the use of water for washing or shaving, fat and oil being superior substitutes. Personal cleanliness is, in fact, at a discount, but health does not suffer, for the dry cold air is healthy and germ-free.

Occasionally the peace of the anticyclones is broken by the invasion of cyclonic storms, in rear of which are northerly winds which may blow with velocities of 50 or 60 m.p.h. and at temperatures of 20° or 30° below zero, laden with powdery snow or sharp ice particles. Such are the *buran* of Siberia and the *blizzard* of Canada, especially formidable on the open steppe lands where there is no wind-break; from them man must find shelter or perish. When the cyclone has passed and the gale has 'blown itself out' the anticyclone resumes its sway; the temperature generally falls still lower, but is no longer unbearable, as the air is calm again.

Before the I.G.Y. resided throughout the year in Antarctica the lowest temperatures in the world are recorded in this zone, not in the Arctic climates to the north; the mean at Verkhoyansk, 'the pole of cold', is nearly —60°, and —90° has been recorded. Canada, owing to its smaller size, is less extreme, but means of —30° and extremes of —70° occur in the Mackenzie Valley.

The higher angle of the sun in February quickly makes itself felt in a rise of temperature. We have seen that along the western margin February is as cold as, or colder than, January, and this is true also of most stations round the Baltic; but at Helsinki February is 0·5° warmer than January, at Leningrad 1·5°, at Tobolsk 7° and at Olekminsk 13°. Temperature rises rapidly in March and April, but does not reach temperatures of growth before May, when winter suddenly leaps into summer. The ice and snow melt rapidly, the surface of the ground is thawed and converted into an impassable morass, the ice breaks on the rivers and, with the rains that occur at this season, causes widespread floods, especially since the large rivers of the zone, the Yukon, Mackenzie, Ob, Yenisei and Lena, flow from south to north and their upper reaches are set free while the mouths are still ice-bound.

Winter transport is by sledge on the universal highway provided by the snow-covered ground and frozen river, lake and even sea; summer transport on the ground is by road. Between the two is the period known as 'Rasputitsa' in Siberia, when for some weeks flood and mud bring all transport to a standstill. But the rapid rise of temperature as the sun climbs higher soon dries the ground and the melting of the river mouths allows the floods to subside.

The summers are really hot, the June mean at Tobolsk is 59° and July 65·7°—warmer than a London July—while maxima above 90°, very rare at London, are recorded almost every year. Added to these genial temperatures is the benefit of long hours of sunlight, and the effect on vegetation is surprising.

In August the temperature begins to fall markedly, in September night frosts occur and by the middle of October the land is fast in the iron grip of the winter frost.

Rainfall. So effective is the barrier of the western mountains that the continental type hardly anywhere receives more than 30 inches of rain and in most places has less than 20 inches. By further decreasing rain the climate degenerates eastward into steppe and, in Central Asia, actual desert. Cyclonic and relief rain decreases in importance inland and convectional rain increases proportionately. As the following table shows, there is at Helsinki still a trace of marine influence in the autumn maximum of rain, but by Leningrad summer is the wettest season with nearly two-fifths of the yearly total and by Tobolsk the summer share has increased to half the total.

The air of winter is so cold as to be incapable of holding much moisture, and the winter anticyclone supplies conditions unfavourable for precipitation; it will be noticed that the drought of winter becomes more complete as the winter temperatures fall and as the anticyclone becomes more intense.

The winter precipitation is, of course, in the form of snow which in

	Total Rain	Percentage Rainfall occurring in			
		Spring M. A. M.	Summer J. J. A.	Autumn S. O. N.	Winter D. J. F.
Helsinki .	24	19	29	31	21
Leningrad .	19	19	39	27	15
Tobolsk .	19	15	51	23	11

spite of the dry winters falls surprisingly frequently. There are 65 days with snow in the year at Irkutsk, though the six winter months during which mean temperatures are below freezing are credited with a total precipitation of only 3 inches. The snow cover in regions such as this is not always continuous, since occasional strong winds strip off the protecting mantle, piling the powdery snow into drifts and exposing large areas of bare ground which are thus robbed of the moisture which the melting of this snow would have provided for the spring-sown crops. The prevention of this drift by autumn ploughing is one of the practices of dry farming; the snow collects in the furrows and so is conserved.

VEGETATION AND CULTIVATION

Two main types of vegetation characterize the cold climates; where the rainfall is adequate the coniferous forest reigns supreme, where it is deficient grassland takes its place. The sub-arctic forest stretches continuously across Eurasia and North America and abuts on its southern edge against the temperate forest, which differs from it in the greater abundance—in the Old World the predominance—of deciduous types and in the possession of larger species, e.g. red pine, white pine and Douglas fir. But in the interior of continents the great grassland formations (prairie and steppe) replace the temperate forest and penetrate into the sub-arctic zone (see Fig. 30 on p. 87).

The boundary line between grassland and desert, both in temperate and sub-arctic zones, is difficult to demarcate. In a region where the climatic control is so clearly transitional it is left to local conditions to determine the plant associations. Along the river banks tongues of forest, nourished partly by underground water, penetrate far into the grasslands, e.g. along the valley of the Dniester and the Bug; where there is shelter from desiccating winds outlying patches of forest are established or have survived. Doubtless the hand of man, particularly aided by fire, has brought about considerable readjustments, his activities tending in general to drive the forest northwards and replace it by grass, sometimes by design and with a view to cultivation, more often unintentionally.

Strong winds, and especially salt-laden winds, are a factor inhibiting forest growth, for which reason the extreme western margins of the zone (e.g. in Norway), though abundantly watered, have trees only in sheltered places, the open country being bleak moorland.

The Coniferous Forest. These comprise the 'Northern Coniferous Forest' of the 'sub-arctic zone' in Canada and Alaska and the typical 'Taiga' of Siberia. The actual amount of rain that this forest demands is very small, 10 inches being sufficient if, as is usually the case, it is concentrated into the season of active growth. Economically the sub-arctic forest is of less importance than the temperate forest, for the trees are smaller in size, becoming more and more stunted to the north towards the limit of tree-growth; the commercially valuable species of the temperate forests, too, such as the white pine and red pine of eastern Canada and the Douglas fir of British Columbia, do not occur. A few hardy deciduous trees persist, e.g. poplar, larch, birch and willow, some of them extending actually to the limit of tree-growth, but the commercial hardwoods of the temperate forest have been eliminated. Spruce is an important tree which extends as far as the limit of tree-growth, and since it provides good pulp it will probably have an important bearing on the future development of these forests. At present the forest is of little economic importance, trading in little else but furs and skins.

Grasslands. Grassland is often the dominant association even where climatic conditions favour forest, but except in continental interiors it can always be ascribed to inhibitions, e.g. strong winds or fires. It has been claimed that even parts of the prairie and steppe could support forest, but that the trees have been destroyed by fire and bison. Along the forest margin the grass is fairly rich, but in the direction of greater aridity the cover becomes less continuous, open patches of bare soil appear and widen and there is a steady degeneration towards desert. At first the stronghold of pastoral nomadism, these steppes and prairies have been recently brought under cultivation, especially where good communications allow extensive cereal growing, e.g. along the trans-Siberian and the Canadian trans-continental railways. The trend of economic development and the great increase of population during the last century has focused attention on the problem of cereal food supply, and it is particularly in these sub-arctic grasslands that the greatest advances have been made. Several qualities of the climate combine to give them great advantages for cereal growing: the early summer incidence of rainfall, the rapid rise of temperature in May and June, the hot summers with July means of 65°–70° and mean maxima of 80°–90°, the long hours of daylight, the sunny skies of summer, and the dry air of autumn. The aridity, the length and severity of the winter, and the shortness of the summer, despite its warmth, were elements which did not offer great possibilities for wheat cultivation under the conditions that obtained fifty years ago, but the almost yearly improvement in quick-growing and drought-resistant varieties promotes an ever-increasing area to the position of potential wheatland.

REGIONAL TYPES: ALASKA AND CANADA

The marine type of cold climate has only a very small development in North America, namely, along the Alaskan coastal strip from Sitka to Cook Inlet. The rainfall of the Pacific slopes is prodigious (Latouche records 172 inches), but the decrease across the ranges is striking and the interior of Alaska has only 10 inches in places. The mountain barrier is more formidable and complete than in Europe, and this tends to make the transition to the continental type more sudden, but as far as temperature is concerned the effect is offset by the adiabatic warming of the westerly winds in their descent of the eastern slopes. Thus Calgary is 15° warmer than Winnipeg in the same latitude, although it stands nearly 3,000 feet higher. The high plains at the foot of the Rockies, both in Canada and U.S.A., owe their comparatively mild winters to the frequency of the *Chinook* winds, giving spells with temperatures well above freezing. The Chinook typically occurs when pressure is high over the plateaux to the south and when a depression is crossing Canada—a common combination of conditions during winter and spring. The resultant wind is generally south-westerly, but like the foehn wind, is much modified by local topography. Being the result of a passing cyclone it shows no preference for any time of the day or night. Under Chinook conditions heavy rain falls on the western slopes, the liberation of heat by condensation checking the fall in temperature as the air rises; a bank of black cloud appears on the crest of the mountains and the Chinook descends rapidly on to the plains. The rapidity with which the temperature rises is astounding, a change of 30° or 40° in a quarter of an hour is not unusual and in extreme cases the thermometer has been known to rise 30° in three minutes. Although 40° is seldom exceeded, the warmth seems like midsummer by comparison with the anticyclonic cold that has gone before.

The economic results are of the first importance, the dryness and warmth cause the snow (never very deep in these somewhat arid regions) to disappear as if by magic. Thus winter pasture is available almost throughout the year, especially farther south in Montana—an asset of great value in an essentially pastoral region.

Farther east and north the winter climate is more severe, Winnipeg has two months below zero, while Dawson City has four months, a January mean of —24° and extremes as low as —70°. Anticyclonic conditions are more persistent here and in the still, clear air radiation reduces the temperatures to very low figures. But the anticyclone is

neither so intense nor so persistent as in Siberia, and while the temperatures are not so extreme there is greater risk from blizzards and sudden
changes of temperature in which comparatively mild periods are
followed by intense cold and icy winds.

East of Winnipeg the winters are not quite so cold, the moderating
influence of the Great Lakes begins to make itself felt and, farther east,
the effect of the sea. It has been pointed out (p. 213) that the Great
Lakes and the St. Lawrence lowlands constitute the path by which
most of the cyclones leave North America and these cyclones bring
heavy winter snow and rain to the eastern parts of Canada, which has
precipitation all the year round, thus sharing the régime of the eastern
states of U.S.A. (p. 216).

The climate of Newfoundland and Labrador shows evidence of the
cold Labrador current. This has little effect in winter, when Labrador
is actually warmer than the interior of Canada (i.e. the current is actually warming), but it keeps the summers unseasonably cool; the warmest month at St. John's, in the latitude of Paris, does not reach 60°,
while at Nain, in the same latitude as Edinburgh, it is only 47°. The
cold current carries the July isotherm of 50° as far south as Hamilton
Inlet, in the latitude of Dublin, north of this the coast consists of barren
tundra land. Newfoundland thus shows a degree of similarity with
Norway, 15° or 20° farther north, in the following respects:

1. The cool and somewhat cheerless summer.
2. The relatively mild but damp and chilly winter.
3. The delay of minimum temperatures to February and of maxima
to August.
4. The heavy precipitation distributed evenly over the whole year.
5. The frequency of fogs, especially in winter.

SCANDINAVIA

The most noteworthy feature of the Scandinavian climate is the
extraordinary mildness of the coastal zone; Tromsö (70°N.) has a
January mean higher than Bucharest, which is 25° nearer the Equator,
while Vardö, in the extreme north of Norway, has never experienced
temperatures as low as the extremes at Paris and Berlin. The North
Atlantic drift, which is responsible for this phenomenal warmth, permits free navigation up and down the coast and sea-fishing is pursued
throughout the winter months. Inland the temperature falls with great
rapidity, for the continental anticyclone attempts with some degree
of success to establish outflowing winds which restrict the sphere
of marine influence. Along the Norwegian coast south-easterly or
southerly winds are normal and the mixing of these cold winds with the
warm air over the sea causes frequent 'smoke fogs'. These outflowing
winds are frequently replaced by south-westerly winds, often of gale

force, associated with passing depressions; three or four such gales may be expected along the coast during each winter month. February is colder than January, partly because of marine influence, partly because the anticyclone intensifies and so brings to bear a more continuous continental influence.

Rain and snow are frequent; in the north snow falls on about 100 days in the year, but in the south, where precipitation may take the form of rain even in midwinter, on only 30 or 40 days.

Sweden has a much more severe climate; the Gulf of Bothnia is frozen over every year, and sledges can sometimes pass from Finland to Sweden across the frozen Åland Sea and islands; pack-ice is met with in most years north of Öland and Gothland. Haparanda is closed to shipping from November to the end of May, and Stockholm is only kept open by the use of ice-breakers. Winter here is very dry, nowhere in Sweden having as much as 10 inches in the six winter months. This applies still more to late winter and spring when depressions keep almost exclusively to the west coast route, and the Baltic lands are fine, dry and sunny.

In summer the prevailing wind, drawn into the continental low, is westerly and generally light; the fjord inlets often modify the direction of the wind which usually blows up them, especially during the day. Depressions are few and shallow and rainfall along the coast is less than in winter, but inland convectional rain begins to contribute a preponderating share; in fact summer rain exceeds winter rain immediately the mountain divide is crossed.

SIBERIA

In consequence of its great distance from the sea and of its segregation by mountain barriers from the warmer influences to the south, the heart of Siberia presents the extreme case of the continental type. Verkhoyansk (Jan. —59°) has the reputation of being the coldest spot on earth and appears to be colder than the North Pole itself. Yet at the other extreme the July mean is 60° and forests flourish though the summer is only four months long. Semipalatinsk has a July mean of 70°, although the temperature is below 43° for seven months in winter. Winter is extremely dry (3 inches to 4 inches in the six winter months), a deficiency which brings no hardship as it is far too cold for its utilization in any case, and the rainfall, small as it is, is conveniently concentrated into the summer months. In the south, however, even the small amount required is not available and the vegetation degenerates into poor steppe and desert. The better watered parts of the steppe are, however, potentially very productive, being generally blessed in addition with a rich black soil, ideal for the cultivation of cereals (cf. the prairies).

On approaching the east coast the rainfall increases and the summer maximum becomes very well marked; this is the monsoonal sub-type.

TEMPERATURE

Station	Lat.	Long.	Alt. (ft.)	J	F	M	A	M	J	J	A	S	O
DUTCH HARBOUR	54°N.	167°W.	4	32	32	33	35	41	46	51	51	47	42
KODIAK . . .	58°N.	152°W.	6	29	31	34	36	43	50	55	54	50	42
EAGLE . . .	65°N.	141°W.	4	−15	−4	7	27	45	55	59	53	42	25
DAWSON . .	64°N.	139°W.	1,052	−23	−11	4	29	46	57	59	54	42	25
CALGARY . .	51°N.	114°W.	3,389	12	15	25	40	49	56	61	59	51	42
EDMONTON .	53°N.	114°W.	2,158	5	11	24	41	51	57	61	59	50	41
QU'APPELLE .	51°N.	104°W.	2,115	0	2	16	38	50	59	64	62	52	40
WINNIPEG . .	50°N.	97°W.	760	−4	0	15	38	52	62	66	64	54	41
MARQUETTE .	47°N.	87°W.	734	16	16	25	38	49	59	65	63	57	46
TORONTO . .	44°N.	79°W.	379	22	21	30	42	54	64	69	67	60	49
MONTREAL .	46°N.	76°W.	187	13	15	25	41	55	65	69	67	59	47
ST. JOHN'S .	48°N.	53°W.	125	24	23	28	35	43	51	59	60	54	45
VESTMANNO .	63°N.	20°W.	43	35	35	36	40	45	50	53	52	47	42
THORSHAVN .	62°N.	7°W.	84	38	38	37	41	44	49	51	51	48	44
BERGEN . .	60°N.	5°E.	72	34	34	36	42	49	55	58	57	52	45
TRONDHJEM .	63°N.	10°E.	70	26	26	31	39	46	54	57	56	49	41
BODÓ . . .	67°N.	12°E.	7	30	28	30	36	43	50	55	54	48	40
COPENHAGEN .	56°N.	13°E.	16	32	32	35	42	51	59	62	61	55	47
UPSALA . . .	60°N.	18°E.	4	24	23	27	38	49	57	62	59	50	41
KÖNIGSBERG .	55°N.	21°E.	7	27	27	32	42	51	60	63	62	56	46
HAPARANDA .	66°N.	24°E.	8	12	11	18	29	40	53	59	55	46	35
HELSINKI . .	60°N.	25°E.	37	21	20	25	34	46	57	62	60	52	42
LENINGRAD . .	60°N.	30°E.	16	18	18	24	36	48	59	63	60	51	41
MOSCOW . . .	56°N.	37°E.	480	12	15	24	38	53	62	66	63	52	40
ARCHANGEL .	65°N.	41°E.	22	8	9	18	30	41	53	60	56	46	34
KAZAN . . .	56°N.	49°E.	250	7	10	20	38	54	63	68	63	51	39
ORENBURG .	52°N.	55°E.	360	3	6	17	38	58	66	71	67	55	39
BARNAUL . .	53°N.	84°E.	531	0	3	14	34	52	63	68	62	51	35
URUMTSI . .	43°N.	87°E.	2,875	5	9	23	47	61	67	72	70	59	41
HARBIN . . .	46°N.	127°E.	525	−2	5	24	42	56	66	72	69	58	40
YAKUTSK . .	62°N.	130°E.	330	−46	−35	−10	16	41	59	66	60	42	16
VERKHOYANSK .	68°N.	133°E.	328	−58	−48	−24	9	36	56	60	52	36	6
VLADIVOSTOK .	43°N.	132°E.	50	5	12	26	39	48	57	66	69	61	49
OKHOTSK . .	59°N.	143°E.	30	−11	−7	7	21	35	45	55	55	46	27
NEMURO . .	43°N.	146°E.	80	23	22	32	37	44	50	58	63	59	50

RAINFALL

N	D	Yr.	Ra.	J	F	M	A	M	J	J	A	S	O	N	D	Total
35	32	40	19	5.4	7.1	5.6	3.4	5.0	2.7	2.3	3.1	5.8	8.4	6.8	7.2	62.8
35	30	41	26	4.8	4.6	3.9	3.9	5.3	4.7	3.4	5.4	5.6	7.3	5.8	6.4	61.1
2	−11	24	74	0.5	0.4	0.4	0.4	0.8	1.5	1.8	2.0	1.3	0.8	0.5	0.5	10.9
1	−13	23	82	0.8	0.8	0.5	0.7	0.9	1.3	1.6	1.6	1.7	1.3	1.3	1.1	13.6
28	19	38	49	0.5	0.6	0.7	0.8	2.3	2.9	2.6	2.5	1.3	0.7	0.7	0.5	16.1
25	14	37	56	0.9	0.6	0.7	0.8	0.8	3.2	3.5	2.4	1.4	0.7	0.7	0.8	16.5
22	9	35	64	0.8	0.8	1.0	1.1	2.3	3.5	2.8	2.0	1.6	1.1	0.9	0.8	18.7
21	6	35	70	0.9	0.7	1.2	1.4	2.0	3.1	3.1	2.2	2.2	1.4	1.1	0.9	20.2
33	23	41	49	2.2	1.8	2.1	2.3	3.1	3.5	3.1	2.8	3.2	3.0	3.0	2.5	32.6
37	27	45	49	2.8	2.4	2.4	2.3	2.8	2.7	2.7	2.8	2.7	2.6	2.6	2.5	31.4
33	19	42	56	3.7	3.2	3.7	2.4	3.1	3.5	3.8	3.4	3.5	3.3	3.4	3.7	42.7
37	29	41	37	5.4	5.0	4.6	4.3	3.8	3.6	3.8	3.7	3.8	5.4	6.0	5.4	54.6
37	35	42	18	5.8	4.8	4.4	3.8	3.2	3.3	3.1	3.1	5.7	5.8	5.3	5.5	53.8
41	38	43	13	6.7	5.3	4.9	3.7	3.3	2.6	3.2	3.6	4.7	6.1	6.5	6.6	57.2
39	36	45	24	9.0	6.6	6.2	4.3	4.7	4.1	5.7	7.8	9.2	9.3	8.5	8.9	84.3
34	28	41	31	4.3	3.0	3.4	2.5	2.2	1.9	2.8	3.4	4.4	5.0	3.9	3.4	40.2
34	29	40	25	3.3	2.7	2.3	2.0	2.0	2.1	2.6	3.0	4.5	4.0	4.0	3.1	35.5
39	34	46	30	1.5	1.3	1.4	1.4	1.5	2.0	2.4	2.6	2.1	2.2	1.9	1.7	22.0
32	25	41	39	1.3	1.1	1.2	1.2	1.7	2.0	2.7	2.8	2.0	2.1	1.7	1.6	21.4
35	29	44	36	1.7	1.4	1.5	1.5	2.0	2.4	3.4	3.5	3.0	2.4	2.3	2.3	27.4
23	15	33	47	1.5	1.1	1.0	1.0	1.2	1.5	1.8	2.1	2.4	2.2	2.0	1.4	19.2
32	25	40	41	1.8	1.4	1.4	1.4	1.8	1.8	2.2	2.9	2.5	2.6	2.5	2.4	24.7
30	22	39	45	1.0	0.9	0.9	1.0	1.6	2.0	2.5	2.8	2.1	1.8	1.4	1.2	19.3
28	17	39	54	1.1	0.9	1.2	1.5	1.9	2.0	2.8	2.9	2.2	1.4	1.6	1.5	21.0
22	12	33	52	0.9	0.7	0.8	0.7	1.2	1.8	2.4	2.4	2.2	1.6	1.2	0.9	16.8
25	11	37	61	0.5	0.4	0.6	0.9	1.6	2.2	2.4	2.4	1.6	1.1	1.0	0.7	15.4
24	11	38	68	1.1	0.8	1.0	0 9	1.4	2.0	1.7	1.3	1.3	1.2	1.2	1.2	15.2
17	6	33	68	0.8	0.6	0.6	0.6	1.3	1.7	2.2	1.8	1.1	1.3	1.1	1.1	14.2
22	10	40	67	0.5	0.3	0.5	1.5	1.1	1.5	0.7	1.0	0.6	1.6	1.6	0.4	11.3
21	3	38	74	0.1	0.2	0.4	0.9	1.7	3.8	4.5	4.1	1.8	1.3	0.3	0.2	19.3
−21	−41	12	112	0.9	0.2	0.4	0.6	1.1	2.1	1.7	2.6	1.2	1.4	0.6	0.9	13.7
−34	−51	3	118	0.2	0.1	0.1	0.2	0.3	0.9	1.0	1.0	0.5	0.4	0.3	0.1	5.0
30	14	40	64	0.1	0.2	0.3	1.2	1.3	1.5	2.2	3.5	2.4	1.6	0.5	0.2	14.7
6	−8	22	66	0.1	0.1	0.1	0.2	0.5	1.1	0.5	1.8	2.1	0.7	0.2	0.2	7.5
39	29	42	40	1.3	1.0	2.2	2.9	3.7	3.7	3.8	4.3	5.6	3.8	3.3	2.3	37.9

MONSOONAL SUB-TYPE

In winter the monsoonal variety differs little from the continental in temperature and rainfall, for the prevailing influence is continental, but in summer the differences are considerable, for the prevailing influence is now marine. In winter Siberian cold extends to the shores of the Pacific where some of the lowest mean temperatures for their latitude in the world are recorded, but in summer the south-east monsoon tempers the heat, while in Siberia the thermometer rises unchecked; the July isotherm of 70° passes near Vladivostok in latitude 43°N. and then swings due north to Olekminsk in 60°N. Again, Manchuria is almost as dry in winter as Siberia, but its summer rainfall is more than double the Siberian. The essence of the Manchurian climate is, then, the alternation of a continental winter and a maritime summer and especially the great contrast in humidity between the seasons.

Winter. The Manchurian winter is very severe, the rivers are frozen for five or six months, snow lies everywhere, sledges are the regular means of transport, skins and quilted clothing are worn. Everywhere north of Vladivostok and the great bend of the Sungari the January mean is below zero and temperatures of —40° occur in northern Manchuria. To this bitter cold must be added the discomfort of the strong wind, often of gale force, of the winter monsoon.

Summer. In April comes the thaw, the Siberian high is yielding, and by May warm winds are coming in from the sea, bringing the first of the summer rains. Crops, wheat or barley, are sown in early April as soon as the frost is out of the surface of the ground, and the further thaw provides water for the germinating seed. Temperature rises rapidly and, aided by the rains, the plant makes rapid growth; by June the earliest of the cereals are ripened and a second crop (buckwheat) may be snatched before October frosts put an end to the growing season.

The July isotherm of 70° encloses Manchuria, though since much of the land is high the actual mean temperatures do not reach this figure. The monsoon rains last from May to September, reaching a maximum in July, the amount decreasing from 40 inches near the coast to about 20 inches or less at the scarp of the Great Khingan, beyond which the influence of the monsoon is scarcely felt.

SUGGESTIONS FOR FURTHER READING

Many of the references quoted at the end of the last chapter deal also with these climates. See also *Atlas de Climat de Norvège*, nouvelle édition par A. Graarud et K. Ingens, Geofis. Pub. II, 7, Kristiania, 1922; 'Le Climat de la Sibérie Orientale', A. Woeikof, *An. de Géog.*, 1897; E. M. Filton, 'The Climate of Alaska', *M.W.R.*, 1930. For conditions of erosion and deposition under arid conditions see C. A. Cotton, *Climatic Accidents*, 1942.

XIII

ARCTIC CLIMATES

The isopleth of 3 months with 43°, which has been adopted as the lower limit of these climates, probably follows a regular course in the southern hemisphere, encircling the globe at about 55° south latitude, but in the northern hemisphere its course is much less regular. Representing as it does a summer value, it would be expected to extend polewards over the land and equatorwards over the sea; actually it reaches well into the Arctic Circle in Alaska and Siberia, but is carried by the cold currents down the coast of Labrador almost as far south as Belle Isle Strait and through the Bering Strait as far south as the Aleutian Islands.

The climates included within this zone might be divided by the 43° isotherm for the warmest month into two types, namely:

1. *Tundra climates* with a summer, however short, during which the ground is free from snow for a sufficient period to allow the growth of the typical tundra vegetation.

2. *Perpetual frost climates* in which, even at midsummer, there is insufficient warmth to melt the snow and to permit the growth of even the hardiest plants.

The former are habitable and though scantily populated supply some interesting examples of climatic control of occupation and habits; the latter can never be inhabited and are of little geographical interest, though they present many interesting meteorological problems and are, of course, of vital importance in polar exploration. Little detail was known of the climates of these regions until recently, but the increasing importance of long-range flying, at present strategic but later perhaps commercial, has stimulated an interest in these lands. Many new meteorological stations have been established and many scientific expeditions fitted out, some of them wintering in the Arctic, and the experience of the International Geophysical Year, with winter residence at the South Pole, has wiped out Verkhoyansk's record of cold; $-126.9°$F. is the lowest temperature recorded—on 24th of August, 1960, at Vostok, the Russian base near the Pole of Inaccessibility.

Pressure and Winds. The poleward decrease of pressure from the 'horse latitude' highs is not continuous as far as the poles, but appears to reach a minimum at about 60°S. in the southern ocean and along a more irregular and sinuous line in the north polar regions. This girdle of lowest pressure represents the most frequented track of the cyclonic storm centres which, it has been suggested (p. 72), occur along the polar front, or line of convergence of the polar and westerly winds. Beyond this low-pressure trough pressure increases again to reach a maximum which in the Antarctic is probably in the neighbourhood of the pole, but which in the less regular Arctic appears to lie between Alaska and Greenland. Out-blowing winds from these high-pressure centres take on, as a result of deflection by the earth's rotation, an easterly direction which has been found to be prevalent wherever the polar anticyclone is well established; their direction is more regular in the Antarctic than in the Arctic. The Greenland ice-cap produces its own anticyclonic circulation which is only rarely invaded by winds from without. Nearly 80 per cent of the winds on the western side blow from the east or south-east, and nearly 70 per cent of the winds on the eastern side blow from the north or north-west.

The anticyclones are most likely the direct results of the chilling of the lower air layers by the polar cold and there is a great deal of evidence to show that their extent is affected from year to year by the extension of the pack-ice, the limits of the anticyclone being more or less coincident with the limits of ice. They are comparatively shallow phenomena and are replaced above by a general cyclonic circulation with westerly winds, as is shown by the drift of high clouds and of the smoke of Mt. Erebus. From these anticyclonic centres the surface winds gravitate outwards, aided in some cases, and especially in Antarctica, by the creep of cold, heavy air down the slopes of the ice-cap.

The downward creep is sometimes converted into a downward rush of air whenever the pressure gradient is increased, as, for example, by the passage of a deep depression along the edge of the polar anticyclone. Then the air may be strongly warmed by compression and arrives at the coast as a wind of foehn character; such winds are met in Greenland blowing down the fjords, especially of the east coast, and also along the coast of Antarctica. In general the air movement in anticyclonic areas is characteristically light and the polar anticyclones are no exceptions, calms being frequent, but on the other hand severe blizzards, sometimes with wind velocities of 150 m.p.h., are distressingly frequent, especially in certain localities. It is a curious fact that while at certain stations blizzards are comparatively rare and innocuous, at others (e.g. Adelie Land and Cape Evans) they are numerous and very severe. During Sir Douglas Mawson's expedition of 1912 the wind velocity in Adelie Land averaged 50 m.p.h. for the whole year, 85 m.p.h. for 24 hours and 107 m.p.h. for 8 hours. Adelie Land has well earned the name 'the home of the blizzard', but it is an exceptional station; at Amundsen's base at Framheim the average wind velocity from April 1911 to January 1912 was only 10 m.p.h. The cause of these contrasts is apparently to be found in the local configuration, especially in the existence of large areas of high ground and of steep slopes which add a large gravitational force to the pressure gradient. The winds of gale force which accompany the blizzard break down inversions of temperature and by causing the mixing of air layers usually bring about a considerable rise in temperature, further increased by foehn effects. Physiologically, however, the still, anticyclonic weather that precedes is much to be preferred in spite of the intense cold.

Cyclonic storms are unable to invade the polar high pressures and are concentrated round the margins, especially in the southern ocean where there is a constant procession of depressions which make these latitudes the stormiest on earth. But in the north polar regions the depressions travel far into the Arctic Circle along the gulf of warmth and associated low pressure of the Norwegian Sea; Spitzbergen experiences as many westerly as easterly winds, for the depressions pass chiefly to the north of the islands.

Temperature. The inequality of the length of day and night reaches its maximum at the poles, where there is a six-month day and a six-month night, while everywhere within the polar circles enjoys at least one day on which the sun does not set. This mathematical factor introduces a new conception in climates, for diurnal range has now little meaning, since insolation is absent in midwinter and continuous at midsummer, however low the angle of the sun. The annual march of temperature, too, is profoundly altered, for temperature represents the balance of heat received over heat radiated away; elsewhere this generally reaches a minimum in January, after which the higher angle of the sun and the greater length of day cause the temperature to rise, but at the North Pole the sun does not rise until the spring equinox and loss by radiation goes on continuously up to this point, where the minimum is reached. In lower latitudes where the sun rises earlier, the minimum is reached earlier, but everywhere in the zone the rise of temperature in early spring is slow. The sun is low in the sky and its feeble rays, largely reflected from the white surface, have little power to melt the snow. Most of the heat available is used up in this process, for the specific heat and latent heat of snow are high, and little is available to raise the air temperature.

The ground, deeply frozen in the long cold winter, thaws only to a depth of a few inches in the short cool summer. The result is a surface layer of mud, saturated with ice-cold water, overlying an impervious hard-frozen subsoil (permafrost). In the tundra zone of the northern hemisphere the temperature does not rise above freezing-point until June, the mean of the hottest month does not reach 50°, and by September the thermometer is below freezing again and winter has taken hold.

Although air temperature is low in summer the sun's rays are not without power; where they shine on solid objects they may raise their temperature to 60° or more and they give a feeling of warmth and comfort to the body; the black-bulb temperatures, in fact, may exceed 100°, though the air temperature is below freezing. The average maximum temperature reaches about 60° and occasionally exceeds 70°, and thanks to the anticyclonic conditions and the dryness of the air the sky is generally clear and a high percentage of the possible hours of sunshine is attained.

In the warmth of the sun and aided by the long hours of daylight plant-growth is fairly vigorous, the most favourable conditions being realized where the slope of the ground is such that the rays of the sun meet the surface at a high angle, for these conditions give both warmth and good drainage. Plants, especially flowering plants, thrive in such spots with a success that is surprising.

Autumn is warmer than spring, for the snow cover which delayed the advent of the latter is no longer operative, or at least not so continuous. September, especially in marine climates, is often as warm as June (e.g. Jan Mayen, June, 36·9°; Sept., 37·4°). The winter temperature depends partly on the distribution of land and water; Jan Mayen feels the influence of the North Atlantic drift and has a minimum (in March) only 8° below freezing, Spitzbergen (in February) of −2·4°. More continental stations have lower minima, e.g. Sagastyr, on the Siberian coast, −36°. At oceanic stations the annual range of temperature is considerably diminished (cf. Sagastyr 70° and Jan Mayen 11°). That the winter temperatures are but little below freezing-point is, however, of small account, for they are too low for plant-growth in any case; it is the low mean temperature of summer and the absence of really warm days that is the serious drawback. The mean temperature of the warmest month at Spitzbergen does not reach 42° and the highest temperature recorded is 58°; actually, then, these climates, in spite of their mild winters, are bleak and inhospitable in the extreme.

The permanently frozen seas behave more or less as continental areas, and have a larger though not extreme range, for sea ice cannot become very cold since the temperature of the water below is about 29° and ice is a good conductor of heat. Absolute minima are very low, most stations having recorded 40° or 50° below zero, but only in the extreme continental conditions of Antarctica and perhaps on the Greenland ice-cap do temperatures rival that of Verkhoyansk. Winter is generally calm and the cold is by no means unbearable; but the occasional blizzard is fatal unless shelter can be reached, and the darkness, added to the numbing cold, plays havoc with health and spirits.

Precipitation. Though rain sometimes occurs in the Arctic summer, the usual form of precipitation is snow. The actual amount which falls is difficult to measure because of drifting, especially since much of the snowfall is connected with blizzards. In fact it is often impossible to tell whether snow is falling at all, for the wind drives along a flurry of snow and small crystals of ice which may be falling from the clouds or may be swept up from the ground. What is quite clear, however, is that the amount of precipitation is very slight, probably 10 inches or 12 inches in most places. This is quite in accordance with expectations; in the first place the anticyclonic conditions are unfavourable for precipitation, the air being dried by its descent and further dried by the adiabatic warming due to gravitational descent from the high plateaux (especially in Greenland and Antarctica); in the second place the air, being cold, can hold but little moisture, even when saturated; and in the third place there is a general absence of thunderstorms and other convectional effects which are such a fruitful source of precipitation in other zones.

Under the circumstances it is at first sight surprising to find a permanent snow cover and such huge glaciers and ice-caps as exist in Antarctica and Greenland; but if the accumulation is slow the dispersal is even slower, and the dimensions of the glaciers and ice-caps are to be ascribed mainly to the slow rate of melting. The air temperature is never much above freezing-point—in Antarctica hardly ever—and snow disappears by sublimation rather than melting. In any case the cold air has little capacity to take up water vapour. 'Absorption thawing' occurs round solid objects on the ice, e.g. morainic debris, but is not an important agent in disposing of snow and ice on a large scale.

It is likely that high plateau ice-caps recruit their moisture from the upper layers of air into which they rise; the cyclonic westerly winds of this upper circulation (see p. 242) must import considerable moisture which could be precipitated as hoar frost by contact with the cold mountain-tops and ice-caps. In this connection it is interesting to note that neither the summit of the Greenland ice-cap of today, nor the summit of the Scandinavian ice-cap of the Pleistocene coincides with the highest point of the underlying land.

Twelve inches is only exceeded where cyclonic storms invade the polar zones, e.g. Jan Mayen (14 inches), the east coast of Greenland (Angmagsalik, 36 inches), Kerguelen (33 inches), South Georgia (35 inches). Cyclonic winds sometimes break through the Greenland anti-cyclone, blowing from coast to coast, and bring heavy falls of snow.

In regions such as these the maximum fall is in winter when cyclones are deepest and most frequent, but in regions influenced by the anti-cyclones alone there is a tendency to a maximum in summer when the air carries most moisture.

Fog occurs frequently, both in the Arctic and Antarctic, especially where warm water meets cold water or cold land, e.g. Labrador and the South Orkneys.

THE ARCTIC AND ANTARCTIC COMPARED

Whether from coincidence or for some fundamental reason connected with the structure of the earth, the Arctic is an almost land-locked sea, while the Antarctic is a sea-girt land. From this there result some important and interesting climatic differences. The seas which surround Antarctica, being continuous, are uniform in temperature, in which respect they differ strongly from the independent oceans and semi-independent seas which surround the north polar regions. As a result of its greater regularity, of its uniform enclosure by sea, and of the uniformity of that sea, the Antarctic shows a greater uniformity of climate (see Fig. 72).

Through the only considerable breach in the land girdle of the Arctic Ocean, marine influence, strengthened by the warmth of the North Atlantic drift, penetrates deeply, allowing the zero isotherm of

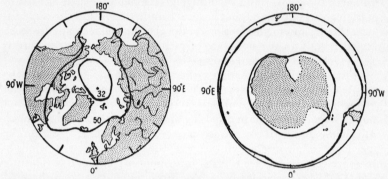

Fig. 72.—Approximate Positions of the Summer Isotherms of 32° and 50° in the Polar Regions

January to spread to the tip of Spitzbergen in latitude 77°N. Antarctica, on the other hand, is strongly held by the polar anticyclone and warm influences are unable to penetrate; nowhere does the mean of the warmest month rise above freezing-point and only in favourable sites on the tip of Grahamland have flowering plants been found. In the north polar regions, on the contrary, with the exception of the Greenland ice-cap and some of the islands, the land has everywhere summer temperatures adequate for lowly plant-growth and the polar limit of the tundra climates is set, not by cold, but by the sea. Thus the mainland of Antarctica belongs almost entirely to the 'perpetual frost climates' and the 'tundra climates' are virtually confined to the northern hemisphere.

LIFE IN ARCTIC CLIMATES

Plants. The tundra vegetation has to contend with extremely adverse circumstances and the number of species that can endure the conditions is small. The rhythm of the seasons compels a strongly rhythmic habit in the plant, but the rhythm is very unevenly balanced; a long phase of inactivity is followed by a short season in which life must be lived at high pressure. The growing season is short, only two or three months perhaps, and the plant must pass through its life cycle with all possible speed, in which it is aided by the more or less continuous

daylight. The seeding habit is generally discarded in favour of vegetative reproduction, the fruits being often barren; the plant grows
actively, almost feverishly, until overtaken by the winter cold, and when
it is nipped by the winter frosts it bears the last of its flowers and fruit
still unready on its branches.

The second disadvantage with which tundra vegetation has to contend is the physiological drought. Though its roots may be in a waterlogged soil the plant may suffer from drought, for the soil-water may
be ice-cold, the soil may be sour and acid with decomposing vegetation,
while drying wind and strong sunshine may be demanding rapid
transpiration from its leaves. Against this risk of drought tundra plants
employ all the usual adaptations, especially an extended root system
(but not extended downwards because the lower layers of soil are frozen)
and a dwarfed sub-aerial growth (the cushion habit being characteristic).

The drainage is generally bad, for the soil is permanently frozen
below a shallow surface layer, there is surface water for only a few
months and under these conditions erosion is slow and ineffective; the
normal geographical cycle is thus scarcely operative. The chief agents
in fashioning relief are landslip and the downhill creep of half-thawed
ground on a frozen subsoil. Pools and marshes of stagnant water lie
throughout the summer, waterlogged areas growing only sphagnum,
lichen and sedges. Where drainage is better the tundra is a dreary
stretch of lichen and coarse grass with occasional hummocks of a
brighter green where the home of an arctic fox or snowy owl enriches
the soil with its refuse. Dwarf willows, birches and alders occur in
sheltered hollows which are better drained and protected from the
wind; they seldom grow more than two feet high, but their shoots and
branches, rich in protein, provide nourishing food for herbivorous
creatures. But it is on the sun-warmed southern slopes, when the sun
melts early and the ice-cold water drains away from the roots, that the
flower carpet of the tundra brightens an otherwise monotonous landscape. These are veritable garden plots bright with the blossom of
campion, rock rose, monkshood, purple saxifrage, Iceland poppy,
forget-me-not, thrift, yarrow, geum, willow-herb and numerous other
flowers with most of which we are familiar in our own latitudes.

Animals. The rhythm of the seasons impresses itself on the animal
life as on the plants; most of the birds leave at the end of summer for a
warmer land, the reindeer retreats to the margin of the taiga, the arctic
wolf follows, and other animals make long journeys in search of winter
food supply. Hibernation, a common escape from winter scarcity, is
not practised, the winter is too long and the summer too short to build
up the necessary reserves of fat or food supply. A surprisingly large
number of animals, including the musk-ox, arctic hare and lemming,
brave the rigour of winter, seeking their food in the vegetation (e.g.
reindeer moss), which is buried under only a thin mantle of snow, since

TEMPERATURE

Station	Lat.	Long.	Alt. (ft.)	J	F	M	A	M	J	J	A	S	O
BARROW POINT	71°N.	156°W.	20	−19	−13	−14	−2	21	35	40	39	31	16
HEBRON .	58°N.	63°W.	49	−6	−5	6	18	32	40	47	48	41	31
RAMAH .	59°N.	63°W.	16	−4	−9	3	19	36	44	48	46	42	33
UPERNIVIK	73°N.	56°W.	65	−8	−9	−6	25	35	41	41	34	34	25
ANGMAG-SALIK	66°N.	38°W.	104	17	13	17	24	33	41	44	42	38	30
GRIMSEY .	67°N.	18°W.	4	29	27	27	30	36	43	46	46	43	37
JAN MAYEN	71°N.	8°W.	76	27	26	24	28	31	37	42	43	37	31
BEAR IS. .	74°N.	20°E.	131	15	12	12	18	29	35	40	39	35	29
SPITZ-BERGEN	78°N.	14°E.	37	4	−2	−2	8	23	35	42	40	32	22
VARDO .	70°N.	31°E.	33	22	21	24	30	35	42	48	48	43	35
MALYE KARMAK'Y	72°N.	53°E.	49	2	3	5	14	25	35	44	43	35	26
SAGASTYR .	73°N.	124°E.	11	−34	−36	−30	−7	15	32	41	38	33	6
EVANGEL-ISTAS IS.	52°S.	75°W.	180	48	48	48	47	42	40	39	40	41	42
ANO NUEVO	55°S.	64°W.	174	46	46	45	42	39	37	35	37	39	41
MACQUARIE ISLAND	55°S.	159°E.	13	43	43	42	41	39	37	37	38	39	40
SOUTH ORKNEYS	61°S.	45°W.	23	32	33	31	27	19	15	13	15	20	25
SOUTH GEORGIA	54°S.	36°W.	13	41	42	40	36	32	29	29	29	33	35
MCMURDO SOUND	78°S.	167°E.	Coast	24	16	4	−9	−11	−12	−15	−15	−12	−2
LITTLE AMERICA	78°S.	161°W.	33	19	6	−8	−21	−24	−38	−39	−38	−18	−14

RAINFALL

N	D	Yr.	Ra.	J	F	M	A	M	J	J	A	S	O	N	D	Total
−7	−15	10	59	0·3	0·2	0·2	0·3	0·3	0·3	1·1	0·8	0·5	0·8	0·4	0·4	5·6
20	4	23	54	1·0	0·7	0·9	1·1	1·6	2·2	2·8	2·8	3·4	1·6	1·1	0·6	19·8
25	13	25	57	0·8	1·0	2·5	4·3	1·1	2·4	3·5	1·7	2·0	4·1	5·4	4·2	33·0
14	1	17	49	0·4	0·5	0·7	0·6	0·6	0·5	0·9	0·1	1·1	1·1	1·1	0·5	9·1
23	19	28	29	3·5	1·7	2·2	2·4	2·8	2·1	2·1	2·5	4·0	6·3	3·4	2·7	35·7
32	30	36	17	0·6	0·7	0·6	0·5	0·6	0·7	1·1	1·3	1·3	1·4	1·1	0·7	10·6
28	25	32	15	1·6	1·5	1·2	0·9	0·5	0·6	0·9	0·9	1·9	1·7	1·3	1·2	14·2
21	18	25	28	1·1	1·4	1·2	0·7	0·8	0·8	0·7	0·9	1·6	1·4	1·0	1·2	12·8
11	6	18	44	1·4	1·3	1·1	0·9	0·5	0·4	0·6	0·9	1·0	1·2	1·0	1·5	11·8
28	24	33	27	2·7	2·6	2·1	1·6	1·4	1·5	1·8	2·0	2·4	2·5	2·5	2·6	25·7
13	6	21	42	0·3	0·2	0·3	0·3	0·5	0·6	1·2	1·7	1·6	1·2	0·5	0·4	8·8
−16	−28	1	77	0·1	0·1	—	—	0·2	0·4	0·3	1·4	0·4	0·1	0·1	0·2	3·3
43	46	43	9	12·0	9·7	11·3	11·3	9·4	9·2	9·6	8·6	9·3	9·9	9·9	10·0	119·4
42	44	41	11	2·6	2·5	2·7	2·9	2·8	1·6	1·6	1·5	1·3	1·3	1·8	2·2	24·0
41	42	40	6	5·0	3·5	4·5	4·6	4·1	4·2	3·9	3·8	4·0	4·2	3·4	3·6	48·8
28	31	24	20	1·5	1·5	1·8	1·7	1·3	1·2	1·2	1·4	1·0	1·0	1·4	0·9	15·9
37	39	35	13	3·3	4·1	5·1	5·3	5·5	5·0	5·5	5·1	3·4	2·6	3·4	2·9	51·2
14	25	1	40	—	—	—	—	—	—	—	—	—	—	—	—	—
1	19	−13	58	—	—	—	—	—	—	—	—	—	—	—	—	—

precipitation is everywhere light. The sea is a great source of food, and many animals, e.g. polar bear, derive their winter food supply from this source; even the reindeer is said to make shift with seaweed when nothing else is to be had. But the winter is hard and when the polar night is nearing its end, when food is scarcest and the cold is greatest, they are reduced to a very poor condition and numbers perish.

Insects, like plants, begin their life with the advent of summer, mosquitoes, especially, appearing in swarms to the sore discomfort of man and beast. The reindeer suffers most, for his antlers, at this season in velvet, are a prey to the insect swarms. Because of this the Samoyedes take their herds away from the swamps to higher and drier ground.

Man. Although aircraft, snowcats, weasels and tractors have been very helpful to explorers and scientists the life of resident man is always a struggle against nature; by necessity a food-collector he is often a fisher as well as a hunter, for the summer is too short to store up food against the long winter. Fishing, in fact, plays such an important part in his life that settlement in the higher latitudes is almost restricted to the coast. The seasonal change of habits necessitates a nomadic existence, the summer home is the portable tent of skins, the winter home a more permanent structure of earth or snow. It should be remembered that constructional materials, especially wood, are scarce and the native must utilize what is to hand, hence the 'igloo' of the Eskimo. In summer the rivers and the sea are the chief media of transport (in canoes), for much of the land is swamp; in winter the sledge drawn by dogs or reindeer, for the frost converts river, land and even sea into a uniform medium.

The physical hardships and risks that man is called upon to undergo result in a high death-rate from accident and exposure; frostbite and snow-blindness are maladies directly attributable to climate, but otherwise the polar regions are healthy. Cold is a preservative and putrefactive organisms are inactive at low temperatures, germ-borne diseases are practically negligible risks. Scurvy, once the scourge of the lands of winter night, is due to a lack of the vitamins that fresh food contains, and can be avoided by careful attention to diet.

SUGGESTIONS FOR FURTHER READING

Articles on Arctic climates will be found in most books on the polar regions, such as Rudmose Brown, *The Polar Regions*, 1927; O. Nordenskjold, *Geography of the Polar Regions*, 1928; *Problems of Polar Research*, 1928, published by the American Geographical Society and containing contributions on climate and meteorology by H. H. Clayton and Griffith Taylor; W. H. Hobbs, *Characteristics of Existing Glaciers*, 1911, etc. See also *British Antarctic Expedition*, 1910–13. *Meteorology*, vol. 1, Discussion by G. C. Simpson, Calcutta, 1919. Simpson, 'Antarctic Meteorology', *Geog. Journ.*, 1929. H. H. Lamb, 'Differences in the Meteorology of Northern and Southern Hemispheres', *Met. Mag.*, 1958. Reports of the I.G.Y.

XIV

DESERT CLIMATES

The sole criterion of the desert climate is aridity, but aridity in its turn produces a number of secondary characteristics such as sunshine, temperature range and even relief and soil types which help to conjoin deserts into a single climatic group wherever and for whatever reason they occur.

Meaning of Aridity. Now aridity, while mainly a matter of rainfall, may be qualified by a number of other circumstances which may serve to mitigate or increase it. For example, the tundra lands have seldom more than 10 or 12 inches of rain, yet far from being dry, the soil in summer is waterlogged because run-off is slow (there is none at all for eight months) and there is no loss by downward percolation, the subsoil being frozen and impervious. Again, parts of Western Australia with little more than 10 inches of rain grow good crops of wheat, for the scanty rain falls conveniently at the season when the plant most needs it and when evaporation is least, and furthermore the rainfall possesses a high degree of reliability. Along the banks of water-courses or where underground streams exist vegetation (oases) may thrive with no rain at all. Similarly an impervious layer below the soil will conserve soil moisture, giving the plant a degree of independence of rain—a condition artificially reproduced in dry-farming areas by deep ploughing and the formation of a hard 'plough-sole' at a depth of 9 inches or so.

On the other hand there are climates with 20 inches of rain which falls so extravagantly in thundery showers that most of it is lost by run-off on the hard-baked soil and by evaporation in the hot, dry air and bright sunshine which follow. There are thirsty, porous soils which can absorb almost any amount of rain and have none to give to the growing plant. Numerous other examples might be quoted, but enough has been said to emphasize the complexity of factors at work in determining effective aridity. The real significance of the desert is its unproductiveness and its consequent inability to support settled self-dependent communities. The best test of the desert is, therefore, its vegetation which is clearly not always determined by precipitation alone; but rainfall is the most significant single element in the climate and deserts may be expected wherever the rainfall is less than 15 inches and generally presumed where it falls below 10 inches. A discussion on the limits of desert climates is given on pp. 85 and 86.

The Semi-arid Transition Climates. The tropical climate degenerates polewards into desert beyond the limits of the migration of the equatorial rainfall belt, but retains its summer maximum of rain to the last; the Mediterranean climate degenerates into desert equatorwards beyond the limit of the migration of the stormy westerlies, but retains its winter maximum as far as the rains occur; the trade-wind coasts degenerate into desert westwards beyond the reach of the rains which the trade winds bring; the cool-temperate climates degenerate into desert eastwards as the westerlies lose their moisture; the Mongolian desert lies beyond the reach of the monsoon rains; the Great Basin is robbed of its rains by the mountains which encircle it. In almost every type of climate we may trace a progressive degeneration into desert, but in each case the other climatic characteristics are retained to the last, the essential change being the progressive decrease of rain and the secondary characteristics which this begets.

These transitional semi-arid areas of degenerate climate comprise the scrubland marginal to the great hot deserts of the tropics and the steppes marginal to the great cold deserts of extra-tropical latitudes. Here dwell the nomadic pastoral folk, an unsettled class living under conditions of climatic insecurity, subject to fluctuations of prosperity, and prone, because free and unattached, to great migrations, often to the discomfiture of their neighbours. Under modern economic conditions they may become great agricultural lands, growing millets and cotton in the tropics and cereals, especially wheat, in extra-tropical latitudes. They have been described in the appropriate chapters in connection with the climatic type whose decay they represent.

The Causes of Aridity. The main cause of aridity is distance from marine influence, hence continental interiors are especially dry. But in evaluating distance from marine influence we must take into consideration the prevailing winds, their direction and their constancy; thus the deserts extend down to the western shores in the trade-wind belt, but in the belt of the less constant westerlies they do not extend to the eastern shores. The one exception to this is Patagonia and in this last case, as in the Great Basin, in Peru and in Australia, the degeneration is accelerated by the interposition of a mountain range.

The second great cause of aridity is directly connected with the planetary circulation. High atmospheric pressure is a condition unfavourable for precipitation; regions of seasonal high pressure suffer from seasonal drought (e.g. the winter drought of continental interiors), the permanent areas of high pressure will normally be areas of permanent drought. But it must not be supposed that the converse is true and that deserts are necessarily always areas of high pressure; in fact it often happens that the desert gives rise by excessive convection to intense low pressures, as, for example, in Sind in July and in Northern Australia in January (see pp. 152 and 161).

The greatest deserts in the world occur beneath the 'horse latitude' highs: the Sahara, Arabia, Australia, Kalahari, Atacama. These are the great trade-wind deserts, covering an enormous space of the earth's surface and doomed to backwardness on account of their aridity. Because of the difficulties of crossing them they present one of the most significant barriers of race, creed and civilization to be found on the earth. Because of their unproductiveness and the harshness of the environment they must be inhabited by nomadic peoples of great independence and strength of character, self-reliant, proud and lawless. The desert is the stronghold of Mohammedanism, the simple, rigid and severe creed of a monotonous and stern environment.

Types of Desert. Deserts in all latitudes are united by their aridity, but they must be divided by their temperature. As far as summer temperature is concerned there is little difference between them despite differences of latitude. Luktchun (July 90°) is as hot as most places in the Sahara though 20° farther north, and the highest figure recorded here (118°) compares not unfavourably with Saharan stations. Some of the highest temperatures in the world are recorded in Death Valley (Cal.) in 36°N., and therefore well outside the tropics. But deserts in high latitudes are distinct in the possession of a well-marked winter season which is sometimes very severe; four months at Luktchun, for example, have means below freezing. We may therefore divide deserts into two types (see p. 98):

1. Hot deserts with no cold season.
2. Cold deserts, with a cold season, i.e. with one or more months below 43°.

The annual temperature range in the former is less than 50°, in the latter, except in Patagonia, it generally exceeds that amount. Such a large range is a feature generally associated with extreme continentality, and in fact the cold deserts, with the exception mentioned, are all excessively continental; it is to this that they owe their aridity.

HOT DESERTS

To all intents and purposes the hot deserts are the trade-wind deserts, by far the most extensive being the Sahara, which, with its direct continuation the Arabian desert, covers in all an area of more than three million square miles. Australia and North Africa are unfortunate in having their greatest width in these arid latitudes, but North America is favoured in narrowing rapidly south of 30°N., where the Gulf of Mexico actually functions as a source of supply of humidity to the surrounding lands. South Africa and Australia are unfortunate in presenting their steepest slope to the east and consequently suffering from a long, arid western slope; South America is better arranged, the profile being reversed, and enjoys a long, well-watered eastern slope, its tropical desert being confined to a comparatively narrow strip between mountain and coast in Northern Chile and Peru.

Wherever the trade-wind deserts come down to the western seaboard the coast is washed by a cold current—the Equator-seeking
return current on the east side of each of the oceanic anticyclonic
swirls, reinforced (see p. 47) by the upwelling of cold water from below
as the off-shore winds skim off the surface water. The chilling influence
of these currents has a profound effect on the climate of a somewhat
narrow coastal strip. Here is found a peculiar marine variety of the
desert climate, with cool summers, a marked diminution in temperature range, both annual and diurnal, a greater humidity and more
mist and cloud. It is best seen in the Atacama and Kalahari deserts,
but it may also be clearly recognized in Southern California, Rio de
Oro and perhaps Western Australia. In contrast with this the normal
hot desert is typically continental, with dry air and a huge diurnal
range of temperature and a considerable annual range.

THE MARINE TYPE OF HOT DESERT (PERUVIAN)

Pressure and Winds. It is perhaps surprising, at first sight, that
the cold currents along the coast should exercise so powerful an influence
over regions which, being in the trade-wind belt, should receive their
prevailing wind off the land to the east. On closer inspection, however,
it transpires that the local wind along the coast is an off-sea wind;
thus at Iquique south-west winds make up nearly 80 per cent of the
total, at Walvis Bay 50 per cent, in South California westerly or north-
westerly winds prevail throughout the year, and in Rio de Oro
northerly winds are the most frequent. To some extent these winds are
integral parts of the circulation round the oceanic anticyclones, but
they are more properly local diversions of the planetary winds, due,
in large measure, to the attraction of the heated continental interior
and therefore to be classified as sea-breezes or even monsoons. The
attraction is naturally greatest in summer and particularly during the
heat of the afternoon; consequently it is during summer that marine
influence is strongest, which accounts for the remarkably low summer
temperatures. During winter when pressure is higher over the continental interiors off-shore winds are more frequent; in South-west Africa
these winds, descending from the plateau, have a foehn character so
that, curiously enough, the highest temperatures are recorded during
winter (cf. p. 177, Berg Winds). There are somewhat similar winds
in Chile and California (see pp. 176 and 174).

Temperature. Thanks to the influence of the cold currents these
regions are, for tropical deserts, remarkably cool; the mean of the
hottest month does not far exceed 70° (at Swakopmund it is only 63°),
and that of the coldest month is about 60° (at Swakopmund it is only
55°). Extreme temperatures rarely exceed 80°, never exceed 100°, and
never fall below freezing. The annual range is in the neighbourhood

of 10° (cf. 40° in the Sahara) and the daily range is generally less than 20°. The crest and trough of the temperature curve are delayed until February and August. In fact in all respects except rainfall the climate is typically marine.

The importance of the sea as a control of temperature is demonstrated by the extraordinary uniformity of the means, irrespective of latitude, of all coastal stations, as the following table shows:

Station	Lat.	Hottest Month	Coldest Month	Mean Annual
Callao	12°S.	71	62·5	66·5
Arica	18°S.	71	62·0	66·0
Iquique	20°S.	70·5	61·5	64·8
Antofagasta . . .	23°S.	71	62·0	65·5
Walvis Bay . . .	23°S.	67	57·3	62·5
Luderlitz Bay . . .	27°S.	66·7	56·0	62·5

The isothermal maps bring this point out quite clearly, the isotherms running parallel to the coast through a considerable range of latitude.

Inland the temperature rises fairly rapidly, and except where climate is complicated by relief, there is a steady and somewhat rapid passage to the Saharan type. Windhoek, 400 miles inland, has a mean annual temperature 7° hotter than Swakopmund, although it stands more than 5,000 feet higher, and the hottest month at Windhoek is more than 10° hotter than at Swakopmund. When the wind gets into the east these higher temperatures extend towards the coast and it is under these conditions that the maxima are reached at coastal stations; the temperature at Cape Juby, for example, only exceeds 80° when the harmattan blows off the Sahara.

Thus in general the temperature extremes of the marine type are much less severe than in the continental type shortly to be described, but the higher humidity makes them more uncomfortable to endure; particularly one misses the cool, refreshing nights of the Sahara which are such a pleasant relief after the burning heat of the day.

Humidity and Rainfall. The presence of the cold current serves to increase the aridity since on-shore winds arrive cool, and being warmed by contact with the land, have their moisture capacity increased. In addition, Peru, South-west Africa and South California lie in lee of high land, so that easterly (land) winds are descending winds and therefore dry. These two influences, added to the normal high pressure of the latitude, ensure that virtually no rain falls. The whole of the coast from Arica to Caldera has, on an average, less than

1 inch of rain a year, the mean at Swakopmund is 0·7 inch. Western Australia and Rio de Oro are unsheltered by high land to the east and furthermore the cold currents are less powerful, the aridity is therefore slightly less; the Australian coast records 10 inches as far north as 26° S. (cf. 32° in Chile and South-west Africa).

Very rarely a storm invades the desert and brings a short and sometimes heavy shower of rain; Iquique had 2½ inches in a few hours on 22nd June, 1911. Showers such as this, spread over a number of otherwise rainless years, serve to give Iquique an annual average of 0·05 inch and definitely establish winter as the 'rainy season'.

The scantiness of the rain is the more striking in view of the high humidity which prevails as a result of off-sea winds. At Walvis Bay the mean relative humidity in January is 85 per cent and in July 77 per cent; along the coast of Peru and North Chile it is generally about 70 per cent and seldom falls below 50 per cent; at Cape Juby it is 82 per cent in January and 91 per cent in July. Fog, mist and heavy dew are everywhere characteristic and often persistent features. Iquique in winter is cloudier than England and the seaward slopes are often shrouded for days on end with dense fog through which the sun cannot penetrate. The condensation is often considerable, clothing may be saturated, the ground is wet as after rain and the dried-up branches of the tamarisks drip on to the desert soil.

The mists form over the cold waters near the coast as the air flows landwards and roll up to the coast and up the slopes of the land. Sometimes they lie at sea-level but more often the base of the mist rises on meeting the land and hangs at about 2,000 feet to 5,000 feet above it. In Peru there is a belt of vegetation at 5,000 feet nourished by almost perpetual mist. The level tends to rise by day and sink to sea-level during the night, following the diurnal motion of the air. They do not survive far inland in the face of the desert conditions, in South-west Africa not more than 70 miles; in Peru and Chile they fail to cross the Coast Ranges and from the desert of the longitudinal valley one can watch the heavy vapours rolling over the crest of the ranges but melting away at once in the dry desert air beyond.

In view of the close relationship between these mists and the sea breezes it might be expected that they would be most numerous during summer and during the heat of the day, at which times the sea breeze is strongest; but actually winter is, in most places, the foggiest season and night the foggiest time. The explanation is to be found in the close balance between factors making for condensation and factors causing re-evaporation, and especially the greater heat in summer of the desert interior which is therefore better able to dissolve the mists. In some places the coast may be actually cloudier in summer than in winter (e.g. Callao), but this never applies to the hill slopes, which are comparatively clear at this season. The uncondensed moisture passes over the longitudinal valley to the Cordillera beyond then condenses there; on these slopes summer is the cloudiest season and actual rain may fall at about 8,000 feet. The scanty rainfall, in so far as it is due to the same causes as the mist has, like the mist, a winter maximum along the coast, but a summer maximum on the Cordillera; the latter is a tropical régime and very valuable for irrigation.

THE CONTINENTAL TYPE OF HOT DESERTS (SAHARAN)

The coastal strips just described are only narrow fringes and grade rapidly eastwards into the true hot deserts, the climate becoming more and more extreme into the heart of the continent. The annual range of temperature on the west coast of Australia is less than 20°, in the desert interior it is 30° or more; the coast is frost-free as far south as Geraldton (29°S.), but inland night frost occurs as far north as Alice Springs on the tropic. The annual range at Alexandria is 22°, at Cairo 28°, and at Asyut 32°.

Temperature. The annual range of temperature is not very great (30° or so) and is less significant than the huge diurnal range; so great, in fact, is the difference between night and day that mean values convey a very inadequate picture of the true climatic conditions. The dry air and the cloudless sky offer no obstacle to the passage of the sun's rays, which make themselves felt as soon as the sun shows above the horizon, which it does with almost startling suddenness. As it climbs higher into the sky the heat becomes rapidly more intense until by noon the thermometer stands, in summer, above the hundred mark; in the afternoon it often rises above 120° and sometimes above 130°. After sunset the heat is quickly lost by radiation and the temperature falls as rapidly now as it rose in the morning. The nights, even in summer, are distinctly cool and in winter frosts occur. Thus the daily range of temperature is great (25° or 30°) and often excessive. In Death Valley, California, the mean diurnal range for August, 1891, an exceptional month, was 64° and the greatest diurnal range in that month was 74°. Thus the inhabitants must be prepared for great extremes; the Arabs wear heavy clothing as a protection against the chilly nights and coverings for the head and face to protect them from the heat, glare and dust of the daytime.

The rays of the midday sun, beating on the barren ground, make the sand and rock so hot that they seem to scorch the feet through the shoes. The air above them is heated by conduction and a shimmering heat haze is set up, the differential refraction in the heated layers resulting in the *mirage*. The heat gives rise to strong convection currents resulting in strong, though variable, winds which whip up the dust and bear it along in clouds. These may be only local swirls (*dust devils*), but sometimes they are connected with cyclonic storms passing to the north and are on a much greater scale. The *simoom*, fairly frequent in the northern Sahara during the hottest months, is the most fearsome type—a swirling rush of scorching air (120°–135°), laden with dense clouds of blistering sand through which it is impossible to see more than a few yards.

Humidity and Rainfall. The prevailing wind, having come over wide stretches of land, and in some cases high land and mountain ranges, is normally very dry; at Aswan the mean relative humidity for January is 46 per cent and for July 30 per cent, while the afternoon mean in May is only 15 per cent. Under these conditions evaporation is often twenty times the rainfall. To some extent the humidity varies according to the wind direction, relative humidity at Cairo is often more than 80 per cent with northerly winds, but when the *Khamsin* blows off the desert it is often below 25 per cent and has been known to fall as low as 2 per cent. The temperature at the time is well over 100° and the heat and dryness are most distressing; skin and finger-nails crack, the hair is brittle and electric, and the whole body seems desiccated. It was the great dryness of the air which allowed the ancient Egyptians to mummify their dead. Scarcely a cloud appears in the sky which is sometimes intensely blue, but more often hazy with dust. The night sky is of a transparent clearness through which the moon and stars shine with a matchless brilliance. It is not surprising that the ancient Egyptians and the Chaldeans were well versed in astronomy, that the sun was worshipped as a god and that the star and crescent have been adopted as emblems on their flag.

At night the air often cools below the dew-point and copious dew may be deposited, but the dew is quickly re-evaporated in the morning sun.

Such is the dryness of the air that rain has little chance of reaching the ground; sometimes clouds may be seen from which rain is falling but it does not penetrate far in the thirsty air. The chief rain occurs in the form of heavy storms which break with great violence and suddenness, giving, perhaps, an inch or more of rain in an hour. There is no vegetation to check the run-off and in a few minutes the wadis fill with rushing water and farther down unsuspecting travellers may be overwhelmed by the rapidity of the flood.

Higher land is better watered, forming sometimes altitude oases, such as the Ahaggar and Tibesti highlands, and the Macdonnell Range. The rainfall régimes of these altitude oases often afford an interesting glimpse of the conditions up aloft. It has been noted (p. 18) that in the deserts of the high-pressure belt the poleward margins get their rainfall from the westerlies, the equatorward margins from the migration of the equatorial rain belt. Farther into the heart of the desert all rain disappears, but the highlands allow us to trace the two types towards their meeting-point. For example, the southern ranges of the

Sahara, the Air and the Tibesti, are watered by rain-bearing winds from the south-west, but farther north, in the Ahaggar, the two régimes overlap and rain falls from both sources, winter rains from the westerlies and summer rains from the S.W. monsoon.

COLD DESERTS

Only Eurasia and North America are wide enough in intermediate latitudes to bring about deserts, but Patagonia, in spite of the narrowness of South America, is virtually desert because of the completeness of the screen which the Andes provide against the Brave West Winds. The last has certain maritime traits which call for separate treatment, but the other two have strong points of resemblance and can be considered together.

Three contributory factors account for their aridity: (1) The distance from the sea. (2) The basin form with surrounding highlands. (3) The intensity of the anticyclone that covers them in winter. In consequence few rain-bearing winds penetrate and these must descend into the basins and are thus warmed and dried.

The desert stretches from the Caspian Sea to the Khingan scarp, but it is not continuous since numerous mountain groups form altitude oases which break the desert up into a number of more or less isolated basins, usually basins of inland drainage, often at considerable altitude above sea-level. In America the same climate appears in the Great Basin, where the relief type is reproduced with great fidelity.

Such rain as falls round the Caspian and Aral seas is winter rain (Merv 92 per cent, Teheran 86 per cent, Quetta 82 per cent), showing that this region is a degenerate Mediterranean climate; such rain as falls in Gobi and Takla Makan is summer rain (Urga 88 per cent, Kashgar 77 per cent), showing that this is a degenerate cool-temperate continental climate.

Winter. The dominant influence at this season is the continental anticyclone; in the heart of Asia its reign is absolute and winter is a dry season; Kashgar has less than 1 inch in the six winter months. But to the south and west it is disturbed by the passage of occasional cyclonic storms bringing a little rain and snow to Russian Turkestan and the high plateaux of Persia and Afghanistan where snow lies throughout most of the year. Here the rain persists into spring, in fact the spring months are the wettest; Merv has 63 per cent of its rain in the three months February to April, Teheran 46 per cent and Quetta 50 per cent. Thus the rainy season coincides with the melting of the snows of the mountains and the two combined cause sudden and considerable floods, e.g. the floods of the Euphrates and Tigris on which the ancient irrigation empires depended.

In the Great Basin the rainfall is more evenly distributed, but spring is here also the rainy season, 37 per cent of the yearly total falling at Salt Lake City in the three months March, April and May.

In Turan the mean temperature of January is in the neighbourhood of freezing-point (Samarkand 32°) and the thermometer not infrequently falls below zero; across the Pamirs in the Tarim desert it is colder still (Kashgar 22°). Partly this is due to greater altitude, but mainly to the greater continentality; in point of fact greater altitude is not altogether a disadvantage and it is at the lower stations where, owing to the drainage and settling of cold stagnant air, the lowest temperatures, both mean and extreme, are recorded. Luktchun, in the Turfan depression, 50 feet below sea-level, has a January mean of 13°, i.e. 9° colder than Kashgar, although it stands more than 4,000 feet lower. Warmer weather can in fact be found, as a rule, by ascending the slopes, a fact well known to the natives who take advantage of it in many ways.

The duration of winter increases to north and east; in Mongolia there are six months below freezing, in Eastern Turkestan two or three, in northern Turan one or two, and in southern Turan none. The arrival of spring is everywhere very sudden, at Samarkand April is 11° warmer than March and May 11° warmer than April, at Kashgar April is 14° warmer than March, at Luktchun 21°. The daily range is enormous, often as much as 90°, and daily maxima are high; by June the shaded thermometer is exceeding 100° in the heat of the day. The suddenness of this rise of temperature is the result of the prevailing dryness. In other climates much of the spring warmth is expended in melting snow, warming water and drying the soil, but here there is little snow to melt and virtually no water to warm and evaporate. The result is that spring is actually warmer than autumn; April is 3° hotter than October at Samarkand, 6° at Kashgar, and 11° at Luktchun.

In the Great Basin the continentality is less extreme and autumn is warmer than spring, but the difference is only very slight; at Boise City, for example, the April and October temperatures are almost identical.

Summer is almost as hot as in the Sahara; Petro-Alexandrovsk has a July mean of 83°, Tashkent of 81° and Luktchun exceeds 90°. In the hottest hours of July days the thermometer rises above 110° and Luktchun has recorded 118°. At the last-named station local peculiarities of the site undoubtedly add to the heat; the absence of cloud facilitates insolation, the bare, dry ground is strongly heated and warms the overlying air, and further, owing to its position in a deep depression, ascending air must be replaced by air already heated by contact with the slopes and further heated by compression as it sinks into the denser air layers. Higher altitudes are, however, little cooler; even Lhasa (11,600 feet) has a July mean of over 70° and the temperature in the sun is sometimes more than 140°.

The continental lows are now centred over these arid interiors and air movement is somewhat variable, but as in the hot deserts strong winds from almost any point of the compass spring up by day, carrying clouds of dust and sand which darken the sky. The *Karaburan* is of this type, blowing strongly from the north-east in the Tarim Basin.

PATAGONIA

Patagonia, transitional between steppe and desert, differs from other cold deserts in that it is not typically continental, for on account of the narrowness of the continent and the force and regularity of the west winds there is always a certain marine influence which prevents the winter temperature from falling very low or the summer temperature from rising very high; there is, moreover, at the foot of the Andes a 'foehn' effect which increases the warmth and aridity, so that the isotherm of 32° nowhere enters the continent. Owing to the dryness of the air there is, however, a considerable diurnal range of temperature—about 15° in winter and 20° in summer—so that temperatures below zero are not unknown in the south and temperatures above 100° in the north. The strong wind, a characteristic feature of the climates of these latitudes everywhere in the southern hemisphere, increases the aridity and aggravates its ill-effects by raising storms of dust. Rainfall increases to the north and east, Neuquen has 5 inches, Bahia Blanca 21 inches, and there is a gradual passage to the subtropical climates of the better watered parts of the Pampa.

DESERT LAND FORMS

The individuality of the desert environment is made up of a subtle mixture of climate and relief, but the relief of arid lands presents such a consistent and peculiar association of forms that it might almost be considered as an element of the climate. Exfoliation, frost-action and mechanical erosion by wind-blown sand, typical desert agents, produce a hard, clear profile and jagged lines in exposed rocks, exaggerated, in the absence of a softening mantle of a vegetation cover, by the bareness of the rock edge seen in a transparent atmosphere against a brazen sky. One misses, too, the graded line, the steady fall of the ground towards base-level, the ineluctable waste of land and the transport of the debris to the sea. The absence of continuous slopes and the ever-changing landscape of the sand-dunes give an air of restlessness to desert scenery that contrasts strongly with the well-ordered, purposeful lines of the normal fluviatile relief.

Watercourses (wadis or arroyos) that originate in the desert contain water only once in a while and do not often escape from the desert, but are lost in sand and stones. Those starting beyond the desert, rising, perhaps, in the melting snows of mountains, dwindle as they make the passage of the desert. Their thalweg is graded, but not their valley sides; they flow in cañons with steep, rugged, irregular sides sculptured crudely along the lines of weakness in the rocks. The Nile alone escapes from the Sahara, the Orange and the Kunene traverse the Kalahari, the Snake River and the Colorado escape to north and south of the Great Basin, but more often than not the rivers end in the desert in basins of inland drainage. In many cases it is doubtless true that the basin is tectonic and the inland drainage is not the result of the desert, but in any other climate the rivers would have drained the interior basins. The very structures that produce inland drainage tend to perpetuate it by erecting mountain screens to keep out rain-bearing winds.

Aided by the rivers which sometimes rise on the mountain slopes which project into the higher rain-bearing layers of air, the waste of the land creeps slowly down and buries the hills and mountains, little by little, in their own refuse. Screes and alluvial fans are built out-wards and unite to form belts of piedmont gravel which may spread across the valley, so that the slopes rise suddenly out of the flat aggraded floor. The water escapes as springs, perhaps, along the foot of the gravel belt and here is found a fringe of cultivable land, e.g. the line of oases in the Tarim Basin, joined by the trade routes which pass through the basin on their way to Ferghana and Western Turkestan.

In the absence of wood, stone or adobe form the prevalent building material for the houses, which are flat-roofed and often arranged to collect and conserve what little rain falls; in this way they are the exact opposite of the houses of humid climates whose sloping roofs are designed to disperse the excessive rain.

Evaporation everywhere exceeds precipitation and lakes are highly saline, e.g. the Great Salt Lake (18 per cent salinity), the Dead Sea (24 per cent) and Lake Van (33 per cent). Their shores are thickly encrusted with salt, indicating a once greater extension, and in some cases the lake has altogether disappeared, leaving a *salina* or *salt-pan*.

VEGETATION AND CULTIVATION

Without the aid of irrigation only the most xerophilous plants can keep alive in the desert, which they do by means of the usual devices for reducing transpiration to a minimum and by extending the range of their root system to a maximum. Other plants, less xerophilous, have the power of remaining for many months and even years in the resting state, waiting for the occasional rains to waken them to active life. The incidence of rainfall in the desert is highly irregular and there is therefore nothing in the nature of a seasonal rhythm to be followed; the desert plant, like many desert peoples, is an opportunist, always prepared to take advantage of what fortune provides. The rate of growth of these plants, once germinated, is phenomenal, and an ephemeral carpet of blossom follows the shower with surprising rapidity. Where the rainfall régime is more regular, especially on the borders of the desert, there may be regular seasonal pasture, or even crops, e.g. the alfa grass of North Africa, and it often happens that the seasonal supply of pasture induces a seasonal migration of pastoral tribes. The irregular and usually torrential showers in the desert proper do more harm than good; they are not sufficiently reliable to be put to use and the sudden floods to which they give rise are a source of much damage and danger. In Peru and Chile, where the rivers support a narrow ribbon of irrigated cultivation, the trail of sand, gravel and boulders which these floods leave in their wake may smother a whole season's carefully tended crops.

Where the desert approaches the sea, e.g. in Peru and South-west Africa, there may be available an additional water supply in the mist and fog which sometimes pass inland. Certain desert plants possess the power of absorbing moisture from these mists and dews through their leaves, and some exude hygroscopic salts on their leaf surfaces which are thus salt-encrusted during the day; the salts withdraw moisture from the cool night air and deliquesce. The mists which daily rise from the Red Sea and bathe the slopes of Yemen provide not only moisture for the famous coffee-trees, but also shade from the heat of the midday sun.

The small amount of rain, or even dew, with which crops can be grown on desert margins is surprising, but broadly considered the cultivation of the desert is mainly a matter of irrigation, either from streams or from underground sources. Streams that originate outside the confines and lose themselves in the desert give rise to marginal and terminal oases, but even when the water has passed below the surface it is not lost. It may be brought to the surface as springs by irregularities of the ground or of the impervious floor, or it may be tapped by wells, often hundreds of feet deep and sometimes, as in Australia, artesian. The resultant oases are among the most isolated of settlements, but they tend to become the foci of caravan routes and markets for exchange. One of the most important and certainly the most characteristic product is the date palm which finds in the hot desert oases its optimum conditions. It is greedy for water at its roots, but cannot stand rain which interferes with pollination at this early stage and spoils the fruit later. If it is to bear good fruit it demands a long hot season (six months above 64° and a large accumulated temperature above this zero point); this last requirement is seldom available in cold deserts.

Maize, beans and millets are grown beneath the shade of the date palm upon whose stems are often trained the creeping branches of the

TEMPERATURE

Station	Lat.	Long.	Alt. (ft.)	J	F	M	A	M	J	J	A	S	O
SAN DIEGO	33°N.	117°W.	87	54	55	57	58	61	64	67	68	67	63
YUMA	33°N.	115°W.	141	55	59	65	70	77	85	91	90	84	72
CAPE JUBY	28°N.	13°W.	Coast	61	61	63	64	65	67	68	68	69	68
INSALAH	27°N.	2°E.	920	55	59	68	76	86	94	99	97	92	80
ASWAN	24°N.	33°E.	326	59	63	70	78	85	90	91	90	88	82
CAIRO	30°N.	31°E.	380	55	57	63	70	76	80	82	82	78	74
MASSAWA	16°N.	39°E.	64	78	79	81	84	88	92	95	94	92	89
ADEN	13°N.	45°E.	94	76	77	79	83	87	89	88	87	88	84
BAGHDAD	33°N.	44°E.	125	49	54	61	71	81	90	95	94	88	80
JASK	26°N.	58°E.	13	67	68	73	80	85	90	91	89	87	83
KARACHI	25°N.	67°E.	13	65	68	75	81	85	87	84	82	82	80
JACOBABAD	28°N.	68°E.	186	57	62	75	86	92	95	96	92	89	79
IQUIQUE	20°S.	70°W.	30	71	71	69	65	63	62	60	61	63	64
ANTOFAGASTA	24°S.	70°W.	13	72	70	70	67	64	63	62	60	61	62
WALVIS BAY	23°S.	14°E.	10	65	66	66	65	62	60	59	57	58	60
PORT NOLLOTH	29°S.	17°E.	16	60	60	59	58	57	55	55	54	55	58
CARNARVON	25°S.	114°E.	15	80	81	79	75	68	63	61	63	66	69
EUCLA	32°S.	129°E.	30	71	71	69	66	61	56	54	56	59	63
ALICE SPRINGS	24°S.	134°E.	2,000	84	82	77	68	60	54	52	58	66	74
*KAMLOOPS	51°N.	120°W.	1,193	23	26	38	50	58	64	70	68	58	48
ASTRAKHAN	46°N.	48°E.	−46	19	21	32	48	64	73	77	74	63	50
TEHERAN	36°N.	51°E.	4,002	34	42	48	61	71	80	85	83	77	66
IRGIS	49°N.	61°E.	360	3	4	19	44	63	72	76	73	59	42
PETRO-ALEX-ANDROVSK	41°N.	61°E.	295	23	29	43	58	73	80	83	79	67	52
KASHGAR	39°N.	76°E.	4,255	22	34	47	61	70	77	80	76	69	56
LUKTCHUN	43°N.	90°E.	−56	13	27	46	66	75	85	90	85	74	56
*URGA	48°N.	107°E.	3,740	−16	−4	13	34	48	58	63	59	48	30
SARMIENTO	46°S.	69°W.	813	65	63	59	51	44	38	38	42	46	53
SANTA CRUZ	50°S.	69°W.	85	61	57	54	48	41	34	33	38	43	48

* Kamloops $\left(\dfrac{T}{R} = 4\cdot7\right)$ and Urga $\left(\dfrac{T}{R} = 3\cdot7\right)$ lie just beyond the limit of deserts as defined on p. 98.

RAINFALL

N	D	Yr.	Ra.	J	F	M	A	M	J	J	A	S	O	N	D	Total
59	56	61	14	1·8	1·9	1·5	0·6	0·3	0·1	0·1	0·1	0·1	0·4	0·9	1·8	9·6
61	56	72	36	0·4	0·5	0·4	0·1	—	—	0·1	0·5	0·2	0·2	0·3	0·4	3·1
65	62	65	8	0·5	0·5	0·5	—	—	—	—	0·5	0·5	0·5	0·5	1·0	4·5
68	58	78	44	Practically Nil												
72	62	77	32	Practically Nil												
65	58	70	27	0·4	0·2	0·2	0·2	—	—	—	—	—	—	0·1	0·2	1·3
86	81	87	17	1·5	0·6	0·6	0·8	—	—	—	—	—	0·3	0·7	1·4	5·9
80	77	83	13	0·3	0·2	0·5	0·2	0·1	0·1	—	0·1	0·1	0·1	0·1	0·1	1·9
63	35	73	46	1·2	1·3	1·3	0·9	0·2	—	—	—	—	0·1	0·8	1·2	7·0
76	70	80	24	1·1	0·9	0·8	0·2	—	0·1	—	—	—	—	0·3	1·1	4·5
74	67	78	22	0·5	0·5	0·4	0·2	0·1	0·9	2·9	1·5	0·5	—	0·1	0·1	7·6
68	59	79	41	0·3	0·3	0·3	0·2	0·1	0·2	1·0	1·1	0·3	—	0·1	0·1	4·0
67	69	66	11	Practically Nil												
64	68	67	12	Practically Nil												
61	64	62	9	Practically Nil												
59	60	58	6	—	0·1	0·2	0·2	0·4	0·3	0·2	0·4	0·3	—	0·2	0·1	2·3
73	77	71	20	0·3	0·9	0·5	0·6	1·5	2·8	1·7	0·7	0·3	0·1	—	0·1	9·5
66	69	64	17	0·7	0·5	0·9	1·2	1·2	1·1	0·9	1·0	0·8	0·7	0·7	0·4	10·1
80	82	70	32	1·8	1·7	1·3	0·9	0·6	0·6	0·4	0·4	0·4	0·7	0·9	1·3	11·1
35	28	47	47	1·0	0·8	0·3	0·4	0·9	1·2	1·1	1·1	0·8	0·6	1·0	0·9	10·1
37	26	49	58	0·5	0·3	0·4	0·5	0·7	0·7	0·5	0·5	0·5	0·4	0·4	0·5	5·9
51	42	62	51	1·6	1·0	1·9	1·4	0·5	0·1	0·2	—	0·1	0·3	1·0	1·3	9·3
26	11	41	73	0·6	0·3	0·5	0·7	0·8	0·9	0·6	0·4	0·5	0·5	0·4	0·7	6·9
39	30	55	60	0·2	0·4	0·5	0·6	0·2	—	—	0·1	—	0·1	0·1	0·1	2·4
40	26	55	58	0·3	—	0·2	0·2	0·8	0·4	0·3	0·7	0·3	—	—	0·2	3·5
33	18	56	77	—	—	—	—	—	—	—	—	—	—	—	—	—
8	−7	28	79	—	0·1	—	—	0·3	1·7	2·6	2·1	0·5	0·1	0·1	0·1	7·6
57	61	51	27	0·1	0·3	0·5	0·4	0·8	0·6	0·8	0·4	0·4	0·4	0·2	0·2	5·1
55	55	47	28	0·5	0·3	0·2	0·6	0·7	0·4	1·1	0·4	0·2	0·4	0·4	1·0	6·0

vine. The desert gardens of Persia and Ferghana grow figs, pome-
granates, mulberries, vines, apricots, tobacco, opium, cotton, wheat,
maize, barley, lucerne, melons, etc.

Desert soils, when irrigated, are proverbially fertile, for the long
ages of drought have allowed the accumulation of soluble salts which
the plant requires. On the other hand, the salts which rise by capillarity
and effloresce at the surface are often a hindrance to cultivation (e.g.
the 'reh' soils of India), either because they are of a kind inimical to

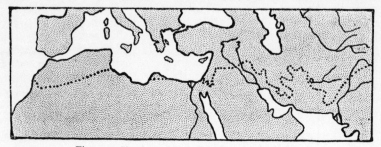

Fig. 73.—Northern Limit of Productive Date Palm

plant life or because they are present in much too great concentration.
In many places the concentration of surface salts is sufficient to form
the basis of mining industries, e.g. the borax and nitre of Chile, the
phosphates of Tunis and the salt that results from the evaporation of
lakes—often a government monopoly and sometimes a medium of
exchange. Also preserved by the aridity, though not due to it, is the
guano of Chincha Islands. Most of these salts are valuable mainly as
fertilizers and there is something rather ironical in the spectacle of the
world's chief fertilizer being obtained from the world's most complete
deserts.

SUGGESTIONS FOR FURTHER READING

Besides Knox, *Climate of Africa*; Ward, *Climate of the United States*; and Taylor,
Australian Meteorology, see H. Schirmer, *Le Sahara*, Paris, 1893; Woeikof, *Le
Turkestan Russe*, Paris, 1914; Lespagnol, *Sur le caractère désertique de l'Australie
intérieure*, 1898; Lyons, *Physiography of the Nile Basin*. Desert and semi-desert
vegetation (as well as other types) are related to climate by Livingstone and
Shreve in *The Distribution of Vegetation in the United States*; Carnegie Inst. Pub.
No. 284, 1921. U.N.E.S.C.O. has a Commission on Arid Lands, their reports
are full of information For conditions of erosion and deposition under arid
conditions see C. A. Cotton, *Climatic Accidents*, 1942.

XV

MOUNTAIN CLIMATES

Increasing altitude brings about a series of quite well-defined changes in the values of climatic elements which have been discussed in a general way on pp. 35 to 40. At first the effect serves only to introduce minor modifications of the normal sea-level climate, such as the altitude variants of the equatorial and tropical climates described on pp. 107 and 126, but ultimately the alteration becomes so profound that we find ourselves in a climate possessing such a number of individual peculiarities as to require separate treatment. But the mountain climate, being the direct result of great contrasts of relief, is compounded of a great variety of contradictory types; of dry burning heat on exposed slopes, of stifling heat in enclosed valleys, of bitter cold in the same valleys on winter nights, of fog and sunshine, of torrential rain and great aridity.

As in polar and desert lands the study of climate in mountain regions is handicapped by the paucity of regular observations and we are driven again to seek information, especially for higher levels, from the all too irregular visits of climbers. The number of stations is, however, increasing and for the European and, to some extent, the North American mountains there is a fair quantity of material now available.

Pressure (see also p. 35). The decreased atmospheric pressure is not of very great importance *per se*; there are, for example, human settlements on the Andes and in Tibet where the pressure is only about two-fifths of that at sea-level, but it is of the greatest importance in that most of the phenomena of mountain climates are the direct results of the rarefaction of the air. In consequence of the fall of temperature which results from this rarefaction the same temperature zones are passed through on the slopes of a mountain up to the snow-line as on a journey into high latitudes at sea-level as far as the line of permanent snow. But the conditions in the corresponding temperature zones are widely different; the characteristic feature of insolation in the polar zone is the long summer day of oblique sunshine through a great thickness of atmosphere, the feature of the insolation in alpine climates is the intensity of sunshine through a thin and often transparently clear atmosphere. The far-reaching results of this difference will be seen later when the relation of climate to vegetation and the habitability of the zones are discussed.

Mountain and Valley Winds. Since mountains occur in each and all of the great planetary wind belts it is clearly impossible to generalize on the major winds of mountain regions, except to say that on exposed slopes and summits, where the conditions approximate to those of the free atmosphere, friction is reduced to a minimum and winds are much stronger in proportion to the barometric gradient than at low levels. But there are a number of local winds, characteristic of mountains and due to their relief, which are of considerable importance. Where the general air movement is not too strong to overcome local influences there is a tendency to a diurnal reversal of wind direction analogous to land and sea breezes, the wind blowing up the slopes during the day and down the slopes at night. These winds are concentrated by the form of the land into the valleys, where they are felt at their strongest. The down-valley wind is the direct result of cooling by nocturnal radiation, the cold air sliding down the slopes and into the valleys, but the cause of the up-valley wind of the daytime is less apparent. Insolation, strongest in calm weather and with a cloudless sky, causes an expansion of air on the slopes which tilts the isobars towards the mountain, and starts a flow of air which meets the slope and continues up the mountain. Often these winds blow with considerable force, especially the night wind; besides, being colder, it is more generally noticed by mountain dwellers and visitors. In the neighbourhood of large snow-fields and glaciers the force of the night wind is strengthened by the cooling of the air in contact with the ice and snow; in fact this effect is in places so profound as to overcome the day wind and to give rise to a fairly regular and persistent system of cold downcast winds, e.g. the *Nevados*, an unpleasant feature of the climate of the higher valleys of Ecuador.

The Foehn and Allied Winds. In mountain districts the horizontal movement of the air is interfered with by the relief and air is compelled to ascend and descend. In doing so it undergoes changes of pressure which result in changes of temperature which, in turn, result in changes of relative humidity, the ascending winds tending to bring rain, the descending winds tending to cause evaporation. These winds are well known in all mountain districts, but the best known is the *foehn* whose name has been adopted as a general term for winds of this type. It is an excessively hot, dry wind occurring typically in the valleys of the northern alpine slopes, and though the wind itself is not due to the mountains it owes its properties to the relief. It occurs when a depression moving to the north of the crest of the Alps draws air from higher pressures lying to the south across the divide. In ascending the southern slopes the air is cooled below its dew-point, clouds are formed and heavy rain occurs, the liberation of heat by the condensation checking the fall of temperature, so that by the time it reaches the divide it has lost most of its moisture but retained much of its warmth. In its descent of the northern slopes it is adiabatically warmed (in descending 5,000 feet it would be warmed about 25°) and blows as a hot, dry wind down the valleys. The foehn is announced by a crest of cloud along the mountains to the south while the air is still, heavy and oppressive, yet so clear that the distant mountains seem very near and tinged with a bluish colour. Then come a few cold puffs of wind, then ominous calm, then the hot blast of the foehn. The thermometer may rise 18° or 20°, the snow vanishes in the hot, thirsty air and everything becomes tinder dry. The wooden chalets are ready to catch fire at a spark, but all fires are put out for safety's sake as soon as the foehn appears.

The average yearly frequency of days with foehn winds in northern Switzerland is as follows: spring, 17·3; summer, 4·9: autumn, 9·6; winter, 9·1; year, 40·9. This is the same as the frequency of cyclones, i.e. a spring maximum (cf. the *Chinook*), a fortunate circumstance as its effect in freeing fields and pastures from their winter covering of snow accelerates the beginning of the agricultural year. But its occurrence in autumn and winter have important consequences, the former helping to ripen the harvest, especially the grape harvest which in some places is dependent on the foehn for its success.

In general it is a southerly wind—it used to be thought that it came from the Sahara, but this is neither true nor necessary—but like most mountain winds it follows the relief lines, blowing most strongly down those valleys whose courses coincide with the pressure gradient, i.e. the north and south valleys such as the Rhine from Chur to Lake Constance, the Aar to Lake Brienz, and the Rhône from Martigny to Lake Geneva, but not in the Rhône furrow from the Furka to Martigny, nor in the Vorder Rhein which lie at right angles to the line of flow.

Similar winds to the foehn occur in all mountain districts where cyclonic storms occur. Though rarer than on the northern slopes it is by no means unknown in the Italian Alps in the valleys of the Ticino and Toce; the *Chinook* (see p. 235) is exactly similar, so are the *Samun* of Persia, descending from the mountains of Kurdistan, the *Nor'-Westers* of New Zealand off the New Zealand Alps, and many others.

Insolation and Temperature. The thin air of mountains, almost free from dust and moisture particles, is extremely ineffective in absorbing the sun's radiation, which is thus allowed to pass through with little loss of intensity. The short wave-length violet and ultraviolet rays are absorbed first, those of the red end of the spectrum penetrating the atmosphere most successfully. Elster and Geitel showed that 40 per cent of the ultra-violet rays penetrate as far as the summit of Sonnblick (10,000 feet), only 31 per cent as far as Kolm Saigurn (5,250 feet) and only 16 per cent reach sea-level. Insolation on mountains is thus not only very powerful but there is also a markedly high proportion of the more refrangible ultra-violet rays. Now these rays are of great value to plant life and are stimulating to animals and man, they are powerful to burn the skin and, combined with the reflection from a white snow surface, quickly produce a deep tan. The bright light, especially when reflected from the snow surface, is trying to the eyes and may cause snow-blindness unless dark glasses are worn.

It is the avidity with which the lower layers of the atmosphere absorb the sun's radiation and especially the infra-red rays that accounts for their rapid warming up, and conversely it is the transparency of the upper layers which diminishes their rise in temperature in spite of the power of the sun's rays. Rocks, stones and soil, on the other hand, absorb heat readily and their temperature rises rapidly in the sun. The result is a surprising difference between the temperature of the ground and the temperature of the air near it, between temperature in the sun and temperature in the shade. Hann quotes an example from Leh (11,500 feet) where, on 11th August, 1867, the shade temperature was 70°, but the black bulb *in vacuo* was registering 214°, which is 22° above the boiling-point for this altitude. As the sun sinks in the sky and as insolation decreases and ceases the ground rapidly loses its heat by radiation through the thin air, especially as the convex form of the mountain presents a large surface for radiation, and night temperatures of the ground surface fall to a very low figure. Thus the diurnal changes of surface temperature are very severe, an important factor in the disintegration of rocks and in the production of the typical jagged peaks and 'aiguilles' of mountain scenery. But the air is as slow to lose its heat at night as it was to absorb it by day and the air temperature lags far above the ground temperature. Furthermore, the air is usually in motion and fresh supplies are imported before the air in contact with the mountain can be chilled. The diurnal motion of the air operates in the same direction; the ascending currents of day-time, cooled by expansion, moderate the daily maximum; the descending currents of the night-time, warmed by compression, moderate the nocturnal minimum. Thus a variety of factors combine to reduce the daily range of air temperature which, as in marine climates, is very small; Mont Blanc has a mean diurnal range in July of only 6·3° as compared with 10° at Geneva.

Effect of Local Relief on Temperature. Up to this point we have considered only the changes of temperature on continuous mountain slopes and have seen that in general high-level stations enjoy a diminished range of temperature both diurnal and annual, but the form of the ground and the exposure of the site and a number of other purely local factors modify or even reverse the generalizations we have made. Exposure with respect to the midday sun makes a very considerable difference to surface temperature and has a profound effect on the habitability of a slope outside the tropics. Owing to the transparency of the air the shaded slope lies in heavy shadow while the sunny slope is bathed in brilliant light and warmth. The significance of the contrast is witnessed by the distinguishing names which these receive: in German *sonnenseite* and *schattenseite*, in French *l'adret* and *l'ubac*, in Italian *adretto* and *opaco*. The exposure may cause curious features in the daily march of temperature; an east-facing slope will have warm mornings and cool afternoons, while the opposite will be the case on a west-facing slope.

Just as the convex form of mountain peaks is a condition disposing to moderate temperatures, so the concave form of the valley disposes to extremes. Valley air is warmed by day from three sides, by contact with the bottom and both slopes, by night it is cooled by radiation from three sides. By night, also, the air, chilled by radiation, flows down the slopes and settles in the hollows, being displaced on the slopes by warmer air. Thus in calm weather, when the air layers are not disturbed by winds, the stagnant air in the hollows may be many degrees colder than at stations thousands of feet higher. It is under these conditions that the lowest temperatures occur—the lowest figure on record in U.S.A. is −65°, reached at Miles City, Montana (2,371 feet) which lies in a deep hollow on the Great Plains; Pike's Peak, though 11,000 feet higher, has never recorded a temperature below −40°. Hot by day and cold by night, valley bottoms have a very considerable diurnal range.

Intermediate in form between the convex mountain and the concave valley is the flat plateau. Here also we find the contrast between surface temperature and air temperature, but the range of air temperature is greater than on mountain slopes, though less than in valley bottoms.

The yearly march of temperature in mountain climates closely resembles the diurnal march, being characterized by:

1. A progressive decrease of mean temperature with altitude.
2. A small annual range on peaks and slopes.
3. A large annual range in valley bottoms.
4. A delayed minimum and maximum.
5. Autumn much warmer than spring.

The last two features are particularly noticeable where there is a heavy winter snow-cover whose melting absorbs the heat of the spring sun. This is paralleled in cold climates and in marine climates, the delay in the latter case being due to the expenditure of heat required for warming the sea.

Humidity and Cloudiness. Most of the water vapour in the atmosphere is concentrated in the lower layers (at 6,500 feet the amount is reduced by a half), thus the air of the mountain-tops is excessively dry and the low pressure further increases physiological dryness by allowing water vapour to be disseminated rapidly into the air. Evaporation is rapid, perspiration disappears at once, the skin is parched, face, hands and lips become cracked, and thirst is increased. On the high plateaux of Peru the Incas were able to preserve their dead by allowing them to mummify naturally.

The daily march of humidity is fairly regular and is intimately connected with the daily movement of the air. The day wind carries moisture upwards, cumulus and strato-cumulus clouds form, their flat bases marking the level at which the dew-point is reached; the night wind carries the moisture down again, the summits are dry and clear, and from aloft one may look down on the billowy surface of the fog that fills the valleys, condensed by the cold stagnant air that has drained into them during the night. Thus at high levels the nights are fine and dry and the best views are obtained in the early morning, there is an afternoon maximum of humidity and of rain and there is a tendency to afternoon thunderstorms; at low levels the nights are often cloudy and foggy, but the mists melt before the morning sun and rise again up the mountain-sides. The limits of the migration of this zone of maximum cloudiness vary with the season, rising in summer and falling in winter. In winter the alpine peaks and the higher valleys, where are all the famous centres for winter sports, rise into brilliant sunshine out of a sea of cloud that enshrouds the lower slopes. In summer these high valleys are cloudier and the lower levels are comparatively free; on the Swiss plateau at 1,500 feet winter is by far the cloudiest season; on the summits of the passes at 7,000 feet spring, and on the mountain-tops at 10,000 to 15,000 feet, summer.

Precipitation. As has been described on p. 39, mountains, by forcing the ascent of horizontally moving air, bring about increased precipitation up to an indefinite level of maximum precipitation, above which, owing to increasing dryness of the air, there is a decrease of precipitation. This greater raininess of high levels is of the greatest importance in arid lands since here alone is water available; the rainfall of the high sierras above the deserts of Peru nourishes the irrigated strips that run as green ribbons through the brown desert sands; the Tibesti and Ahaggar highlands are well-watered oases in the midst of the Saharan desolation; the mountains that surround the Tarim Basin provide a zone of sorely needed pasture between 10,000 and 14,000 feet. But the distribution of rain in mountain regions is extremely complex because of the complexity of relief, as an examination of a rainfall map of, say, the Alps will show in no uncertain manner. In general the heights stand out as areas of heavy precipitation and the valleys as drier strips; thus the St. Gotthard pass has more than 80 inches, but places in the Valais have less than 25 inches.

The daily and annual march of rainfall accompanies the march of cloudiness, rising in summer and in the heat of the day, falling at night and in winter. These tendencies, superimposed on the normal régime for the climate in which the mountains are situated, produce an incidence of rainfall which defies generalization.

Snowfall and the Snow-line (see also p. 37). The position of the limit of permanent snow depends on a number of factors both regional and local and only broad generalizations can be made. The level of snow at any particular time clearly represents the line of balance between accumulation and dispersal, i.e. between the amount of precipitation in the form of snow on the one hand and the rate of melting and evaporation on the other. But both of these depend in turn on the aspect with respect to the snow-bearing winds and to the sun, on the steepness of the slope and other local influences. The line, of course, migrates up and down the slopes in summer and winter. Fig. 74 shows the migration of the line on the two slopes of the Inn Valley in the Northern Tyrol and illustrates the magnitude of the difference that aspect makes. The term 'snow-line' refers, however, to the line above which the snow does not melt in summer, i.e. to the

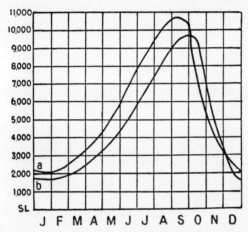

Fig. 74.—Mean Height of Snow-level in the Inn Valley in Northern Tyrol

a. On south-facing slope
b. On north-facing slope

crests of the curves in Fig. 74, and is determined mainly by two values, namely: the winter precipitation which determines the rate of accumulation and the summer temperature which determines the rate of ablation. It is a line of some biological significance, for plants can grow where the snow disappears in summer (cf. the tundra), no independent life being possible above this line.

VEGETATION AND CULTIVATION

The variations of temperature and rainfall which mountains bring about are sudden and abrupt, so that the climatic optimum may be displaced within a very short distance from one vegetation type to another. Consequently there is a great variety of plant associations within a small space and a corresponding variety of cultivation and of human occupations. It is quite common for one family in the French

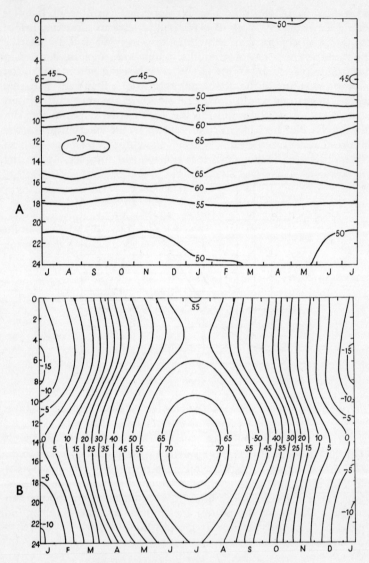

Fig. 75.—Mean monthly temperature at each hour of the day.
(A) Quito. Lat. 0°. Alt. 9,350 ft.
(B) Verkhoyansk. Lat. 68° N. Alt. 328 ft.

Alps to cultivate vineyards on the lower slopes in an almost sub-tropical climate and wheat or barley in a temperate climate at higher levels, to cut their wood from the forest zone, to pasture their cattle on the alpine meadows and to act as guides among the glaciers and peaks. The short distance between the zones encourages seasonal migration from one to another and the rigour of the higher altitudes in winter is easily escaped (contrast the tundra dweller). In general there is a movement of animals and man up the slopes in spring following the melting snows and down again in autumn; but sometimes the movement is in the reverse direction, e.g. in Central Asia where the Kirghiz in winter drive their flocks and herds above the level of the clouds into bright sunlight that bathes the high pastures watered by the rains of the summer season.

There is, of course, a vast seasonal difference between the 'Alpine' climates of mountain tops in fairly low latitudes and the tundra climates in very high latitudes; a contrast of seasonal régimes. Winter nights are the times of strong radiation in the thin, dry air and freezing temperatures occur every night, but daytime is a period of strong insolation, more or less from directly above, and the strength of the sun gives a marked temperature maximum, especially in the surface soil, and the upward movement of the air carries that warmth to the air.

The temperature régime is markedly diurnal in low latitude mountains and strongly seasonal in high latitudes with a long cold winter, an annual régime. Nor is there such a marked contrast of aspect in the tropics, for most slopes receive their insolation more evenly from North and South, though the seasonal differences naturally increase as latitude increases.

Whatever the vegetation zone in which the base of the mountain stands the zone of maximum precipitation is nearly always forest, deciduous below, passing upwards into coniferous. Between this forest and the snow-line, in the region of steadily decreasing precipitation, stretches an 'alpine' zone of grassland, narrow in the Alps where the snows are not far above the upper limit of the forest, but very wide in the tropics, becoming drier and drier towards its upper limit (see p. 126). On the plateaux of Arizona and New Mexico, a region which would be desert but for its altitude, the forest extends from 6,500 to 8,000 feet and an alpine flora follows above this level. The low temperature and the high evaporation aided by low pressure and strong winds here demand a xerophytic vegetation. Alpine pastures are renowned for the bright colours of their flowers and for the tender succulence of their grasses, the basis of the Swiss dairying industry. These properties they owe to the bright sunlight, the warm soil and cool air, the fertile glacier silt, the adequate rainfall, and the supplies of moisture from melting snows. The climatic conditions here are in strong contrast with those experienced by the corresponding vegetation in Arctic climates, where the sunlight from the low-angled sun is weak, where the soil is poor and acid, ill-drained and saturated with ice-cold water.

The effect of decreased temperature at higher altitudes is to delay the function of plants by an amount which in the Eastern Alps amounts to about ten days per 1,000 feet of altitude, as is shown by the accompanying diagram based on the work of the brothers Schlagintweit, quoted by Hann. Thus at higher levels the season available for growth becomes shorter and shorter and crop after crop is eliminated. The limit of grain is reached in the Alps at 4,000 to 6,000 feet according to local conditions of rainfall, exposure, etc. On north-facing slopes the vegetation is on an average about a fortnight behind that on

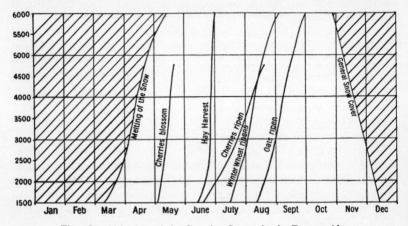

Fig. 76.—Altitude and the Growing Season in the Eastern Alps

south-facing slopes. Boissier has shown that on the southern slopes of the Sierra Nevada in Spain, where conditions are African, the limit of the vine is 7,000 feet higher than on the northern slopes where the conditions are more like those of the Meseta, that of the olive 1,700 feet and of rye 600 feet higher.

The phenomenon of temperature inversion has some important effects on vegetation and on settlement in valleys and basins. Houses and farms avoid the valley bottoms which are cold and foggy and subject to severe cold snaps, preferring the slopes and the alluvial fans raised above the valley floor, which occur where tributary streams debouch into the main valley. The liability to late and early frosts in such 'frost holes' or 'frost pockets' shortens the growing season and prohibits the cultivation of delicate plants. The coffee plantations of São Paulo, for example, are always situated on hill-sides for the drainage of cold air. So sharp is the line of division between air above and air below freezing-point in some cases that the bottom of a shrub may be nipped by the frost, but its crown may be green, healthy and unaffected. The so-called *thermal belt* of some of the Appalachian valleys is a zone about 300 feet above the floor of the valley untouched by frost, which retains its verdant green through the winter, separated by a perfectly horizontal line from the blackened, frosted vegetation of the valley bottom.

The Mountain Environment tends to isolation because of difficulties of transport and intercommunication and so to self-reliance but also to insularity and suspicion; the climate disposes towards versatility since the inhabitant gains experience of several climatic zones and usually spends a part of the year in each. He is agriculturist, pastoralist, forester and indoor worker in one. Physically he is adapted in some degree to the diminished pressure in which he lives; a larger lung development and wider chests begin to be noticeable above about 5,000 feet, and at greater heights there is a marked increase in the number of red corpuscles (see p. 36). Usually the physique is good, e.g. the people of the South American 'Puna' and some of the Kirghiz, but elsewhere there is a marked degeneration, e.g. the Mexican Indians. Cretinism, prevalent in certain valleys, has sometimes been attributed in part to the darkness of deep north-facing depressions which never see the sunlight.

REGIONAL TYPES

Out of the great number of regions with mountain climates it will be sufficient to describe three only:

1. The Alps, an example of mountain climates rising from the cool-temperate zone.
2. The High Andes, an example in tropical latitudes; and
3. Tibet, rising from the arid interior steppes.

THE ALPS

Chiefly because they are best known and because they are well supplied with meteorological stations the Alps have already been used to illustrate many of the general points of the mountain type of climate in the foregoing pages; the more particular characters which they owe to their position and surroundings will be discussed below.

The most arresting feature of the winter pressure distribution in Europe is the ridge of high pressure—the climatic backbone of Europe —an offshoot of the Asiatic high which extends from the Black Sea to the Meseta of Spain and continues to the Azores. It is this high pressure which gives Switzerland its fine winters, with clear, crisp, cold air and cloudless skies. It allows local factors full play, permitting mountain and valley winds to develop freely and encouraging inversions of temperature and other climatic phenomena.

In summer, too, the Azores high sends a ridge-like extension eastwards into Central Europe, its axis again coinciding with the axis of
the Alps; again local conditions are favourable for the free development of the mountain climate in its ideal form. Thus by virtue of
their position in the planetary wind system, as well as by virtue of
their relief, the Alps function as a climatic divide, separating the
cloudy cyclonic climates of Central Europe from the sunny Mediterranean province. One may recognize an altitude variety of the latter,
with the characteristic tendency to summer drought over most of the
southern slopes in Provence and in the Maritime Alps, but the greater
part of the Alps shares the Central European régime with summer
rain.

The climate becomes more extreme to the east in proportion to the
distance from marine influences, the change being more than normally
rapid since the relief accelerates aridity on the leeward slopes. There
is a decrease of rainfall eastwards and an increase in the annual range
of temperature; some of the enclosed valleys of the Eastern Alps in
Austria are bitterly cold in winter, especially during the night and
during anticyclonic weather. Klagenfurt has a mean January temperature of 20°, the same as on the Obir, more than 5,000 feet higher,
and frequently the lower station is the colder by some 20°.

THE HIGH ANDES

The succession of temperature and vegetation zones, the *tierra
caliente*, *tierra templada* and *tierra fria*, has been described on pp. 126
and 128; the first two are scarcely more than altitude variants of the
tropical climate, but by the level of the 'tierra fria' we have reached
the land of hot sun and cold shadow—the mountain type of climate.
The temperatures are reduced by the altitude to those of an English
spring (Quito 54°–55°, La Paz 44°–53°), but the conditions are in no
way similar. In contrast with the Alps, aspect has here very little
significance, nor have the seasonal variations of temperature; the annual
range at Quito is less than 1°, at Bogotá less than 2°, at La Paz less
than 9°. The diurnal range is much greater (about 25°) and of much
greater significance than the annual, and the daily procession of
weather is strikingly regular. During the rainy season the mornings
dawn unpleasantly chilly, but soon warm up, the latter part of the
morning being very pleasant with a warm sun and a transparently
clear air. But by midday clouds appear and cover the sky throughout
the afternoon; by two or three o'clock the daily storm breaks, often
with thunder and hail. There is no real dry season on the high Andes
in Colombia and Ecuador, but southwards into Peru and Bolivia it
increases in length and at La Paz there are five completely dry months.
It is during this dry season that the greatest extremes of temperature

are recorded; the thermometer generally falls below freezing at night and by day the sun scorches and blisters through the thin air.

The conditions of the 'tierra fria' are exaggerated in the *Puna* between 10,000 and 13,000 feet; the nights are bitterly cold, even in summer, while in winter the wind is so cold that it sears the skin. Woollen masks are used to protect the face and the heaviest of ponchos and shawls are worn. The sunlight is intensely brilliant and scorching hot, the shadow is dark purple and icy cold, for the clear air can neither hold the heat nor disperse the light. The natives avoid the shadow and even do their cooking out of doors in the sunlight. The daily winds blow with great force, raising clouds of dust; in winter they are of a penetrating cold that blows through almost any clothing and brings a crop of ailments, especially to the lungs—at Oruro they call them the 'Harvest of Death'. The rains last only about three months and during this season the lakes and rivers expand and rise, but dwindle again in the dry season to shallow salt-pans and dry stony courses which is their normal nature for the rest of the year.

The xerophytic vegetation consists only of coarse tufted grass and cushion-like plants with small leathery leaves and buds sealed in hard cases. Agriculture is limited to the hardiest of crops; potatoes, beans and barley are grown wherever water is available. The staple industry is pastoral, depending on the mountain wool-bearing llama, alpaca and vicuña.

Above the puna and up to the snow-line, intruded by the tongues of glaciers, stretches the bleak, uncultivable moorland known as the *paramos,* a region of still more stunted vegetation, lichens and mosses with, in favoured spots, mats of plants bearing large bright blue and purple flowers, but barren and inhospitable in the extreme.

TIBET AND THE PLATEAUX OF CENTRAL ASIA

The Himalaya Mountains, like the Alps, function as a climatic divide, barring the way against the summer monsoon purely by virtue of their relief. To the north of them lies the lofty plateau of Tibet whose northern rim is the Kuen Lun, beyond which lie the desert depressions of Eastern Turkestan and the Tsaidam. The last vestiges of the monsoon climate are recognizable in the south-east of Tibet, in the deeply trenched valleys of the upper courses of the Tsanpo, Salween, Mekong and Yangtse, by way of which the moonsoon currents find access to the plateaux and bring as much as 40 inches of rain to Lhasa. This is the most favoured part of Tibet, fortunate in possessing an Alpine type of relief, with fertile aggraded valleys and hill-sides which are densely forested but can be cleared for the cultivation of maize, millets and wheat. It is the most densely populated part, and the only towns of any size are to be found here.

TEMPERATURE

Station	Lat.	Long.	Alt. (ft.)	J	F	M	A	M	J	J	A	S	O	N
BROCKEN . .	52°N.	11°E.	3,773	24	23	26	33	42	47	50	49	45	38	30
SCHNEEKOPPE .	51°N.	16°E.	5,300	19	18	22	29	38	44	47	46	41	34	26
OBIR . . .	47°N.	14°E.	6,643	20	21	23	29	37	43	48	48	43	35	27
SONNBLICK . .	47°N.	13°E.	10,097	9	8	11	15	24	30	34	34	30	23	16
SÁNTIS . . .	47°N.	9°E.	8,202	16	16	17	24	31	37	41	41	37	30	23
ANDERMATT .	47°N.	9°E.	4,741	20	24	28	36	43	50	53	52	47	39	30
DAVOS . . .	47°N.	10°E.	5,121	19	23	27	36	41	50	54	52	47	38	30
BEVERS . .	47°N.	10°E.	5,610	14	20	24	33	42	49	53	51	46	36	26
PUY DE DÔME .	46°N.	3°E.	4,780	28	29	30	35	41	48	52	52	48	40	34
PIC DU MIDI .	43°N.	1°W.	9,380	18	18	19	22	29	37	44	44	38	30	24
SITNIA KOWO .	42°N.	24°E.	5,709	23	25	27	36	45	50	55	55	48	41	32
PIKE'S PEAK .	39°N.	105°W.	14,111	2	4	8	13	23	33	40	39	32	22	11
LEH . . .	34°N.	78°E.	11,503	17	19	31	43	50	58	63	61	54	43	32
AREQUIPA . .	16°S.	72°W.	8,041	58	58	58	58	58	57	57	57	58	58	58
USPALLATA .	33°S.	70°W.	9,335	53	52	51	46	38	34	36	38	39	42	48

RAINFALL

D	Yr.	Ra.	J	F	M	A	M	J	J	A	S	O	N	D	Total
25	36	27	—	—	—	—	—	—	—	—	—	—	—	—	—
20	32	29	2·2	2·0	2·2	2·3	3·0	4·0	4·7	3·5	3·6	2·8	2·5	2·5	35·3
22	33	28	3·1	3·4	4·6	5·3	5·4	6·6	6·7	6·4	5·5	5·9	4·1	3·4	60·4
11	20	25	4·9	4·9	6·3	6·6	6·2	5·5	5·6	5·1	4·6	5·1	4·6	5·3	64·7
17	27	25	5·7	6·7	6·7	8·1	7·8	11·2	12·3	10·8	8·3	7·2	4·8	6·1	95·7
22	37	33	3·8	4·2	3·5	3·2	3·4	3·5	4·3	4·7	5·5	5·6	3·1	2·9	47·7
21	37	35	1·8	2·2	2·2	2·2	2·3	4·0	4·9	5·0	3·7	2·7	2·2	2·5	35·7
16	34	39	1·4	1·0	1·6	2·2	2·6	3·4	4·3	4·3	4·2	3·5	2·4	1·8	32·7
29	39	23	6·5	6·0	6·5	5·5	4·8	5·5	4·7	5·3	5·3	5·4	5·1	5·6	66·2
19	29	26	5·3	6·6	6·4	6·8	5·7	5·1	2·8	2·9	4·4	4·6	5·5	6·5	62·6
27	40	24	2·2	2·2	3·2	3·2	4·2	4·4	3·4	3·2	3·2	2·3	3·6	1·6	36·7
6	19	38	1·6	1·5	2·0	3·5	3·8	1·6	4·2	3·8	1·7	1·4	1·9	2·6	29·6
22	41	46	0·4	0·3	0·3	0·2	0·2	0·2	0·5	0·5	0·3	0·2	—	0·2	3·2
58	58	1	1·2	1·7	0·6	0·2	—	—	—	—	—	—	0·1	0·4	4·2
53	44	19	0·8	1·2	0·2	—	0·1	0·6	—	1·6	0·2	1·3	0·8	0·2	7·0

The rest of Tibet is an arid plateau, virtually desert, most of it with less than 10 inches of rain or snow, its aridity increased by the diminished pressure and high winds. About 2 inches of snow falls in the south-west, winter precipitation related to the cyclonic storms which bring rainfall to the Punjab and Kashmir (see p. 140). The Pamirs and the Tian Shan form altitude oases with 25 inches to 30 inches (mostly summer rain) and separate the deserts of Eastern from those of Western Turkestan, but the rest of the plateau of Central Asia is so dry that the snow-line is some 4,000 feet higher than on the south face of the Himalaya (here at 16,000 feet) and the streams flow only in summer when the snows are melting. The lakes are mostly saline and there are expanses of dazzling white salt deposits that mark the site of others that have vanished.

Winter is rigorous, the excessive cold (readings of $-30°$ and $-40°$ appear to be not uncommon) being aggravated by high winds during the day and sometimes by blizzards of great violence. Summer is short, the growing season lasting only from April to September, and no month is free from the risk of frost owing to the large diurnal range of temperature. The nights are generally calm and clear and the early mornings are generally the pleasantest time of the day, for strong winds which spring up in the morning are still, as in winter, one of the most unwelcome features of the climate, bringing clouds of penetrating dust and sometimes even snow. They are not, however, without their uses, for they help to sweep away the snow from the pastures, especially in spring.

SUGGESTIONS FOR FURTHER READING

The influence of altitude on the various elements of climate are fully treated in Hann's *Handbook of Climatology* (Ward's translation); the climate of Switzerland in J. Maurer et al., *Das Klima der Schweitz*, 1864–1900, 2 vols., Frauenfeld, 1905. A good account is also given by E. De Martonne in *Les Alpes*, 1926. The influence of insolation and aspect are interestingly treated by Alice Garnett in Publication No. 5 of the Institute of British Geographers, 1937, *Insolation and Relief*.

XVI

CHANGES OF CLIMATE

No branch of climatology has received so much attention during recent years as that of climatic change; geology, botany, zoology, anthropology, meteorology, astronomy and other kindred sciences are daily supplying an almost overwhelming mass of fresh evidence bearing on the subject; theories as to its causes are being acclaimed, rejected, revived and modified from day to day; few are without serious objections and none find universal acceptance. It will only be possible here to give an outline of the better established facts relating to the climates of the past and to discuss briefly the means by which they may have been brought about.

So long as the Nebular Hypothesis of the origin of the solar system held the field, it seemed natural to suppose a certain amount of progressive cooling of climates on the earth's surface, and there appeared to be much confirmatory evidence in the geological record for warmer climates in the past; but unfortunately for this view some of the oldest rocks known contain records of an Ice Age in what are now temperate latitudes, and it is clear that from very early times there have been climates of a nature comparable with those of today. Life, evidenced by fossils, has existed on the earth from before the Cambrian and probably from a very much earlier period, from which it is apparent at least that the somewhat narrow temperature limits of life have not been overstepped at any time since then.

Evidence for changes of Climate in Geological Time. In the more remote periods the nature of the fossils is not of great assistance in identifying climatic types, for most of them are aquatic forms showing very little adaptation to climatic conditions, but with the greater specialization of forms it becomes possible, by Mesozoic times, to recognize differences between boreal and tropical types. Still later, with the evolution of more specialized land fauna and flora, much information can be deduced from fossils as to the climatic conditions of their habitats, e.g. the mammoth, the sabre-toothed tiger, the arctic willow, etc.

The evidence provided by the nature of the rocks is more satisfactory, e.g. boulder clays indicating glacial climates, salt and gypsum deposits indicating aridity, coral limestones indicating warm seas, coal seams which, from the absence of annual rings of growth in the trees, have been interpreted as tropical forest vegetation. Other plants (see Fig. 77) did develop rings, demonstrating the contemporary existence in other parts of the world of seasonal changes, which are further proved by alternations of coarse and fine sediments, resulting from seasonal flood and low water (see p. 298), and alternations of rock salt and gypsum. Rock salt is more soluble in warm water than in cold, gypsum more soluble in cold water than in warm; seasonal changes of temperature result, therefore, in the crystallization, from a saturated solution, of salt in winter and gypsum in summer. Judged on such criteria as these the British Isles appear to have passed through many climatic vicissitudes, including tropical rain forest in the Upper Carboniferous, sub-tropical arid in the Triassic, cool-temperate in the Cretaceous, warm-temperate in the Eocene, and arctic in the Quaternary.

Evidence of Climatic Change in Historical Time. There can be no reasonable doubt that though the climates of geological time differed in no fundamental respects from those of the present time, the distribution of the climatic zones and their limits have not always coincided with those of today. But with regard to historic time (using historic in its widest sense to include the record of the last 7,000 years for which we have written records, however scanty) there is less, though increasing, agreement among authors. The evidence is collected from a great variety of sources, few of them absolutely convincing in themselves, but all pointing with such a high degree of accord in the one direction that it seems impossible to escape the conclusion. Furthermore, as will be shown later (p. 301), modern research tends to show that much of the later 'geological time', for which climatic oscillations are definitely proved, was contemporary with the earlier 'historical time' elsewhere.

Early efforts to demonstrate these changes were stultified by attempting to prove a gradual and progressive desiccation—a wrong conception and one easily defeated. At the present moment the prevalent hypothesis, well supported by the facts, is that of oscillation of climate above and below a certain mean; periods of warm climate alternating with cooler, periods of dry climate alternating with wetter. The evidence includes:

Rainfall and other climatic records such as the meteorological register kept by Claudius Ptolemaeus at Alexandria in the first century and that of Tycho Brahe at Uranienborg in the sixteenth.

Records of floods and droughts.

Records of dates of sowing and harvesting of crops (there are records of the wine harvest in parts of Europe since 1400).

Records of dates of freezing of harbours and rivers (winters with ice off the Danish coast have been recorded since 1350).

Descriptions and comments on weather in contemporary literature; usually relating only to phenomenal weather.

Legends, e.g. 'The Flood' and 'The Twilight of the Gods'.

Variations in width of spacing between the annual rings of growth in trees, especially in the sequoias of California, some of which are more than 3,000 years old.

Past distributions of plants sensitive to climatic limitations, e.g. the date palm and the vine.

Dead forests in lands with rainfall now inadequate for forest growth; peat bogs in lands too dry at present.

Evidence of settlement (e.g. ruined cities) where settlement is now impossible because of climate, e.g. Palmyra in the Syrian desert which, it is claimed, must have had a population of more than 100,000 but now has not enough water supply for 1,000.

Evidence of agriculture (e.g. wine presses and threshing floors) where agriculture is now impossible.

Roads round dry lake basins and bridges over watercourses now dry.

Irrigation works where rainfall is now adequate or old conduits from sources now dry.

Records of lake levels, e.g. the Caspian Sea and Victoria Nyanza.

Old strand lines of lakes; dried-up lakes and salt-pans.

Advance and retreat of glaciers.

Burials in Greenland: Coffins have been excavated from soil which is now permanently frozen. The coffins were penetrated by plant roots showing that the summer thaw must have penetrated much more deeply at the time of burial.

Migrations of peoples on a large scale which may be ascribed to, or at least correlated with, increasing drought of their home region.

All this material must be treated with great care and circumspection, for there is a grave risk of discovering the changes of climate for which one is seeking in phenomena really due to other causes. Thus misrule may lead to the decay of irrigation works and so to the abandonment of settlements; great invasions and migrations may have their cause in personal ambition; improvement of varieties may allow the extension of cultivation into areas previously considered unprofitable; irrigation works may alter the level of lakes and rivers; new crops may put old ones out of use; wind-blown sand may destroy oases; or a few barren years may result in the evacuation of irrigated lands which the desert then reclaims.

CAUSES OF CLIMATIC CHANGE

Any change in the distribution of the climatic elements presupposes a change in climatic factors, nearly all of which have, at one time or another, been held entirely responsible for Ice Ages; but so complicated are the facts that it is unlikely that it will ever be possible to explain them all on the supposition of a single variable. Most of the earlier hypotheses assumed a change of solar climate, a change in the amount of the heat received from the sun or in its seasonal distribution.

Croll's Theory. Thus Croll based his hypothesis on changes in the eccentricity of the earth's orbit and supposed that at periods of high eccentricity the long winter of the hemisphere in aphelion would not be balanced by the short hot summer in perihelion. Ice would accumulate in winter and not be dispersed in summer and would ultimately grow into ice-sheets. The precession of the equinoxes, which depends on the eccentricity of the orbit, has a period of 25,000 years, and one would expect alternate glaciations in each hemisphere recurring with that frequency. There is, however, a further effect, produced by the bunching of the planets, which further increases the eccentricity and gives a period of 100,000 years. The hypothesis does not fit the facts as they are now known; the glaciations in the two hemispheres were synchronous, not alternating, their period is neither 25,000 nor 100,000 years and is not even regular; there have been oscillations of climate between glacial periods for which the theory offers no explanation and finally the close of the Ice Age appears to have been only about 15,000 years ago, compared with Croll's calculation of 80,000 years.

Drayson's Theory was based on another variable, the obliquity of the plane of the ecliptic which, at present, is $23\frac{1}{2}°$. The amount by which this has varied is not agreed upon; Stockwell and Milankovitch give the figure as $22°$ as a minimum and $24\frac{1}{2}°$ as a maximum, Lagrange gives $21°$ and $28°$, but Drayson required $11°$ and $35°$—an amount no astronomer is prepared to allow. It is claimed that at maximum obliquity the polar regions, large areas of which would experience the long polar night, would be glaciated. Like other mathematically controlled theories it demands regular, recurrent glaciations, the interval being about 20,000 years and the last retreat being placed 7,000 years ago. The facts, however, are otherwise.

Tyndall, Chamberlain, Humphrey, Frech and others. There is a group of theories which seek to explain Ice Ages and other climatic changes by variation in the amount of carbon dioxide, volcanic dust and other impurities in the air, which, by checking outward radiation, help to maintain the earth's temperature. Careful inquiry and experiment show that they are quite inadequate alone to produce the known effects; furthermore, Ice Ages do not follow great volcanic outbursts such as might increase the carbon dioxide or volcanic dust content of the air.

The Sunspot Cycle. Huntington and Visher have elaborated a theory which is partly based on the variation in intensity of solar radiation with the development of sunspots. It is a curious paradox that sunspot maxima, at which time the sun's energy is greatest, are correlated with low temperatures on the earth. They are also correlated with an intensification of pressure gradients (the permanent highs are higher and the permanent lows are deeper), a greater storminess, especially in certain belts, and a slight redistribution of rainfall. It is probably the greater storminess which accounts for the lowering of temperature, heat being lost by convection and ventilation. The sun-spot cycle has a period of about 11 years, a figure which emerges in many meteorological phenomena, e.g. in the oscillations in level of the African lakes, in the spacing of the annual rings in the Californian sequoias and in the annual laminæ in the 'varve' clays of North America during the ice retreat at the end of the last glacial period.

The Brückner Cycle. There are other frequencies which emerge from time to time in an empirical study of the periodicity of phenomena, but whose causation remains enigmatical. The best known of these is the 35-year or Brückner Cycle, found by Brückner to recur, albeit very irregularly, in a variety of phenomena including the advance and retreat of the Alpine glaciers, variations in the dates of opening and closing of Russian rivers, the level of the Caspian Sea and the rivers emptying into it, the date of the grape harvest and the price of grain. There would appear to be irregular alternations of cold, wet periods with hot, dry periods especially in the continental cool-temperate climates, which might be regarded as fluctuations in the distance of penetration of marine influence. In hot, dry years the marine influence is restricted to a narrow western zone and the continental type of climate spreads westwards, rain decreases by about 20 per cent in Germany and 30 per cent in Russia. Consequently harvests suffer in the East (where normal rainfall is on the margin of sufficiency) and benefit in the West (where rainfall is usually excessive and sunshine deficient). There is generally a small lag of dry seasons after warm and of wet seasons after cold, as the following figures show:

Warm, 1746–1755, 1791–1805, 1821–1835, 1851–1870.
Dry, 1756–1770, 1781–1805, 1826–1840, 1856–1870.
Cold, 1731–1745, 1756–1790, 1806–1820, 1836–1850, 1871–1885.
Wet, 1736–1755, 1771–1780, 1806–1825, 1841–1855, 1871–1885.

Kreichgauer and Wegener. The most difficult single fact to explain in all the history of climate is the glaciation of tropical latitudes in Permo-Carboniferous times. A courageous attempt to solve this was made by Kreichgauer, who supposed that the position of the poles had varied through geological times. The same idea is incorporated by Wegener in his ingenious theory of Continental Drift and applied by Köppen and Wegener to the climates of the past. The theory is carefully worked out and succeeds in explaining many hitherto irreconcilable anomalies of past climates and the past distribution of plants and animals. Briefly the conception is one of rigid blocks of

continental material floating in a more or less fluid substratum and
free to move westwards and equatorwards in obedience to forces which,
it must be confessed, are generally agreed to be inadequate to overcome the resistance. The land masses of the world are conceived to

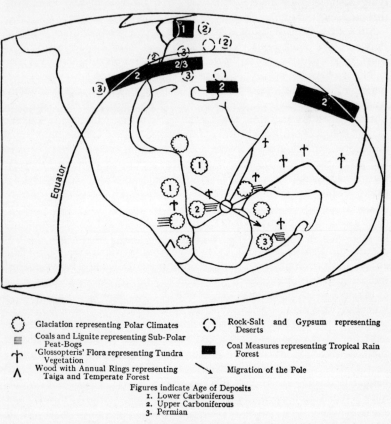

Fig. 77.—Evidence for Climatic Zones in Permo-Carboniferous Times
(*After* Wegener)

have been more or less united in Carboniferous times into one large
continent, Pangaea, which contained the pole of the period in what is
now South Africa, but with which Brazil, India and Western Australia
were in close juxtaposition. Even this compression is inadequate to
bring together within reasonable compass the great expanses in these
areas which were glaciated at the time and it is further suggested that
the pole migrated slowly south-eastwards from Brazil to Australia
between early Carboniferous and Permian times, and that the
glaciated area and presumably all the climatic belts migrated slowly

in company with it (see Fig. 77). Fig. 77 reproduces Wegener's diagram summarizing the evidence for the distribution of climatic zones at this time; a concentric arrangement round a pole in South Africa is clearly brought out, the equatorial forest, represented by the coal measures, lying along a great circle exactly 90° from the Pole and extending through North America, across Europe and Asia to China.

Subsequently rifts appeared in this great continental mass of Pangaea, the component parts spreading and opening fanwise to the south to allow the birth of the Indian and Atlantic oceans.

The palaeoclimatology of other periods is worked out with equal ingenuity, but it is impossible here to enter into a description of them. At first sight the theory appears to supply the key to all past climates, but on closer inquiry difficulties appear, many of which are so grave as to compel postponement of its acceptance. The Quaternary glaciation, for example, is a very recent episode, yet Wegener requires the postglacial separation of North-west Europe and North America, thus demanding a westward migration of America at a phenomenal rate. Again, the 'squantum tillite' in the Carboniferous of Boston, Mass., a boulder-clay of obvious glacial origin, associated with 'varve' clays (see p. 298), occurs in close proximity, both in space and time, to coal-beds which are considered to represent the tropical forest. Again, the British Isles during the Cretaceous are placed in latitude 20° N., but the chalk in our islands contains far-travelled erratic pebbles, indicating transport by shore ice or icebergs and therefore the proximity of glaciers reaching sea-level.

Relief and Glaciation. It must be remembered that altitude, as well as latitude, brings about a fall in temperature and that the sequence of climatic zones up a slope is similar in many respects to that along a meridian. Altitude has been invoked to explain the Permo-Carboniferous glaciation of the tropics, but it is difficult to believe that such a huge area can have risen above the snow-line which must have lain at least 8,000 feet up if conditions were in any way like those of today. The altitude of the land may well have been one of the contributory causes of glaciation [it can be shown, for example, by means of a buried channel of the St. Lawrence, that the Great Lakes region of North America stood some 2,000 feet above its present altitude (related to sea-level) during the Pleistocene], but it can hardly be held accountable for glaciation on a widespread scale.

Continentality and Climatic Changes. But the indirect results of high relief are more far-reaching and may have a profound effect on remote areas by modifying the movements of air and of ocean currents—factors which are powerful enough to overcome the effects of latitude, as the 'Gulf of Warmth' demonstrates in the North Atlantic. It is a significant fact, to be dealt with in more detail below, that glacial periods are invariably associated in geological history with

strong relief, and Ramsay has carefully developed this conception of
intimate relationship between mountain-building and climate.

The heat received from the sun in the tropics is so far in excess of
that in high latitudes that the tropics would be unbearably hot and
high latitudes so cold as to be uninhabitable were it not for the
redistribution brought about by winds and currents. Clearly any
interference with the freedom of this circulation will induce extremes
of climate, while any change which encourages the circulation and
increases its efficiency will result in a more even distribution of temperature over the earth's surface. By analogy with the conditions in
the North Atlantic and the Barents Sea we may imagine what would
be the result of opening wide the Bering Strait and giving free access
for the warm North Pacific drift into the Arctic Ocean. The climate
of Alaska would resemble that of Norway and there would be open
water in winter perhaps as far as the mouth of the Mackenzie River.
Conversely, the closing of the North Atlantic by the emergence of the
Scoto-Icelandic Rise would be followed by the refrigeration of the
Norwegian Sea and the rapid growth of ice-caps in Norway and
Sweden which, after all, are in the same latitude as South Greenland.
The indirect results on pressure and winds would be very complicated,
but not impossible to reconstruct, and the changes of climate would
be widespread and profound.

The 'Change of Level' Hypothesis. By a careful and detailed
application of such principles Brooks has given precision to this 'change
of level' hypothesis which had previously suffered from vagueness and
loose treatment. His method is to work out a basal temperature for
each latitude (i.e. the temperature found near the centre of an ocean
in that latitude) and to reconstruct the summer and winter temperatures for any given period by adding to or subtracting from that basal
temperature figures which represent the effect of continentality. In
the case of the Pleistocene and later changes two wind systems only
are concerned—the westerlies and, north of 70°N., the polar easterly
winds. In the zone of influence of the former the effect of land to the
west is to lower the winter temperature (vide Manchuria), while
the effect of land to the east is negligible (vide Norway); in summer the
effect of land either to east or west is to raise the temperature (vide
Siberia); while at any season the effect of ice is to lower the temperature.
The amount of lowering or raising of temperature is worked out from
present conditions and applied to well-established palaeogeographical
reconstructions of the distribution of land and sea; the values arrived
at are checked over with values estimated by other authors on independent bases and show, on the whole, quite good agreement. A warning
may be sounded here that there is a risk that the degree of accuracy
apparently obtained may be misleading, for our knowledge of past
geographies at any precise time is not very reliable.

The reconstruction of precipitation is a more complicated and difficult matter, but much can be done by applying certain principles, e.g.:

1. That convectional rainfall is mainly equatorial and continental, the amount of the latter type being proportional to the distance from the moisture supply.

2. Orographic rainfall occurs mainly on windward coasts, the amount being proportional to the humidity of the air and the relief of the coast.

3. Cyclonic rainfall depends on the facilities offered for the passage of depressions; i.e. the storms are generated over oceans, pass eastward, following water areas and low land, die out over land and avoid anticyclonic areas.

Each of the above hypotheses seeks to show how the variation of one or more climatic factors might bring about variations in the climatic elements, culminating in Ice Ages. None of them can be rejected as having no influence at all, though many can be dismissed as primary causes. We shall see later that changes of relief and level offer the only really satisfactory explanation of the major events, but that it is likely that some of the others must be invoked to account for minor oscillations.

THE EVOLUTION OF CLIMATE

Mild Geological Climates. Glaciation is the exception rather than the rule in the climates of the past, the more normal conditions, typified by the Silurian, the Lower Carboniferous, the Jurassic and the Eocene, have been characterized by the mildness and a uniformity of conditions which is surprising to us who live in what is still a glacial epoch, though apparently on the wane. Reef-building corals, which at present inhabit seas with a mean annual temperature above 68°, occur fossil into latitudes where now the mean temperature is below 50°, where long, severe frosts occur and where there is floating ice in the seas. Sub-tropical plants such as magnolias, sequoias, etc., occur fossil in the Eocene of the Arctic, the range of palms and cycads was very widely extended. In short, the floristic and faunistic (and presumably the climatic) distinctions between intermediate and high latitudes were much less clearly marked than today, or, in other words, the temperature belts appear to have extended polewards far beyond their present limits.

Causes of Mild Climates. These long periods of uniformly mild conditions correspond to periods of peneplanation and uniformly low relief. Denudation and deposition, working slowly but incessantly, had reduced the mountains to stumps and base-levelled the continents to flat and monotonous plains. The seas, partially filled with sediment from the wastage of the land, had spread in shallow sheets over the low flat margins of the continents. The widespread seas allowed their equalizing influence to spread unimpeded over the whole earth, perhaps even to the poles. They provided a large surface for evaporation, the air was humid and the high water vapour minimized the escape of heat out into space, especially in high latitudes whose atmosphere at the present day can hold very little because of the cold. The low relief offered the minimum surface for radiation (mountains are very wasteful of heat in this respect) and the absence of mountain barriers permitted moisture-laden currents to enter far into the interior of the continents, which were small and low, the wide sluggish rivers and swampy flat plains constantly renewing their humidity. The ice-sheets and glaciers of the last mountain epoch had disappeared, gradually retreating as the relief was lowered, and their chilling influence was withdrawn. The sea-bottom temperatures were higher and there was little cold upwelling along the leeward shores of land masses; warm water, on account of its lightness, covered the surface of the ocean and there were no cold surface currents, such as the Labrador current, which depend for their persistence as surface currents today on the continual supply of fresh (and therefore light) cold water from the melting of ice. In the absence of polar ice-caps the polar anticyclones had ceased to exist and the warm westerly winds extended to the poles. With the vanishing of the polar anticyclones passed the 'polar front' and with it most of the storms of this now turbulent zone.

Glacial Climates. This placid and uneventful régime was brought to an end by the next orogenic revolution. The land began to emerge from the sea, hollows were formed in the sea-bed into which the waters withdrew, the sea shrank back laying bare the margins of the lands. The air became drier, the climate more continental. New land masses began to interfere with the circulation of the oceans, new mountain chains excluded the moderating influence of the sea from the hearts of the continents, where deserts came into existence. The polar regions, robbed of warming currents, grew steadily colder; ice began to form in winter; from year to year it appeared earlier and lingered later; finally it survived the summer and ice-caps formed and grew. For their nourishment water was withdrawn from the oceans whose level fell farther and whose influence was further restricted. As the poles grew colder the temperature gradient was increased between them and the Equator, the polar anticyclone came into existence, the 'polar front' was born and storminess increased. Floating icebergs carried arctic cold into lower and lower latitudes, the melting of the glaciers and ice-caps in high latitudes yielded a constant supply of cold water which slowly sank and crept along the ocean floor to the Equator, chilling the rest of the ocean by contact and mixture.

Now an ice-cap, once formed, is difficult to dislodge, and the larger it grows the more firmly is it established. Its white surface reflects much of the solar radiation and in warm weather a mist tends to form which protects it from the sun's rays. As it grows larger it develops its own anticyclone which forces cyclonic storms to pass along its equatorward margin, supplying it with fresh moisture and helping it to grow by marginal accretion. As the anticyclone extends, the polar limit of the westerlies is driven nearer and nearer to the limit of the trades, the pressure gradient becomes steeper, and storminess increases further.

Crises of Evolution. Thus from a single cause there follow a great variety of results, which, in turn, set in motion other causes, all of which operate in the same direction. The new relief calls into existence a variety of climatic types, often sharply demarcated by mountain boundaries; new climatic environments are created and to the stimulus of the new demands nature responds with marvellous adaptability. New species arise, narrowly confined within climatic zones; there is a wholesale extinction of unadaptable forms. These are the crises of evolution out of which have arisen new virile groups, stimulated by hardship and triumphing over difficulties, to become the dominant stock of the following age. Out of the Hercynian revolution and the climatic stress that followed arose the reptiles who dominated the Mesozoic land fauna, from the Alpine revolution and the climatic changes that followed man received the impulse that has brought him to his present state.

Frequency of Glaciation. The geological record shows that there have been at least four periods in the earth's history when true glacial conditions existed and these follow close after four of the greatest mountain-building eras. These may be dated as follows, the earlier ones by means of the 'radio-active clock', the last by other more accurate means (see p. 298):

Quaternary, 700,000–20,000 years ago, following the Alpine orogeny.
Permo-Carboniferous, 260,000,000 years ago, following the Hercynian orogeny.
Lower Cambrian, 500,000,000 years ago, following the Late Proterozoic orogeny.
Lower Proterozoic, 750,000,000 years ago, following the Early Proterozoic orogeny.

Thus there is a periodicity of about 250,000,000 years governing these major revolutions, a cause for which has been suggested by Joly and Holmes as being due to the periodic melting and resolidification of a basaltic substratum depending upon the accumulation of heat generated by the disintegration of radio-active minerals. In addition to these there have been minor revolutions each followed by minor refrigerations, but not, so far as is known, culminating in Ice Ages.

It may be significant that the most important of these, the Caledonian (Silurian to Devonian) and the Laramide (Cretaceous) gave birth to mountain ranges oriented more or less meridionally and therefore interfering less with the circulation between Equator and poles.

It is at first difficult to see why there should be such a considerable lag of the cold climates after the orogenic maxima; the Alpine maximum occurred in the Miocene, yet warm climates persisted through the

Fig. 78.—Land and Sea during the Eocene

early Pliocene and glaciation did not reach a maximum until the Pleistocene. But the maximum of relief does not coincide with the maximum orogeny; isostatic readjustments result in steady uplift for some time afterwards, and there are other factors, as we shall see later, which cause a lag of the glaciation behind the highest relief.

Tertiary Climates. The early Tertiary illustrates the conditions that have already been described as 'mild geological climates'. Fig. 78

Fig. 79.—Land and Sea at Maximum Glaciation
More important ice-caps shown shaded

shows the distribution of land and sea in Eocene times, from which it is seen that there are four wide and independent lines of sea communication between the Arctic and the seas to the south through which there was a free exchange of water between the Tropics and the Arctic. Then came the Alpine crustal movements, culminating in the Miocene, but followed by a long period of readjustments and vertical movements, the highest relief being probably reached early in the Pleistocene, since when erosion has probably succeeded in lowering it to some extent. The resultant refrigeration begins to be noticeable in the Later Pliocene rocks of East Anglia, towards the end of which period a prevailing north-easterly wind banked up shells against the coast. This north-east wind may be the first indication of a clockwise swirl round the glacial anticyclone forming over Scandinavia. Fig. 79 shows the advanced stages of the process of elevation which finally established the Ice Age.

Glacial and Interglacial Periods. At this point it may be mentioned that the Pleistocene glaciation was by no means a simple and single event, nor, it may be argued by analogy, were previous glaciations. There is abundant evidence that there were 'interglacial' periods within the Ice Age when temporarily the climate improved and at times was milder than at the present day. The evidence for this consists of:

1. Alternations of glacial with non-glacial deposits, e.g. boulder-clay with fluvioglacial gravels or even with sediments laid down in a fairly warm sea.

2. Alternations of arctic and temperate floras and faunas. The typical glacial fauna of North-west Europe includes the mammoth, the reindeer, the musk-ox, the woolly rhinoceros, the variable hare, the Arctic lemming, etc., the typical interglacial fauna includes the cave-lion, cave-hyena, the straight-tusked elephant, etc.

3. Oscillations in the position of the ice front with important results in river drainage; advance and retreat of glaciers.

4. Alternation of glacial and fluviatile relief types in the evolution of valley profiles and other scenic forms.

5. Formations and dissection of river terraces probably due to changes in the régime and load of rivers.

Penck and Brückner established the succession of four glacial periods in the Alps separated by three interglacials, but these cannot always be recognized on the lower ground. In Britain only two major glacial advances are recognized, the *Older* and the *Newer drift* and in the North German plain there are three terminal moraines representing major periods of glaciation. In North America five are accepted; the general correlation is as follows:

Alps	North German plain	Britain	North America	Date
Gunz	—	} Older Drift	Nebraskan or Jerseyan	—
Mindel	Elster		Kansan	430–370,000
Riss	Saale	} Newer Drift	Illinoian	130–100,000
Wurm	Weichsel		Wisconsin	40–18,000

Causes of Interglacial Periods. To understand these oscillations of climate it is necessary to examine carefully the sequence of events which may be expected in an area covered by an ice-cap and in adjacent areas. We have seen (p. 292) that as the ice-cap grows by the congealing of precipitation the level of the sea sinks and that of the land rises relatively; on a conservative estimate of the dimensions of the ice-sheets this depression of the sea-level has been estimated at about 400 feet. After a time, however, the growth of the ice-cap constitutes an excess load on the surface and the glaciated area begins to sink. If isostatic compensation were perfect the amount of depression would be about one-third of the thickness of the ice accumulated (for the ice has about one-third of the density of the substratum), but because of the rigidity of the crust the compensation is not complete and there is a considerable lag behind the theoretical amount. The result of this sinking of the ice-caps was to bring them below the level of maximum precipitation and in some cases to bring them below the snow-line so that ablation began to gain on accumulation. Furthermore, the depression of the whole area permitted a partial return to freer circulation of currents and winds, an effect which was rapidly increased when the melting of the ice-caps returned water to the oceans and raised the sea-level. There was, in fact, a temporary return to non-glacial conditions.

But the depressed areas, relieved of their load of ice, began to rise again, eventually rising above the snow-line; accumulation of snow began to exceed ablation and glacial conditions returned. Thus as a result of the lag of compensation there would be oscillations on either side of the stable condition before equilibrium was reached. But it is not, of course, a perpetual motion cycle, the amplitude of the oscillation must be smaller each time and stability must eventually result unless and until factors external to this process undergo a change.

THE CLIMATIC BELTS AT THE TIME OF MAXIMUM GLACIATION

The glacial maxima were by no means of equal severity, nor were the climatic conditions within or without the glaciated area the same during each, but with each advance there occurred certain migrations of climatic belts, of which the following pages give a generalized account.

Europe. Along the margins of the ice the climate was of the *tundra type*, though possibly the summers were warmer than in the tundra of today. Survivors of this tundra flora are found today at the higher levels in Scotland, Scandinavia and the Alps, whither they retreated as the climate improved. The tundra zone did not occupy all the European plain; in the east there was a *steppe climate* of an extreme continental type dominated by the east winds which prevailed on the southern edge of the glacial anticyclone. Steppe animals and plants have been found fossil as far west as France, from which it appears that the influence of the western ocean was not felt far inland.

Pluvial Periods in the Great Basin. The southward spread of the ice-sheets was accompanied by an extension of the polar anticyclones with their prevailing east winds, and the sphere of influence of the westerlies with their cyclonic storms retreated before their advance. Hence on the equatorward margin of their present limit pluvial periods correspond to glacial advances. The Great Basin received a much increased rainfall and large lakes formed against the boundary scarps; the Great Salt Lake represents the dwindled survivor of a once extensive lake, known as *Lake Bonneville*, whose shore-line may still be seen about 1,000 feet above the present level of the lake. Numerous small salt lakes in Western Nevada, Humboldt, Pyramid, Walker, Winnemucca, Honey, and the Carson lakes are the remnants of another known as *Lake Lahontan*.

Pluvial Periods in North Africa. The Mediterranean became a much frequented track for cyclones and its shores were well-watered, in summer as well as in winter. Even in the Northern Sahara pluvial periods were contemporary with the glacial periods in higher latitudes. Where now there are dry wadis, or short streams losing themselves in the sand, there were then continuous rivers whose graded courses (in contrast with the ungraded wadis) and normal fluviatile form with interlocking spurs may still be recognized.

The Birth of the Nile. Today northerly winds prevail over Egypt for most of the year, but above them are westerly winds which may be met on Mount Sinai. During the pluvial periods these westerly winds were the surface winds and brought heavy rain to the highlands which lie to the east of the Nile. Down these there flowed torrential rivers which carried volumes of silt and debris across what is now the valley of the Nile. But the Nile did not exist then. At the present day it suc-ceeds in crossing the desert mainly by the aid of the Blue Nile flood waters from the Abyssinian monsoon. The monsoon depends on the heating of the Asiatic continent, but during glacial times a colder Asia attracted a less powerful monsoon and Abyssinia had dry trade winds. The White Nile lost itself in the desert and probably did not succeed in crossing it to the Mediterranean until postglacial times, about 14,000 years ago, when the monsoon began to reach the Abyssinian mountains.

China. But if the summer monsoon was weak the winter monsoon was doubly strong, since the increased cold added to the intensity of the continental high. Thus in China the glacial periods of North-west Europe are represented by periods of increased aridity and periods of heavy aeolian aggradation, when the continental influence was redoubled and the winter winds transported still greater quantities of loess from the arid interior on to the plains.

The Sudan. On the equatorward side of the trade-wind deserts the climatic belts appear also to have shrunk towards the Equator, so that here the desert gained on the savanna. Lake Chad, for example, although it was once much larger, was also once much smaller. The shrinkage was probably 'glacial', the expansion 'interglacial'; at the present day the lake is increasing in size and in freshness—a postglacial recovery towards more pluvial conditions. Similarly Lake Titicaca, occupying an analogous position in 16°S., is at present increasing in size and drowning the valleys of the rivers that empty into it. There are 'fossil ergs', or vegetation-covered dunes from Lake Chad, along the upper course of the Niger, to the sea in Senegal, which testify to a southward encroachment of the desert, presumably during glacial maxima, and a northward postglacial retreat.

The Equatorial Regions. The concentration of the pressure belts doubtless strengthened the circulation of the trade winds and gave rise to a deep and stormy equatorial low with increased rainfall. The equatorial lakes of Africa swelled to twice their present size—Lake Victoria was continuous with Lake Kioga whose level was 600 feet above the present. The glaciers on Ruwenzori and Kilimanjaro descended by 8,000 or 9,000 feet to within 5,000 feet of sea-level, but the snow-line only descended about 3,000 feet, a clear indication that the cause was increased rainfall and not decreased temperature.

THE CLIMATIC BELTS DURING INTERGLACIAL PERIODS

During interglacial periods the climate reverted towards conditions more or less similar to those of today, but in Central Europe and North America the effect of the glacial anticyclones to the north was felt as strong, cold, dry winds and an unpleasant variety of steppe climate prevailed. These strong winds carried clouds of dust derived partly from the moraines uncovered by the retreat, partly from silt beds left by floods in basins of inland drainage during the preceding glacial period. This dust now forms the thick fertile cover of loess so plentifully distributed about the limits of glaciation. During the Riss-Wurm interglacial loess formation occurred on a large scale in Central and Eastern Europe, and during the corresponding Iowan-Wisconsin interglacial in the plains of the Upper Mississippi.

CLIMATES DURING THE RETREAT OF THE ICE

Chronology. The dating of events in glacial and postglacial time is not always very reliable. Dates are based mainly on the rate of growth of peat bogs and deltas and on the degree of weathering of deposits, but from the last retreat onwards we have an accurate method of computation by the actual counting of the seasons represented by alternate layers of coarser and finer laminae in the 'varve' clays of Lake Ragunda in the south of Sweden. The lake was drained in 1796, so that the top layer of silt represents the sediments laid down in that year. The coarser of the alternating layers beneath are assumed to correspond to the summer floods while the glaciers which discharged into the lake were rapidly melting, the finer layers to winter conditions when melting was slower or ceased. The same method has been employed in other areas, notably by Antevs in North America.

By these means De Geer has estimated that by 10000 B.C. the ice had left the Baltic and reached the present coast of Sweden. If this is so, then the retreat has been surprisingly rapid—too rapid, in fact, to find universal acceptance, but De Geer attributes it to the sudden invasion of the Norwegian Sea by the warm waters of the Gulf Stream Drift. Certainly during the latter part of the retreat there was an open sea, the 'Yoldia Sea', across the Baltic lands. In the comparatively short period since then there have been numerous changes of level of land and sea in North-west Europe, frequent openings and closings of the connections between the Baltic and the North Sea and between the Baltic and the Arctic. Inevitably these brought about oscillations of climate which have not ceased to the present day; during periods of high relief, when marine connections were severed, the climate of North-west Europe was severely continental; during periods of submergence the climate was milder, wetter and, in brief, more marine.

Changes of climate, if long enough maintained, must after some time-lag have brought about a change in the plant cover. Thus the stumps of forest trees, buried in thick deposits of peat, may be interpreted as a deterioration of the climate towards greater cold and wetness. But care must be taken to ensure that the change is more than just a local one. In correlating past floras, and in dating the deposits, much use is now made of pollen analysis. Tree pollen grains survive and can be identified in peats and other Quaternary deposits. The relative frequency of the different genera of trees can be expressed quantitatively as a 'spectrum', and by means of this device a chronology of postglacial colonization stages has been established and a correlation effected over large areas in North-west Europe and in North America, though the correlation must still be regarded as very tentative.

The Continental Phase (Boreal period). It has been shown that the rapidity of the retreat was related to the removal of the barrier between the open ocean and the Baltic, but about 6000 B.C. an elevation of the southern shores of the Baltic restored the barrier and gave rise to an enclosed lake, the 'Ancylus lake'. Marine influence was excluded and Central Europe, swept by dry easterly winds, had a strongly continental steppe climate such as is not found today west of the Urals. The rainfall was less than 20 inches. The Scandinavian anticyclone fended off the cyclonic storms which were concentrated along the western margin or else passed along the Mediterranean. Thus Britain, especially Scotland and Ireland, actually had a heavier rainfall than today, with a damp, chilly climate in which peat bogs grew (the 'Early Peat-bog Phase'). Elsewhere was a pervading aridity from which the people of Asia and Eastern Europe escaped westwards towards the better-watered lands. This is the Neolithic invasion which put an end to the artistic palaeolithic culture of Western Europe. The 'Nordics' came from Central Asia, sweeping westwards to the shores of the Baltic, the 'Alpines' followed a more southerly course into France and to the shores of the Atlantic.

The Maritime Phase (Atlantic period). Meanwhile the melting of the ice-caps was gradually raising the mean sea-level, and about 4000 B.C. an important change came about, namely, the invasion of the Ancylus lake by warm, salt Atlantic water through the Sound and the Belts. An open sea was produced, the 'Littorina Sea', by which marine influence penetrated to the farthest ends of the Baltic and the Gulf of Finland. Everywhere the level of the sea appears to have stood higher than today (the Middle Flandrian transgression) as is shown by raised beaches in Greenland, Spitzbergen, Franz Josef Land, Norway and Scotland (the 25-feet raised beach); the climate was milder,

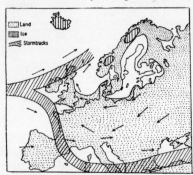
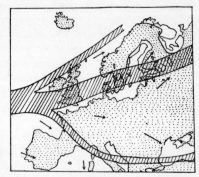

Fig. 80.—Climatic Conditions during   Fig. 81.—Climatic Conditions during
      the Continental Phase   the Maritime Phase

the mean annual temperature of the British Isles being probably 3° or 4° higher than at present. Cyclones began to penetrate along the warm waters of the Littorina Sea, bringing much increased rainfall and much milder conditions to the steppe lands of Russia; the volume of the rivers swelled and the lakes (Caspian and Aral) filled. The steppe dwellers must have prospered and multiplied on the rich pastures which resulted. This period is often referred to as the 'Post-Glacial Optimum of Climate', and, in fact, such mildness has never been repeated since, but Western Europe, and especially the British Isles, suffered from excessive rain, and from the point of view of human settlement the following period was more favourable here.

The Forest Phase (Sub-Boreal period). About 2500 B.C. an elevation of Europe along the central axis caused a restriction of the seas here while the Baltic remained depressed (the Limnaea lake). Much of the Irish Sea and the North Sea were dry land and England was joined to France where now are the English Channel and the Straits of Dover. Cyclones began to follow the more northerly track across the open waters of the Baltic and instead of, as now, crossing Eurasia by the Black Sea and the Siberian lowlands, entered the warmer Arctic by way of the White Sea. The sub-tropical high pressures extended northwards and at least during the summer months covered most of southern Britain, whose climate thus approximated to the Mediterranean type. Ireland, temporarily relieved of the embarrassment of too much rain, blossomed out into the 'Heroic Age', only to decline in the succeeding moister period. The northerly track of the cyclones robbed the steppe lands of their rain, the rivers dwindled and the level of the Caspian fell to a minimum about 2200 B.C. The desiccation of these lands drove out barbarian hordes who overran the better-watered lands of Mesopotamia and the Mediterranean. About 2000 B.C. the Hyksos, or 'shepherd kings' who, as their name implies,

Fig. 82.—Climatic Conditions during   Fig. 83.—Climatic Conditions during
         the Forest Phase   the Peat-Bog Phase

were a steppe people, invaded Egypt, conquering a sedentary people by the greater mobility given by the horse. The Canaanites migrated out of Arabia into Palestine; rather later (between 1600 and 1300 B.C.) the Indo-Europeans entered India by the North-West passes, the Hittites entered Asia Minor and the Achaeans settled in Greece.

The Peat-bog Phase (Sub-Atlantic period). By about 1000 B.C. fresh submergence had set in (the Mya sea) and drowned the marginal forests of the previous period. The encroachment of the sea brought on a moister climate and carried marine influence farther inland. The western margin was now too wet for forests and peat-bogs grew over their decaying stumps. Well-used cyclone tracks carried rainfall to two important areas: to the Steppe lands and via the Gap of Carcassonne to the Mediterranean; the steppe dwellers were settled and contented, the Mediterranean was prosperous and its states ambitious. In fact the misfortunes of this age are attributable to excess of rain rather than deficiency; the Bronze Age culture of Norway perished in the damp cold; the lake dwellings of Central Europe were drowned by the rising of the lake levels. The only migrations of the period are from moister into drier lands—a complete reversal of the usual direction—e.g. the Celtic movements eastwards along the northern foot of the mountains from the Italian Alps to the Caspian Sea. In the Mediterranean, which enjoyed adequate rainfall and a stimulating cyclonic climate, there grew up the great civilizations of Greece, Rome and Carthage, for which the cultivation of grain supplied a sound agricultural basis. Carthage found water supply a greater difficulty than the other two and was compelled to import water by aqueducts. The springs which fed these no longer flow, and Carthage could never have prospered as she did under the climatic conditions of today.

Dry Period (200 B.C.). The climatic arrangement outlined above endured, and with it the prosperity and high culture of the Mediterranean peoples, until the third century A.D., but there was a temporary decrease of rainfall about 200 B.C. which brought about an agricultural crisis in Italy and severely tested the stability of Rome. This dry period was felt also in the steppe and initiated an outward movement. Against this threat of invasion the settled agricultural civilization of China built the Great Wall, but without avail, the 'hunger urge' was stronger than the wall and China was overrun again and again.

Wet Period (100 B.C.). The dry period in the Mediterranean was only a temporary episode. A hundred years later agriculture had revived (though it is true the basis of rural economy was not grain but the more xerophilous olive) and Rome had entered on a period of prosperity with wealth and luxury greater than before. The journal kept by Claudius Ptolemaeus in the first century A.D. at Alexandria throws an interesting light on contemporary climatology. The record describes a high percentage of south and west winds in summer instead of the monotonous north winds of the present; the weather was more changeable and thunder was more frequent. Altogether the records indicate that the cyclonic westerly circulation was a much more important element in the climate of Alexandria than it is at the present day.

Dry Period (A.D. 200–1200). The revival of Rome lasted only about 200 years and then began a period of renewed stress. There was a further decay of agriculture, a shortage of corn and a migration of rural population into the towns; the rivers began to dry up, stagnant water supplied a breeding-ground for mosquitoes and malaria became endemic, sapping the vitality of the people. The steppe tribes, suffering also from drought, became restless; by the end of the first century they were on the Danube and Trajan was forced to conquer and annex Dacia to secure the Danube as a frontier. But there was no stemming the flood; Greece was invaded, Athens fell and then Rome. Out of Arabia came the Mohammedan wave of conquest, actually a religious crusade, but indirectly encouraged by years of drought and want.

Farther north, too, the climate was drying; in Britain the peat bogs ceased to grow and forests took their place. The climate here became more favourable for human occupation, but the outer waves of the disturbance propagated from the steppe prevented the establishment of settled order. The Huns pressed against the Goths, who pressed against the Germanic peoples, who escaped westwards and invaded these islands. The drought of the steppes is proved by the level of the Caspian Sea. In the fifth century the Red Wall was built as a bulwark against the Huns; today that wall extends 18 miles into the lake below the water.

Farther south there is also evidence for the poleward shrinkage of the climatic belts. The Mayan civilization of Yucatan reached its zenith early in the third century and then declined, overwhelmed, it is suggested, by the advance of the tropical forest and the insidious attack of the monotonously hot damp climate. Northern Yucatan apparently lay beyond the invasion of the forest and preserved some vestige of its culture, but the collapse in the south was complete.

The dry warm phase persisted into the tenth and eleventh centuries; Europe was invaded by Magyar horsemen and India by the Mongols. The climate of Northern Europe was so mild that Norway was able to plant and maintain colonies in Greenland where cattle thrived and corn was grown with success.

The Mediaeval Wet Period. The climate swung back in the thirteenth century towards moister and colder conditions. The glories of the Mediterranean were in some degree revived by the city states of Italy and by the high Moslem culture in Spain; the level of the Caspian rose until at the beginning of the fourteenth century it stood nearly 14 feet above the level of today; the steppes were well watered and its people were settled and contented. But the gain of the steppe and the Mediterranean was the loss of North-west Europe which now entered on a wet and stormy cycle. In the thirteenth century the coastal defences of Holland were frequently breached and the polders flooded. Sea-traffic was discouraged; Norway lost touch with her Greenland colonies which were gradually frozen to death.

The Last Centuries. In the sixteenth century, the Elizabethan era of expansion and the age of maritime activity, the curve had swung back to a drier, warmer and less stormy period in North-west Europe and continued in that direction until the end of the eighteenth century. Although the measured values of temperature and precipitation at the few stations with long records suggest that only minor oscillations have occurred since then, the history of glaciers and ice-sheets shows that even small changes can have considerable effects on the volume of ice and the advance and retreat of glaciers. The snouts of most of the larger glaciers in the Alps, for example, have now withdrawn some distance within large terminal moraines that they are known to have been building up quite late in the nineteenth century; outside these lie other moraines of known age and greater antiquity.

The Attainment of Equilibrium. In the process of recovery from the glacial maximum the amplitude of the climatic oscillations has grown progressively less; the earlier changes are beyond reasonable question, but the later changes are so slight as to be within the limits of the present yearly variability. There are no records of weather after the Christian era which might not apply to a single, perhaps very abnormal, year of the present time. Thus allegations of climatic changes in recent times can generally be, and often are, refuted on these grounds. Yet it seems unreasonable to deny that small oscillations may still be occurring, and the balance of evidence and the course of history at least lend some support to this belief.

CIVILIZATION AND CLIMATE

In the outline account of the evolution of climate given in the fore-going pages attention has been focused particularly on North-west Europe. Similar stages might be described for North America, the record of whose glacial climates is fairly complete; but of the history of peoples and of their migrations in the New World we know but little. In Europe and Asia, on the other hand, thanks to many years of patient research into history and pre-history, a wealth of material has been collected, though there still remain many serious gaps to be filled. The Mediterranean region is especially valuable because it has been con-tinuously inhabited from the beginning of civilization by cultured peoples who have bequeathed fairly complete records. Furthermore, its transitional climate makes it peculiarly sensitive to migrations of the climatic belts; it lies perilously near the limit of sufficiency of rain, and if that limit is overstepped disaster results.

From the study of historical changes of climate certain interesting relationships result. The steppes are the sensitive trigger that fires the charge whose echoes are heard over a whole continent. Even more than the Mediterranean the steppe is a region precariously near the limit of rainfall sufficiency; a decrease from 12 inches a year to 10 inches a year may mean a decrease in the supporting power from 100 head of sheep to 10 head. When rainfall decreases pasture must fail and man must go elsewhere or perish. Fortunately for him his mode of life on the steppe makes that departure easy, he has few possessions and no home ties. Outward he spreads, propelling others before him, until the settled peoples of the well-watered agricultural lands beyond feel the shock and perhaps go down before it. He carries with him his language and culture which mingle with those with which he comes into contact. It may be that the Indo-European languages that extend from Western Europe to India and from India to the North Cape disseminated outwards from the European grass-lands, while the Semitic languages, south of the Alpine barrier, spread outwards from the grasslands of Arabia.

The record of these movements seems to show a period of about 600 or 700 years—300 years of unrest followed by 300 years of slow recovery during which new civilizations emerge from the wreckage of the old but rejuvenated by the infusion of vigorous barbarian blood. Looking down the corridor of time we notice that as each disturbance settles down the seat of new power lies northward of the old. We are watching a persistent, though oscillating, recovery from the Great Ice Age which thrust plants, animals and man southwards before the advancing cold. Several writers have stressed the importance of the cyclonic climate with its stimulating variety, its seasonal rhythm and its insistence on constant and unrelaxing effort. It would appear that power has followed the cyclonic belt in its northward retreat.

SUGGESTIONS FOR FURTHER READING

The literature on the subject of climatic change is immense, but two books by C. E. P. Brooks provided, in their day, valuable summaries: *The Evolution of Climate*, Revised edition, 1939, and *Climate through the Ages*, in 1926. *British Floods and Droughts*, 1928, by the same author in collaboration with J. Glasspole, is a treatment of the somewhat elusive material indicated by the title. Geological climates are dealt with by C. Schuchert in *Climates of Geological Time*, Carnegie Inst. of Wash., Pub. 102, 1914; A. P. Coleman's *Ice Ages*, 1926, gives an account of these phenomena for geological times, while the last of them is carefully dealt with by W. B. Wright in *The Quaternary Ice Age*, 1914, and by E. Antevs in *The Last Glaciation*, 1928, No. 7 of the publications of the Amer. Geog. Soc. Research Series; and in *Quaternary Climates*, Carnegie Inst. of Wash., Pub. 352, containing papers by J. C. Jones, E. Antevs, and E. Huntington. Modern techniques and recent discoveries on glaciology are described by H. W:son Ahlmann in *Glaciological Research on the North Atlantic Coast* (R.G.S. Research series, No. 1), 1948; *Glaciers and Climate* (published in Sweden) is a valuable collection of essays on a variety of topics, written by distinguished scholars in honour of Dr. Ahlmann. A readable account of the evidence for climatic change in Central Asia is given by Ellsworth Huntington in *The Pulse of Asia*, 1907, while his 'solar cyclonic hypothesis' is treated together with other interesting material, in *Climatic Changes*, 1922, in collaboration with S. S. Visher. The 'Continental Drift Theory' as affecting past climates is dealt with by W. Köppen and A. Wegener in *Die Klimate der Geologischen Vorzeit*, 1924. Ramsey's theory is most accessible in the *Geol. Mag.*, 1924.

Simpson's theory of the fluctuations of the Solar Constant is available in *Q. J. Roy. Met. Soc.*, 1934. Pleistocene chronology and correlation are fully dealt with in two works by F. E. Zeuner, *Dating the Past* and *The Pleistocene Period*, 1945; the latter has a monumental bibliography.

The following articles, etc., may be consulted:

Antevs: 'Late Glacial and Post-glacial History of the Baltic', *Geog. Rev.*, 1922.
Bishop: 'The Geographical Factor in the Development of Chinese Civilization', *Geog. Rev.*, 1922.

Bovill: 'Desiccation of North Africa in Historic Times', *Antiquity*, 1929.
Brooks: 'World-wide Changes of Temperature', *Geog. Rev.*, 1916.
  'Secular Variations of Climate', *Geog. Rev.*, 1921.
  'The Evolution of Climate in North-west Europe', *Q. J. Roy. Met. Soc.*, 1921.
  'Meteorological Conditions during the (Permo-Carboniferous) Glaciation of the Present Tropics', *Q. J. Roy. Met. Soc.*, 1926.
Butler: 'Desert Syria, the land of a Lost Civilization', *Geog. Rev.*, 1920.
Coching Chu: 'Climatic Pulsations during Historic Times in China', *Geog. Rev.*, 1926.
Curry: 'Climate and Migrations', *Antiquity*, 1928.
De Geer: 'Geochronology', *Antiquity*, 1928.
Douglass: *Climatic Cycles and Tree Growth*, vol. I, 1919; vol. II, 1928, Carnegie Inst.
Godwin, H.: 'Pollen Analysis and Quaternary Geology', *Proc. Geol. Ass.*, 1-11, 4, 1941.
Gregory (J. W.): 'Is the Earth Drying Up?', *Geog. Journ.*, 1914.
Gregory (Sir R.): 'British Climate in Historic Times', *Geog. Teacher*, 1924.
  'Weather Recurrences and Weather Cycles', *Q. J. Roy. Met. Soc.*, 1930.
Hobley: 'The Alleged Desiccation of East Africa', *Geog. Journ.*, 1914.
Hume and Craig: 'The Glacial Period and Climatic Change in North-east Africa', Rep. Brit. Ass., 1911.
Huntington: 'Climatic Changes in America in Historic Times', *S.G.M.*, 1914.
  'Climatic Variation and Economic Cycles', *Geog. Rev.*, 1916.
Lambert, J. M., Jennings, J. N., Smith, C. T., Green, C., Hutchinson, J. N.: 'The Making of the Broads' (R.G.S. Research Series; No. 3). 1960.
Manley, G.: 'The Range of Variation of the British Climate', *Geog. Journ.*, 1951.
Penck: 'The Shifting of the Climatic Belts', *S.G.M.*, 1914.
Shaw: *Manual of Meteorology*, II, pp. 320-5.
Simpson: 'Past Climates', *Q. J. Roy. Met. Soc.*, 1927.
Taylor: 'Climatic Changes and Cycles of Evolution', *Geog. Rev.*, 1919.
Walker: 'World Weather', *Q. J. Roy. Met. Soc.*, 1928.
  Lake Bonneville, U.S. Survey Monograph I.
  Lake Lahontan, U.S. Survey Monograph XI.
  'Report of a Conference on Cycles', *Geog. Rev.*, 1923.

INDEX